America's Most Sustainable Cities and Regions

America's Most Sustainable Cities and Regions

Surviving the 21st Century Megatrends

John W. Day
Charles Hall

With contributions by
Eric Roy
Matthew Moerschbaecher
Christopher D'Elia
David Pimentel
Alejandro Yáñez-Arancibia

 Springer

John W. Day
Department of Oceanography
and Coastal Sciences,
Louisiana State University
Baton Rouge, LA, USA

Charles Hall
College of Environmental Sciences
and Forestry
State University of New York
Polson, MT, USA

ISBN 978-1-4939-3242-9   ISBN 978-1-4939-3243-6   (eBook)
DOI 10.1007/978-1-4939-3243-6

Library of Congress Control Number: 2015952007

Springer New York Heidelberg Dordrecht London

Printed on acid-free paper

Springer Science+Business Media LLC New York is part of Springer Science+Business Media
(www.springer.com)

Preface

According to the U.S. Census Bureau, about 40 million people—or about 14 % of the U.S. population—relocate from one place to another in the course of a year. Of this number, 7.6 million move to a different state. The reasons for moving are many and varied, but often include seeking a better job, to be nearer their work or further from the city where it may be located, to be nearer to family, or to enjoy recreational amenities or a better climate. Increasingly articles have been written suggesting where people might go. For example, *America's Most Livable Cities* ranks such characteristics as unemployment, crime, income growth, cost of living, and arts & leisure. The magazine of the American Association of Retired People (AARP) recently published a review of "The best cities for people over 50" using similar criteria, but even a few related to sustainability or, more precisely, "greenness."

But except for recent media coverage of the long-term potential for drought conditions in California, most people planning to move have no clue about how sustainable different cities or regions will be for themselves or their children in the coming decades, or even what sustainable means. This is a book written by people who have spent their lives examining large forces, or megatrends, at work, which we now apply to the understanding of American regionalism. Whether you want to know how your city or region will fare as these megatrends play out in this century, or gain greater insight into the consequences of moving to a location in the lower 48, this book will empower you with a whole new view of the forces that shaped our development and the American Dream, and will continue to drive our future prospects. The book focuses on the US, but the principles apply globally.

America is a big place stretching nearly 3500 miles from coast to coast. Its landscape is highly varied, with high mountains, broad prairies and forests, enormous deserts, a long and highly variable coastline, and the Mississippi—one of the largest river basins in the world. Superimposed on this natural diversity are a myriad of human systems. Sprawling cities and suburbs extend from coast to coast, and nearly a billion acres of farmland blanket the country. The cities are intimately connected to the natural systems through pathways that are mostly invisible and barely guessed at by those who live in cities.

Enormous changes have taken place in both human and natural systems in America since the arrival of European colonists, invaders some would call them, in the sixteenth century, and the founding of our republic in the late eighteenth century. The impact of Native Americans on the landscape was modest in pre-colonial America. But now, the human imprint is ubiquitous. Photos from space reveal pervasive nighttime light, which visually represent areas of high energy use. Although concentrated in large urban areas, some points of light can be seen almost everywhere.

In 1790, all ten of the largest cities in the new nation were on the east coast. No city had more than 60,000 people. The population of the nation was a scant four million. It has since increased almost 100-fold to nearly 320 million and is projected to be nearly 350 million by mid century. By 2010, almost 50 cities in the U.S. had over a million people. The population has shifted so that seven of the ten most populous cities are now west of the Mississippi. Roads cover the landscape, dams and their reservoirs proliferate everywhere but especially in the west, and most agriculture also takes place west of the Mississippi. Some of the most important farming areas of the country today did not exist, and could not have existed, in 1800, without the monumental re-plumbing of the rivers and draining of aquifers of the west.

What impelled these great changes? Many would argue that it was a vast continent rich in natural resources, a blank canvas, so to speak, and a young nation, newly independent, imbued with a sense of Manifest Destiny. Democratic ideals, hard work, and American ingenuity also comprised central roles in the national ethic. As an aside to this book, the canvas wasn't blank. There were people whose ancestors had lived on this land for more than ten millennia. But this hardly slowed the westward movement.

The word "resources" itself implies that humans know how to make productive use of such materials, and most histories focus on this aspect of human behavior, but societal patterns, changes, and survival are also shaped and determined by the available natural resources themselves and the immutable laws of nature. For example, the rich natural environment played a central role in the populating of the continent in the first part of the nineteenth century. Fertile soils provided bountiful agriculture. Vast old growth forests yielded wood for fuel and building materials and, once cut, additional fertile soils. Waterways supported rich fisheries, energy from hydropower when dammed, and pathways for commerce. Beginning in the second half of the nineteenth century, great stores of fossil fuels, first coal, then oil and natural gas, powered the industrial revolution in America. These fuels underwrote massive water projects in the west that allowed people to prosper in formally inhospitable environments. America took off, along with much of the rest of the world, in an orgy of growth and prosperity.

But the problems associated with all that growth and prosperity are starting to come home to roost, including global scale changes that are impacting society and natural systems alike. Stories fill today's news about climate change, the high cost of energy and impending shortages, degraded natural ecosystems, over-population, and associated problems. Are the fruits of a century of unbridled growth turning sour? We call these problems the megatrends of the twenty-first century. In this book we discuss how these trends will affect cities and regions in different parts of the country. Will some areas be better off than others? How can people in different regions adjust to these problems? Most people are familiar with the drought and fires in California, but the major media have shown you only the tip of the iceberg (to mix our metaphors!).

We take a journey, an odyssey of sorts, across the American landscape to consider the forces that shaped the development of different cities and regions from colonial times to the early twenty-first century, and discuss in detail how the great megatrends will affect the country. Finally, we look at how different cities and regions will fare in dealing with the megatrends. In doing so, we cover topics such as sustainability, the meaning of "green," the value of natural ecosystems, renewable energy sources, the role of technology, and the central role of energy in shaping our society. We show that the laws of nature shape and constrain both humans and nature. In doing so, we use an approach that has been called unified sensibility. This is a holistic perspective that, in the words of Professor Paul Strohm of Columbia University, is a refusal to accept conventional or arbitrary distinctions between different categories of knowledge, be they literary, philosophical, or scientific. Thus, economics is not independent of the natural world and the laws of nature. The result is an alternative history of the United States in terms of physical as well as social forces, and we think the story is all the more exciting and illuminating for it.

Baton Rouge, LA, USA   John W. Day
Polson, MT, USA   Charles Hall

Acknowledgements

In writing this book, we drew on a wide variety of sources.[1] Many individuals provided inspiration, information, insights, and guidance to us. It would be impossible to mention them all. The following people were important in providing specific information, reading parts of the book, helping us in clarifying our thinking, and encouraging us in this effort: Giovanni Abrami, David Batker, James Brown, Robert Burger, Virginia Burkett, Doug Daigle, Herman Daly, David Dismukes, Liviu Giosan, Joel Gunn, Carles Ibañez, John Jacobs, Paul Kemp, Carey King, Jae-Young Ko, Melissa Kopf, Jessica Lambert, Robert Lane, Debora Loader, Sarah Mack, Jay Martin, Michael Mielke, William Mitsch, James Morris, David Murphy, Enrique Reyes, John Rybczyk, Gary Shaffer, Fred Sklar, Joseph Tainter, June Sekera, James Syvitski, and Robert Twilley. Two of us, Hall and Day, wish to acknowledge the incredibly insightful and rigorous training we got from our professor, Dr. Howard Odum (and his brother Eugene Odum), and the wonderful and exciting environment of Chapel Hill in the late 1960s. We especially thank Dr. David Packer of Springer, for his advice, guidance, and keen insights on getting these ideas across. Adrian Wiegman helped tremendously with the logistics of getting the book ready as well as with the

---

[1] We consulted many scientific and trade books, publications in scientific journals, government and private reports and talked to a wide variety of experts on the different subjects that contribute to a comprehensive picture of sustainability. For data and general information such as is included in the narratives on individual cities for such facts as date a city was founded, native American tribes, information about the landscape, economic activity, and the like, we used a number of general references including Wikipedia and an actual paper copy of the Encyclopaedia Britannica published in 1974, as well as reports by the US Census Bureau, the US Dept. of Agriculture and various state agriculture departments, the US Weather Service, and other government reports, as well as newspapers and magazines. These are cited as appropriate.

energy discussion. Finally, we thank our families for their love and support. The ideas and opinions expressed in this book are those of the authors, and do not necessarily reflect those of the individuals mentioned above.

Portions of Chap. 8 were inspired by a separate project focused on examining the biophysical constraints on food localization. This project involved several of the authors of this book (E.R., J.D., D.P., and M.M.) as well as Dr. James Brown and Dr. Joseph Robert Burger at the University of New Mexico. We greatly appreciate the invaluable insights provided by Drs. Brown and Burger.

Contents

Chapter 1

Introduction

In her short story *Silence*, Alice Munro tells us about a woman in British Columbia who loses her husband and her daughter.[1] The husband, a fisherman, heads out one morning when the waters of the bay on which the family lived were hardly choppy. A storm strikes and, by the time the fury ends close to midnight, three boats have been lost, including the husband's. All on board are drowned. Although Munro never gives us the reasons for the daughter's disappearance, we learn that she is still alive and has become a mother. After her husband's death, the wife, the story's main character, moves from the small fishing town in coastal British Columbia to Vancouver. In a short passage, only loosely related to the general progress of the narrative, Munro describes her character's thoughts on this move:

> The cleanness, tidiness, and manageability of city life kept surprising her. This was how people lived where the man's work did not take place out of doors, and where various operations connected with it did not end up indoors. And where the weather might be a factor in your mood but never in your life, where such dire matters as the changing habits and availability of prawns and salmon were merely interesting, or not remarked on at all. The life she had been leading at Whale Bay, such a short time ago, seemed haphazard, cluttered, exhausting, by comparison.

These lines help to capture the wife's response to the tragedy and her vastly shifted personal and economic circumstances, but they also shed light on something equally fundamental: people like us can live tidy, even

---

[1] Alice Monro. 2004. *Runaway – Stories*. Vintage Books, Random House, New York, 335 p.

© Springer Science+Business Media New York 2016
J.W. Day, C. Hall, *America's Most Sustainable Cities and Regions*,
DOI 10.1007/978-1-4939-3243-6_1

sanitized, lives in large towns and cities as long as others labor at occupations such as fishing, farming, forestry, and mining, the often "haphazard, cluttered, exhausting" activities that provide the basis for the everyday necessities of life. These relatively few activities underwrite our whole economy and enable a way of life for most of us. These necessities, and practically everything else that we have come to take largely for granted in the economy, use energy to transform the materials of the natural world (soil, plants, fish, trees, minerals, oil, etc.) into the basic commodities of the economy (food, building materials, gasoline, etc.), and to transport these products to the places where people live, now mostly in cities. This has always been the case since the beginnings of civilization and indeed since the initial separation of home life from work life, perhaps as far back as the first members of our genus *Homo* about 2 million years ago. It is important to understand that while technology may act to make this transformation process more or less efficient and robust, it does not create these basic commodities on which the economy depends.

For most of human history, hunting, fishing, farming, and forestry ran on solar energy that had been captured by plants and stored for a few months to decades, but only in the case of old growth trees for a century or more. Mining was initially carried out at a relatively small scale by people who gained their energy and materials to do so mostly from farming, hunting, fishing, and forestry. The products and energy gained from the natural world are part of what scientists today call ecosystem goods and services, and their exploitation was synonymous with civilization up until the beginning of the industrial revolution in the second half of the eighteenth century. Just consider: everything that dates from before about 200 years ago depended directly and indirectly on solar energy. This was a true solar economy, not the one touted today where fossil fuels are used to make wind turbines and solar panels. Much of what we celebrate as humanity's shared cultural heritage—the paintings and sculpture in the Louvre and other great museums, the pyramids of Egypt and the Mayan world, the Coliseum in Rome, the Taj Mahal, the temples at Ankor Wat, the monasteries of Kyoto, the megaliths of Easter Island, and much, much more—all were built using only energy from the sun.

The majority of the people in the world today live in cities, and the proportion is projected to increase in the coming decades. Demographers call this the urban transition.[2] The world's population increased from about one

---

[2] See David Satterthwaite, Gordon McGranahan, and Cecilia Tacoli. 2010. Urbanization and its implications for food and farming. *Philosophical Transactions of the Royal Society*, 365, 2809–2820. Different references provide slightly different numbers but the general pattern is the same for all of them.

billion at the beginning of the nineteenth century to over seven billion by the end of the first decade of the twenty-first century, and it is projected to grow to more than ten billion by the end of this century. The growth of large cities has been even more dramatic. In 1800, experts estimate that there was only one city in the world with a population of one million or more; Beijing at 1.1 million. By 1900, there were 16 cities larger than one million, the largest being London with 6.5 million. All but two were in what is now called the first or developed world. By 1950, there were 74 cities with more than a million people, most in the developing world, with Tokyo being the largest at 11.3 million. By 2010, there were 442 cities with more than a million people, overwhelmingly in the developing world. The largest five were Tokyo (36.7 million), Delhi (22.2), Sao Paulo (20.3), Mumbai (20.0), and Mexico City (19.5). In 2015, the number is approaching 500. So as world population grew by a factor of about 7 in the 200 years from 1800 to 2000, the growth of large cities of one million people or more grew by a factor or more than 400.

Let's go back to Alice Munro's description of city life versus that of the coastal fishing village. The majority of people in cities are largely unaware of where the stuff they consume every day (gasoline, toilet paper, electricity, water, shrimp, lettuce, plastic bags, Cokes, etc.) originally comes from, or where the waste they produce (water flushed from toilets or what disappears down the kitchen sink, plastic bags, Coke cans, kitchen refuse, waste heat, carbon dioxide, etc.) ultimately ends up. Alice Munro's character was surprised by the cleanliness and manageability of Vancouver, but it is likely that most of her fellow citizens never even thought about it. Vancouver is a modern, first-world city. In teeming third-world cities, an inhabitant may know exactly where waste water goes because it runs down the street in front of the house. Even in the third world, most are only dimly aware of the sources and fates of the stuff of everyday life. However, even if inhabitants of cities are not generally aware of where the stuff comes from, they expect it to be available. This is especially true of the first world.

Greater Vancouver, British Columbia, has a population of 2.3 million—so it is not a large city by world standards today, although it is the third largest metro area in Canada. It is ranked as one of the top ten most livable cities in the world.[3] It is also Canada's most expensive city, and Forbes ranked it as the sixth most overpriced real estate market in the world.[4] It has the largest port in Canada and is a center for the mining and forestry industries

---

[3] See references in Edward Glaeser. 2011. *Triumph of the City*. Penguin Books, New York. 338 p., Wikipedia entry for Vancouver.
[4] Woolsey, Matt (24 August 2007). "World's Most Overpriced Real Estate Markets". *Forbes*. See also links in Wikipedia entry for Vancouver.

(mining and forestry don't take place in the city but most of the people who are rich because of these industries live in the city, or in other large cities). Vancouver is a center for higher education, and is becoming increasingly important for software and video game development (a quintessential twenty-first century activity), biotechnology, aerospace, and the movie industry.

Vancouver is a thoroughly modern city with a lot of high technology and urban amenities. But what do we know about the development of Vancouver? For the third largest urban area in Canada, it was a late bloomer. The first European–American settlement in the area was in 1862. The population of the urban area was about 1000 in 1881, but it grew rapidly to over 20,000 in 1900 and to about 164,000 in 1911. The early growth of the city was fueled by the basic economic activities mentioned above, which were, in turn, supported by the rich natural environment that surrounded the city. A key element in this growth was the Fraser gold rush (mining). The forestry industry also grew rapidly as the old growth forests were cut, although some of these forests still remain in British Columbia. Fishing was and still is a significant economic activity, and farming is still important on reclaimed parts of the Fraser River delta. Because of its location on Puget Sound at the western terminus of the trans-Canadian railway, Vancouver has been and remains a major port city. It wasn't until the second half of the twentieth century that the economy became more "modern" with the growth of high tech and entertainment industries.

People spend the money they earn from working in a variety of ways. Over the world most people pay for their basic necessities first. These include food, housing, clothing, and, if they are lucky, medical care. For a considerable number of people on earth, there is barely enough money or commodities to trade to satisfy these basic needs. But for many others, and, we suspect, most of the readers of this book, there is income left over for spending on things or activities that are not necessary for mere survival. Economists call this excess *discretionary* or disposal income, and over the past century, discretionary income grew dramatically for many people. Most residents of Western countries have considerable discretionary income, and we use it to pay for a great variety of things and activities that, strictly speaking, we do not really need. Among those things are entertainment and video games like those produced in Vancouver, or a trip to Las Vegas to do a bit of gaming (the modern euphemism of the gambling industry for participating in an activity where the odds are stacked against you) or to take in one of the glitzy shows.

Many people in North America have spent some part of their discretionary income on travel to Las Vegas. The city is in fact built on discretionary income. In 2010, the Las Vegas metropolitan area had a population of just over 1.9 million, slightly smaller than greater Vancouver. In contrast, in 1910 there were only 800 people living in Las Vegas, compared to over 100,000 for Vancouver, and the city did not reach 25,000 until mid century. In the post war years, Las Vegas grew explosively, fueled by a tourism industry supported by people spending discretionary income.

Las Vegas came late to a population boom because it sits in the middle of a desert with no farming, fishing, or forestry; in other words, it has a low value of ecosystem goods and services to support a city. Around the world, pre-industrial cities were almost all located in or adjacent to rich natural ecosystems (coasts, forests, rivers with their rich alluvial flood plains). In fact, the first civilizations such as Mesopotamia and the Olmecs in southern Mexico evolved in rich coastal and lower river flood plain areas.[5] A few cities not located in such areas, such as Jericho, prospered as trade centers that were in turn based on products produced in ecologically rich areas. Thus, Las Vegas had to wait until people had enough extra income to begin its growth spurt.

But why did discretionary income grow so dramatically in the twentieth century? As we will see in the succeeding chapters, the underlying reason was primarily the rapid increase in the amount of economic work that was done, which was then fueled by the increase in the consumption of fossil fuels, especially oil, over the past century. In the nineteenth century, coal underwrote the burgeoning industrial revolution. In the twentieth century, the use of all fossil fuels, coal, oil, and natural gas grew rapidly, but oil was the lynchpin of the dramatic economic expansion. For decades, energy was so cheap and abundant that it fueled revolutions in agriculture, transportation, housing, and almost all other aspects of the economy. Provision of the basic necessities of life, particularly in first-world countries, took a decreasing proportion of most family budgets. There was money left over, discretionary income, to do all sorts of things, like travel to Las Vegas in petroleum powered airplanes and cars.

But the world is changing now, and just as dramatically. The great reserves of cheap fossil fuels that seemed limitless for so long have been drawn down. We will see below that "peak oil" is already affecting our economy's ability to grow. Peak oil refers to the time when world oil production will reach its highest level and then decrease inexorably. Many consider that the world is currently near this peak.[6] A future of less abundant and more expensive energy implies that discretionary income will

---

[5] Day, J., J. Gunn, W. Folan, A. Yanez, and B. Horton. 2012. The influence of enhanced post-glacial coastal margin productivity on the emergence of complex societies. *The Journal of Island and Coastal Archaeology*, 7, 23–52.

[6] We will come back to the topic of peak oil in Chap. 7.

stop growing and probably decline. This bodes ill for much of the modern economy, but especially for places like Las Vegas that are almost entirely dependent on people spending excess income. Las Vegas was one of the most hard-hit areas during the recession of 2008, with one of the two highest rates of real estate foreclosures in the nation.[7]

Cities don't produce much at all of the actual stuff that they consume (energy, food, cars, etc.). Cities now "manufacture" (e.g., transform raw materials) and sell finished industrial products into services, culture, ideas, advice. But unlike the protagonist in Alice Munro's Vancouver, most city people are only dimly aware, if at all, of those laboring at the base of the economy. And if they are aware, they are likely to hold these workers in relatively low regard. Wendell Berry, the noted poet and essayist on things cultural and agricultural, wrote "… for an index of our loss of contact with the earth we need only look at the condition of the American farmer – who must enact our society's dependence on the land." Farming is often "considered marginal or incidental to the economy, and farmers, when they are thought of at all, are thought of as hicks and yokels, whose lives do not fit into the modern scene."[8] Edward Glaeser, the author of *Triumph of the City*, denigrated Thoreau and said cities brought civilization to "savage places" inhabited by savages.[9] This statement reflects the lack of appreciation of many city people for how, where, and by whom the stuff they use is produced, how the world functions, and how humans fit into it.

The Future Is Now

Most of us think that the world will muddle on and that the twenty-first century will be more or less like the last century. There will be ups and downs, good times and bad, heat waves and cold snaps, but that water will still flow through the tap at our command because that's just the nature of the world. Most don't expect dramatic and permanent changes. Look at the twentieth century: yes, there were two world wars and many minor ones, the great depression, and natural disasters. But, on average, the economy grew and prosperity spread. We expect the future to be like the past in that we will always be able to buy the stuff we need and want.

---

[7] See Chap. 4.

[8] Wendell Berry. 1972. *A Continuous Harmony – Essays Cultural and Agricultural*. Shoemaker and Hoard, Washington, DC.

[9] Edward Glaeser. 2011. *Triumph of the City*. Penguin Books, New York, 338 p.

What if instead, for most people alive today, the future will be decidedly unlike the recent past? There are great trends afoot in the world that will fundamentally alter the way we live, trends which are already doing so for some. The first decade of the twenty-first century ushered in a time of extraordinary change. In the 1990s, the U.S. economy and that of much of the rest of the world boomed. After the great recession of 2008, however, the U.S. economy—and economies of much of the rest of the world—has struggled with sluggish growth and record deficits. In the U.S., many states have had to slash services to remain solvent, and even deeper financial problems have afflicted most Eurozone countries and Japan. Almost no one, especially economists who should know, foresaw the great recession of 2008. Just as few are able to give a plausible explanation for the weak or absent recovery. In this book you will learn why these events are early evidence of certain major trends that will unfold in the twenty-first century, what are causing these changes, and how they will affect society as a whole and your region in particular. In the coming years and decades, everyone will become aware of how fundamentally important nature, and especially energy, is in underwriting the economy. Parts of North America will share a disproportionate burden of these shifts, and you will want to know how your city or region ranks according to the characteristics that are likely to determine the severity of the impacts.

Chapter 2

Manifest Destiny and the Growth of America: Cheap Energy and Spending Natural Capital

Manifest Destiny is the *sine qua non* of American history. At one time almost every American child was toilet trained on this idea of inevitable expansion of European settlement to blanket the continent. In 1833, the author Horace Greeley is reputed to have said "Go west young man."[1] He expressed the widespread feeling at the time that America was a land of limitless possibilities, that if a person worked hard, especially on the fertile lands of the west (referring to the land of the Ohio Valley and the Midwest) he or she could succeed and prosper.

Indeed, in the late eighteenth century, the central part of the North American continent that would ultimately become the U.S. was a land of practically unlimited resources and possibilities. At the time, the population of non-Native Americans was still largely confined to the east coast, and Jefferson's Lewis and Clark expedition had yet to happen. But after the Louisiana Purchase, the young nation more than doubled in size. Settlers moved west over the Appalachian Mountains to "claim" lands in the Ohio Valley. Flatboats floated down the river to New Orleans. Reports reached the east of vast stands of old growth forests, fertile prairies, mighty rivers, and abundant wildlife. It was all there for the taking, and take it they did.

---

[1] Greeley's actual advice to "any young man" was "Go to the West: there your capabilities are sure to be appreciated and your energy and industry rewarded." Williams, Robert C. 2006. *Horace Greeley: Champion of American Freedom* (Kindle ed.). New York University Press.

© Springer Science+Business Media New York 2016
J.W. Day, C. Hall, *America's Most Sustainable Cities and Regions*,
DOI 10.1007/978-1-4939-3243-6_2

Greeley's words reflected the then popular idea of manifest destiny[2]; Americans were inevitably destined to expand over the continent. Manifest destiny incorporated several beliefs of the time, including that Americans of European descent and their institutions had special virtues, they had a mission to spread these virtues and institutions across the continent, and they had a destiny under a special providence from God to do so. All of this was independent of whether or not the region was already fairly densely populated with Native Americans who had to be killed or displaced for manifest destiny to take place.[2] This idea was a mostly unchallenged component of the social studies curriculum in the US well into the latter half of the twentieth century.

The history of the expansion across North America is bound up in many of our folk tales, songs, and music. Daniel Boone, Davy Crockett, flatboats and steamboats on the Mississippi, the Alamo, wagon trains going west on the Oregon and Santa Fe trails, circling the wagons, the California gold rush, the taming of the west, the Golden Gate; all these contribute to a central theme of our national ethic.

But the idea of manifest destiny was conflicted from the beginning. It was popular during the first half of the nineteenth century when there was growing conflict over slavery and whether it would spread west. Of course there were people there already, Native Americans who had lived on the land for thousands of years and who felt it was theirs. Their communal land ethic was different from the pioneers. They were pushed aside, killed, decimated by disease, and their land largely taken. As the Lakota chief Red Cloud said, "they made promises, more than I can remember. But they never kept but one. They promised to take our land and they took it."[3] Nowadays, there is recognition of the wrongs done to Native Americans, most of which is sincere in spite of a continuing sense of the inevitability of these events. But at the time, the feeling was that most of the continent was sparsely populated and there for the taking if only you were willing to work hard enough.

There is no doubt that the national narrative of America at the time had much to do with the westward expansion. The settlers encountered a land incredibly rich in unexploited natural resources. Vast stands of old growth forests, fertile prairies, rich soils, alluvial plains, abundant wildlife, and rich fisheries awaited them. A network of rivers in the Ohio, Tennessee, and Mississippi valleys provided avenues for waterborne commerce. These rich resources fueled the westward expansion. In the second

[2] Merk, Frederick; Bannister, Lois. 1963. *Manifest destiny and Mission in American History.* Harvard University Press.
[3] http://www.impurplehawk.com/quotes.html.

half of the nineteenth and early twentieth centuries, the burgeoning industrial revolution fueled by coal, oil, and natural gas impelled the expansion forward at an even greater pace.

As the country spread across the continent, it prospered from this abundance. Some areas prospered before others, and there are still large areas in the west where few people live. San Francisco, with access to huge resources through easy ship-based trade, for example, flourished beginning in the middle of the nineteenth century whereas much of Arizona remained sparsely populated until well into the twentieth century. By the second half of the twentieth century, the U.S. had become the richest and most powerful country in the world. A key to understanding the expansion of the U.S. is the enormous amount of natural resources that fed and sustained that growth. These included plentiful water resources in the east, rich natural ecosystems including especially excellent soils located where rain fell reliably, and abundant coal and oil. We don't downplay the importance of human persistence and ingenuity, but natural resources and services were a prerequisite for prosperity and power.

In this chapter, we outline the growth of the United States from the late eighteenth to the early twenty-first centuries with a look at how factors such as natural resources, climate, emerging technology, and energy availability underwrote the country's expansion. In Chap. 4, we will visit 12 characteristic cities and ten regions and show how these iconic examples, including New York, Asheville, Orlando, New Orleans, Houston, Las Vegas, Los Angeles, and Portland developed, and some of the factors that affected their growth. So climb into your wagons, ladies and gentlemen, and get ready for a wild and interesting ride. We are getting ready to travel from sea to shining sea.

Historic Settlement Patterns in the United States

In North America early settlement patterns depended upon areas offering abundant water as rain and in rivers and lakes, including the cheap transportation these resources allowed. Populations initially congregated primarily along rivers and in coastal areas containing abundant and essential natural resources. This abundance provided materials and energy for the production of food, shelter, clothing and tools. Many present day large cities grew into major urban centers as a result of the natural advantages provided by the local environment in pre-industrial times. Later cheap fossil fuels facilitated further and sometimes explosive growth in these resource-rich areas, as well as the development of non-water based transportation systems that played a large part in the growth of urban environments in resource-poor locales. For this reason, energy is sometimes referred to as the "master resource," a topic we will come back to later in the book.

The trend of settling near coasts and along river floodplains and lakes goes back thousands of years prior to European settlement. Some of the largest Native American settlements in the U.S. prior to the arrival of Europeans were located in the Ohio and Mississippi river valleys. Evidence of their cultures and longstanding civilizations dot the landscape throughout these regions in the form of ancient mound formations that still exist today in places like Fort Ancient, Ohio and Poverty Point, Louisiana (recently declared a UNESCO World Heritage Site). Large populations of Native Americans also inhabited the Great Lakes Region and the Northwest. Seasonal floods along the major river valleys maintained fertile soils for agriculture. Lakes and streams provided water sources for game animals and habitat for fish. Coastal areas supplied an abundance of seafood. Settlements on rivers, lakes, and coastal areas also benefited from natural water-based trade routes. Our movie and television view of Native Americans living in the dry and relatively resource-poor West reflects later distributions after most of these tribes were pushed off their lands in the eastern and central U.S. and as the horse (initially from Europe) allowed greater exploitation of the bison.

Early European settlers utilized the wealth of natural resources in areas such as Chesapeake Bay, the Mississippi delta, and other coastal areas that provided easy access to coastal transportation corridors and avenues for entry into the interior of the continent. Settlements such as Jamestown, Boston, New York, Charleston, Savannah, St. Augustine, and New Orleans are but a few of these early settlements located near resource rich coastal ecosystems. In western North America, the north bound Spanish moving from Mexico settled San Juan de los Caballeros near the confluence of the Chama River and the Rio Grande, just north of present day Santa Fe, while others remained closer to the coastal margins in areas such as San Francisco Bay. Meanwhile, the Dutch settled New Amsterdam at the southern tip of Manhattan Island, the English settled in numerous locations along the Atlantic seaboard, and the French settled in the Great Lakes region near present day Detroit, and further south along the Mississippi River in St. Louis, Baton Rouge, and New Orleans. French settlements were also established in the fertile wetlands of Acadiana country in southwest Louisiana and along the Gulf Coast in Mobile and Biloxi.

Waterways served as natural trade corridors for early settlers. Interior settlements of the early colonial period often began as trading outposts that relied on the traffic of traders in fur and other useful products.

A trading outpost was a meeting place for the exchange of goods and information along trade routes. Different ecosystems provided resources to the growing population of settlers allowing for the establishment of a vibrant commercial culture. Traders marketed fur, meat, vegetables, grains, and wood. However, outposts often had the dual purpose of serving as a fort with military garrisons, while also housing tradespeople such as blacksmiths, and mills to process grains or other raw materials. In addition, many coastal areas were especially rich in food resources.

Trends in Urbanization and the Emerging Mega Regions in Population

Trends in urbanization over the past two centuries reflect such things as transportation modes, natural resources, emerging markets, and the growing availability of cheap fossil fuels beginning in the second half of the nineteenth century. In 1800, after the Revolutionary War and the signing of the Constitution, the ten largest cities were located along the east coast (see Table 2.1). These included New York, Philadelphia, Baltimore, Boston, Charleston, Northern Liberties (PA), Southwark (PA), Salem (MA), Providence, and Norfolk. The cities of Northern Liberties and Southwark are now neighborhoods in Philadelphia. If these latter two are included with Philadelphia, it was the largest city in 1800. No city had more than

Table 2.1 Ten largest cities in the U.S. from 1800 to 2000

Date/ rank	1800	1850	1900	1950	2000
1	New York	New York	New York	New York	New York
2	Philadelphia	Baltimore	Chicago	Chicago	Los Angeles
3	Baltimore	Boston	Philadelphia	Philadelphia	Chicago
4	Boston	Philadelphia	St. Louis	Los Angeles	Houston
5	Charleston	New Orleans	Boston	Detroit	Philadelphia
6	Northern Liberties (PA)	Cincinnati	Baltimore	Baltimore	Phoenix
7	Southwark (PA)	Brooklyn	Cleveland	Cleveland	San Diego
8	Salem (MA)	St. Louis	Buffalo	St. Louis	Dallas
9	Providence	Spring Garden (PA)	San Francisco	Washington	San Antonio
10	Norfolk	Albany (NY)	Cincinnati	Boston	Detroit

Note that in 1800, all of the largest cities were coastal
Note: 1800: Northern Liberties and Southwark are now part of Philadelphia
1850: Brooklyn now part of NYC and Spring Garden is part of Philadelphia
New Orleans was not part of U.S. before the Louisiana Purchase in 1803

100,000 people. New York City had a population of 60,515 and Philadelphia, along with its two neighborhoods, had 61,559.

By 1850, two trends are evident. There were high growth rates in a number of cities, and the population was spreading west away from the coast. New York was the largest city, having grown to over 515,000. The next five out of the ten largest cities in 1850 were Baltimore, Boston, Philadelphia, New Orleans, and Cincinnati, all of which had populations greater than 100,000. The final four were Brooklyn, St. Louis, Spring Garden (PA), and Albany, NY. Brooklyn is now part of New York City and Spring Garden is part of Philadelphia.

Four of the largest cities were not along the east coast. Settlers had moved west of the Appalachian Mountains and settled in the Mississippi River basin. Albany is on the Hudson River while New Orleans, Cincinnati, and St. Louis are on the Mississippi and Ohio rivers. By this time, the industrial revolution had begun in earnest. New York, Boston, Baltimore, New Orleans, St. Louis and Cincinnati were all important ports. All of these cities were located on the coast and along rivers where trade was important and their hinterlands were rich natural environments. New Jersey, across the Hudson River from New York and the Schuylkill River from Philadelphia, still proudly wears the official state slogan of the "The Garden State," reflecting its rich agricultural heritage.

In 1900, more large cities were located west of the Appalachians and three cities had more than a million inhabitants (New York, Chicago, and Philadelphia). In order, the top ten cities were New York (with a population over 3.4 million), Chicago, Philadelphia, St. Louis, Boston, Baltimore, Cleveland, Buffalo, San Francisco, and Cincinnati. Chicago owes its rise to second place both to its location as a trade nexus and to the great agricultural expansion in the fertile soils of the Midwest. New Orleans was no longer a part of the top 10 as trade along the Mississippi River was supplemented by freight trains. For the first time, the list of the top ten cities includes San Francisco on the west coast; it first made the top ten in 1870. By 1900, the industrial revolution was well underway and cheap fossil fuels were beginning to transform society. All of the largest cities, however, continued to be located in resource-rich natural environments.

By 1950, however, a number of additional trends become evident. Several cities peaked in population and began to decline. This transition is associated with the growth of suburbs, the expanding Interstate highway system, and the ongoing westward shift of the population. The G.I. bill after World War II enabled veterans to purchase houses in the suburbs with low interest loans. The ten largest cities were New York, Chicago, Philadelphia, Los Angeles, Detroit, Baltimore, Cleveland, St. Louis, Washington, and Boston.

The top five cities each had more than one million people. New York was the largest with almost 7.9 million. Detroit's rapid growth as a major manufacturing hub was the result, first, of the overwhelming demand to supply cars to a population whose major transportation mode had shifted to the personal automobile and, second, access to the iron of Minnesota and coal of Pennsylvania. Low interest loans played an important role in the development of the car culture. Los Angeles was the fourth largest city in 1950 and is a prime example of how cheap fossil fuels allowed the growth of a city in an inhospitable desert environment. The motion picture Chinatown (1972) and the book *Cadillac Desert* (1993) describe the social, political, and economic climate surrounding the massive undertaking of appropriating scarce water resources to build the nation's most populous desert city.

By 2000, further major changes had taken place in U.S. population distribution. More than half of the largest cities were now west of the Mississippi, many in arid and semi-arid areas where the existence of large cities with high-rise buildings and large populations would not have been possible without cheap fossil fuels. New York remained the largest metropolitan area followed by Los Angeles, Chicago, Houston, Philadelphia, Phoenix, San Diego, Dallas, San Antonio, and Detroit. San Jose replaced rust belt Detroit by 2010. The growth of Houston and Dallas reflected the importance of Texas for oil and gas, refining, and petrochemicals as well as the importance of Houston as a major port. Cheap fossil fuels brought roads, airports, and perhaps most importantly, water to the west that allowed cities to grow. The development of affordable air conditioning was instrumental in allowing urban expansion in the hot climates in the south and southwest.

Growing populations of retirees in the sunbelt climates of Florida and Arizona, which are desirable in the winter but often brutally hot without air conditioning in the summer, also played a role in the emergence of large petroleum-based metropolises. Phoenix and cities in the Florida peninsula are major metropolises that have, at least in part, benefitted from the seasonal migration of "snowbirds". Just like real birds, this portion of the populace migrates to the southern climates during the winter months to avoid the harsher winters of northern locales. Snowbirds travel hundreds of miles to reside in second homes that often go uninhabited for the other months of the year. In the Rio Grande Valley of south Texas these snowbirds are called Winter Texans. They more than double the population of some cities in the Valley. There is an active and light-hearted debate as to whether the arrival of Winter Texans raises or lowers the average I.Q. of the Valley. These seasonal populations bring with them disposable income that fuels tourism-dependent economies.

Let's pause for a moment and consider the trends we are discussing and how they influenced the population spreading across the continent. First, until the beginning of the twentieth century, cities like New York, Albany, Chicago, and New Orleans grew up in resource-rich areas and along waterways that provided food and fiber and convenient trade routes. Second, the climate of all of these early cities was moist, providing water for the cities and for agriculture and forestry to feed and fuel them. Finally, energy is a key element that fueled the growth of these cities. Early on, energy came from natural resources that provided food to feed the city inhabitants and wood for heating, cooking, and to fuel their machines. Falling water was another important energy source in these moist climates, powering mills to grind grain, saw wood and do other mechanical work. Many communities were established along the fall line that facilitated hydropower and was the limit of ocean going ships. With the coming of fossil fuels, cities spread into inhospitable environments, especially in the Southwest with the tremendous subsidy of these ancient fuels. Cheap fuels also allowed climate control and easy transportation that connected long distances in the west. Few thought about these fuels ever becoming scarce and expensive.

The 11 Megaregions

The growth patterns described in the previous section led, by the end of the twentieth century, to the majority of the U.S. population becoming concentrated in 11 megaregions that are still growing rapidly according to population and employment statistics (Fig. 2.1).

The sunbelt megaregions consist of regional concentrations of population in Florida, along the Gulf coast from Florida to Texas, the Arizona sun corridor, and southern California.

The most densely populated megaregion is the Northeast, which includes the megalopolis stretching from Washington D.C. to Boston. This megaregion is expected to increase from more than 49 million people in 2000 to over 58 million in 2025.

The Texas triangle megaregion is fueled by the energy sector-based economy of Houston, Dallas, and San Antonio.

The Great Lakes megaregion encompasses the major urban centers of the Midwest, including Chicago, St. Louis, Detroit and Cleveland, and is expected to contain over 62 million inhabitants by 2025.

Less populated megaregions exist out west along the Front Range of the Rocky Mountains near Denver, and in the Northern California megaregion situated around the San Francisco Bay Area.

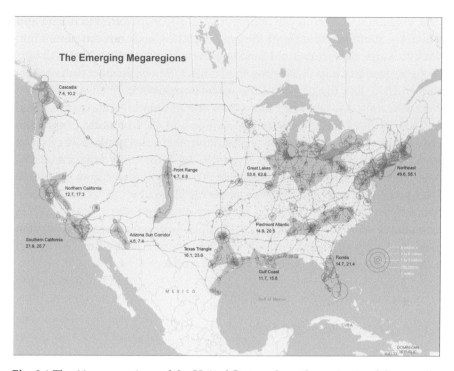

Fig. 2.1 The 11 megaregions of the United States where the majority of the population is located. The two numbers by each megaregion refer to the population in 2000 (left) and the projected population in 2025 (right) (Adapted from Regional Plan Association (RPA) 2008. The Emerging Megaregions. Published on America2050. org. Produced by the RPA)

The Cascadia megaregion is composed of the northwestern population hubs of Portland and Seattle extending north to Vancouver. The Piedmont Atlantic megaregion stretches east from Birmingham to Atlanta and north into the Carolinas.

Since 1790, the U.S. population has increased by a factor of nearly 100, growing from about 4 million to nearly 310 million in 2010. But population grew at a much greater rate in the twentieth century than in the nineteenth. The total population reached 50 million in 1880 and was over 100 million by 1920. It took another 50 years to reach 200 million individuals and approximately 36 years to add the next 100 million. The U.S. is projected to exceed 350 million by mid century.

To a considerable extent, population growth dynamics in the U.S. mirrored that of the globe. World population reached one billion about 1804, two billion in 1927, three billion in 1960, four billion in 1974, five billion in

1987, six billion in 1999, seven billion in 2011, and is projected to reach eight billion by about the middle of the next decade. Check out the population clock to watch the number of humans grow on the counter.[4]

As we saw in Chap. 1, the growth of cities followed an even faster trajectory. In 1800, less than 10 % of the world's population lived in cities and there was only one city with more than one million inhabitants (Beijing).[5] By 1900, 16 cities had more than one million and 13 % of world population lived in urban areas. The number of million plus cities grew to 74 (30 % of world population) in 1950 and over 450 by 2010 (over half the global population).[6] In 2015, there are almost 500 cities of more than a million. Thus growth is centered in urban areas facilitated by the social and economic opportunities of cities and fueled by increased energy use, which in turn allows accelerated exploitation of other essential resources to sustain these urban populations.[7] Growth initially occurred in the first world, but now is overwhelmingly in the developing world.

The American Melting Pot

Changes in immigration policy have affected population and demographics since the early nineteenth century. The earliest records of immigration patterns show an annual increase in immigrants from the late 1840s until the Civil War years when the numbers declined from their peak of 400,000 per year. By 1880, immigrant numbers surpassed the pre-war high of 400,000 per year. More than 200,000 immigrants per year came into the country each year until the initiation of WWI. The period 1905–1914 contained 6 years when more than 1 million immigrants per year arrived in the U.S. Incoming immigrant numbers declined dramatically to less than 100,000/year in 1931 and stayed low until the post WWII years.

---

[4] http://www.census.gov/popclock/.

[5] Chandler, T. 1987. *Four thousand years of urban growth: an historical census*. St. David's University Press.

[6] http://www.un.org/esa/population/publications/WUP2005/2005wup.htm.

[7] Currently, more than half of the world's population lives in cities and as of May 2014, there were 463 cities of a million or more, the vast majority of them in the developing world. Tokyo was still the largest at 36.7 million, but the others in the top five were in developing countries (Delhi 22.2 million, Sao Paulo 20.3, Mumbai 20, and Mexico City 19.5). In 2010, China alone had 89 cities with more than one million people; India had 46, the U.S. 42, Brazil 21, and Mexico 12. So between 1800 and 2010, world population increased sevenfold, but the number of cities populated by over one million people increased by a factor of more than 450.

The majority of immigrants to the U.S. were from Europe up until 1965. The incoming immigrant population increased slowly from 1950 through the mid-1990s when it peaked around 1.8 million annually. It has since hovered between 800,000 and 1.3 million during the first decade of the twenty-first century. Since 1965, the majority of immigrants to the U.S. have been from Latin America and Asia.

The number of foreign-born U.S. residents has increased steadily over time and included approximately 40 million individuals as of 2010. These numbers do not include undocumented immigrants. Immigrants to the U.S. hail from a wide variety of different countries (Fig. 2.2). In 2006, Mexico made up the largest number of immigrants to the U.S. In that year, 31 % of all foreign-born immigrants hailed from south of the border. No other individual country contributed more than 4 % of the foreign-born total in 2006.

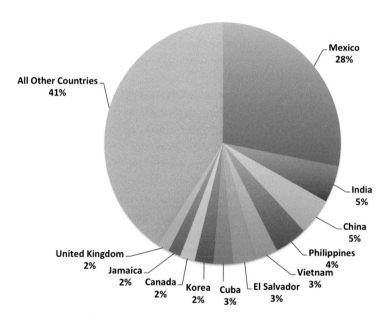

Fig. 2.2 Countries with the largest populations in the United States as percentages of the total foreign-born population: 41,347,945 million in 2013. (Data Source: Migration Policy Institute tabulations of the U.S. Bureau of the Census, American Community Survey and Decennial Census. http://www.migrationpolicy.org/data/state-profiles/state/demographics/US)

The Urban–Rural Imbalance

Current population distribution in the U.S. is heavily concentrated in the large metropolitan areas of the different megaregions. This has not always been the case. Less than 10 % of the population lived in cities until the middle of the nineteenth century, reflecting the concentration of jobs in agriculture. It was not until the first decade of the twentieth century that more people lived in urban than rural areas. In 2010, 55.7 million people lived in rural areas compared to approximately 259 million people (82 % of the total population) living in urban areas.[8]

The trend towards increased urbanization is expected to continue into the near future. By 2030, it is projected that 318 million people—or 87 % of a total population of 366 million—will live in urban areas. This distinction requires a rather precise definition. Urbanized areas are defined as having at least 1000 people per square mile, while urban clusters are areas with at least 500 people per square mile. Areas with population densities less than 500 people per square mile are considered rural. Greater than 80 % of the U.S. population now live in metro areas that occupy less than 20 % of the land area.[9] The suburbanization of the U.S. has contributed greatly to this trend during the latter half of the twentieth century.

Increasing industrialization and the introduction of new technology also enhanced the movement of people into cities. The availability of cheap energy, combined with the increasing affordability of the personal automobile and an ever-increasing consumption based on credit and debt, helped facilitate the suburban housing boom after the Second World War. The personal automobile and the modern mortgage loan were major factors leading to increased suburbanization. Dependence on public transportation was supplanted by the idea of freedom, comfort, and personal mobility that only a car could provide. No longer did railroads and streetcars dictate the worker's place of residence. The Federal Interstate Highway Act also encouraged and aided migration to the suburbs. Cheap gasoline combined with the suburban residences of the commuter class formed the basis of a large part of modern urban U.S. culture. Cheap liquid fuels also allowed the shipment of cheap food to cities. The post WWII consumer culture of the U.S. tended to glorify the urban (and especially suburban) at the expense of the rural. A wider variety of goods and services are available in

[8] U. N. DESA. 2012. *World Urbanization Prospects, the 2011 Revision.* New York: United Nations Department of Economic and Social Affairs of the United Nations Secretariat, New York.
[9] U.S. Census Bureau 2010. https://www.census.gov/geo/reference/ua/urban-rural-2010.html.

urban areas. Rural areas traditionally contained locally owned businesses that have since been supplanted by strip malls and national chain stores. Underwriting this great transition was the abundance of cheap energy, mainly in the form of fossil fuels.

The economies of rural areas are generally based on primary sector economic activities such as agriculture and forestry, fisheries, and mining, including oil and gas extraction. Tourism and retirees also support some rural areas. By contrast, the economies of urban areas tend to be based on secondary and tertiary economic activities such as manufacturing, information, and human services, much of which is dependent on a college-educated workforce. But in the U.S., the economy has shifted away from manufacturing and towards the tertiary and quaternary sectors such as the human services and information sectors, including hospitality, information-based activities like communications and education, and financial services.

This increasing reliance upon the tertiary and quaternary sectors relates to the trend of increasing urbanization and globalization. But this shift in economic activity does not decrease the population's reliance on the raw materials provided by the primary sectors of the economy. Rural areas tend to provide low wages compared to urban areas as technology has supplanted much human labor. Most primary sector jobs do not include an abundance of highly skilled, high wage jobs. Poverty rates in rural natural resource-rich environments are often high where a few landowners tend to control the property and goods produced on these large landholdings. Resident worker populations have little opportunity for economic advancement, and there tends to be less disposable income in rural versus urban areas.

Increasing mechanization of production in the primary and secondary sectors has also shifted the labor force requirements from human-based labor to fossil fuel-based machinery. In addition, the increase in global trade has shifted manufacturing jobs to low-wage developing countries.

Paradoxically, areas rich in natural resources tend to be areas with elevated poverty rates today. This phenomenon is sometimes known as the resource curse or the paradox of plenty. The resource curse concept is evident in the international arena where developing countries that are rich in natural resources or minerals (including agriculture, forestry, fisheries, and mining-especially fossil fuels), tend to be controlled by a few wealthy individuals who profit most by servicing export markets in developed countries. The result is that the country as a whole often remains financially poor and stratified socially.

Perhaps surprisingly, this concept can also be applied to states and counties in the U.S. that are classified as "underperforming" (Fig. 2.3).

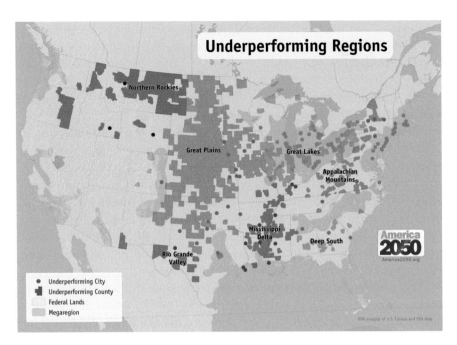

Fig. 2.3 Underperforming regions of the United States by county. These are regions that have not kept pace with national trends over the last three decades in terms of population, employment, and wages. (Figure produced by Regional Plan Association, America 2050: http://www.america2050.org/images/2050_Map_Underperforming_Regions_150.png)

Underperforming regions are those that have not kept pace with national trends over the last three decades in terms of population, employment, and wages. These regions tend to be in agricultural and resource-dependent rural regions, as well as former industrial (manufacturing) regions.[10]

The majority of "underperforming" regions of the country are rural, and in the Midwest and Northeast "Rust Belt", the south, the plains, and the northern Rockies. The America 2050 report goes on to say:

> With the exception of the Great Lakes Megaregion, the identified underperforming counties are overwhelmingly located outside of the megaregions. Most of the 640 underperforming counties are rural counties far from metropolitan centers. Of the 79 identified counties located in one of the 11 megaregions, 75 are in the Great Lakes megaregion. These seventy five counties account for nearly 6 million of the 13.4 million total residents in the underperforming counties.

---

[10] Hagler Y., Yaro R.D., and Ronderos L.N. 2009. *New Strategies for Regional Economic Development*. America 2050 Research Seminar Discussion Papers and Summary. Healdsburg, California – March 29–31, 2009.

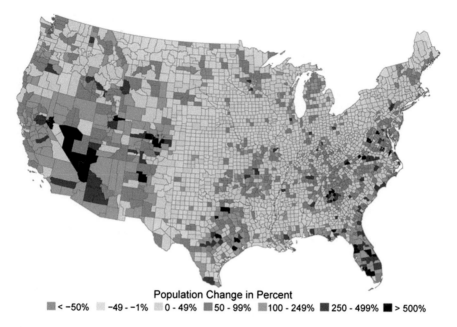

Population Change in Percent
■ < −50% ░ −49 - −1% ░ 0 - 49% ■ 50 - 99% ■ 100 - 249% ■ 250 - 499% ■ > 500%

Fig. 2.4 Percent change in population by county from 1970 to 2008. Many of the areas losing population are the same as the underperforming regions. The regions with the highest growth rates such as the southwest and gulf coast are warmer and require air conditioning. Figure from NOAA and The Census Bureau: https://www. climate.gov/news-features/features/will-hurricanes-change-world-warms

As mentioned in the America 2050 report, many of the areas that are losing population (Fig. 2.4) are the same areas where basic economic activities take place that convert energy and materials of the biosphere (soil, trees, fish, minerals, buried fossil fuels) into products (agricultural products, wood, metal ores, gasoline, natural gas, and coal) that form the base of the economy. In addition, secondary economic sector manufacturing-based economies in the Rust Belt and in the Great Lakes region often underperform as well. This is partially the result of machines replacing human labor, as well as the outsourcing of jobs to lower wage foreign countries. On the contrary, some of these regions are attractive as an escape from industrialization and urbanization, such as for retirees.

The economy in the U.S. is highly dependent on resource consumption and the services associated with consumptive behavior.[11] However, this consumption is itself dependent on the labor of a relatively small percentage of the population who work in the primary sector that provides basic resources to the economy (food, forest products, minerals, energy). People have been departing rural areas throughout the twenti-

---

[11] Burger et al. 2012. The Macroecology of Sustainability. *PLOS Biology*, 10 (6), 1–7.

eth century both as a result of mechanization of agriculture and better opportunities in urban areas. The only time since 1900 that this trend was reversed was during the Great Depression, when job opportunities in urban areas plummeted.[12] Later in the book, we discuss trends in the economy, climate, and energy that may again challenge urban living and portend a new return to rural areas.

---

[12] Source: US Census Bureau 1990 and 2000. Urban/Rural Census Data 1900 to 2000, Online at: http://www.wsdot.wa.gov/planning/wtp/datalibrary/population/PopGrowth SMSA.htm.

Chapter 3

The Myth of Urban Self Sufficiency

Beginning in the mid nineteenth century, the U.S. economy transitioned from a largely rural society, still based predominantly on agriculture and renewable resources, to an urban society dependent to an increasing extent on non-renewable resources and cheap fossil energy, and a globalized industrial agricultural and manufacturing system. The shift from a society based on renewable natural resource flows to non-renewable resource stocks is one of the most important and unprecedented transitions in the history of mankind. This shift allowed for massive human population growth, unprecedented exploitation of the earth's resources, increasing per capita wealth for many, and the dramatic growth of urban areas. Renewable resources are those that can be replaced in a relatively short amount of time such as agricultural crops, trees, and fish. Non-renewable resource stocks are those such as fossil fuels and mineral ores that are not replaced on a meaningful time scale and are forever depleted until they have been practically exhausted. This means that there comes a time when a resource, such as hydrocarbon fuel, is no longer economical to exploit, which occurs long before complete exhaustion. Much of this book is about what happens in different regions when these stocks are depleted and we are forced to depend, once again, on poorer-quality resource stocks as well as daily flows of energy from the sun. Environmentalists and the news media, most of who live and work in urban areas, often speak longingly about a future based on renewable energy. Watch out what you ask for.

Renewable inputs to the economy from farms, forests, and the sea became relatively much less visible to the modern consumer than nonrenewable

© Springer Science+Business Media New York 2016
J.W. Day, C. Hall, *America's Most Sustainable Cities and Regions*,
DOI 10.1007/978-1-4939-3243-6_3

minerals and the materials made from oil (petrochemicals such as plastics) during this transition. In *Drilling Down*, Tainter and Patzek point out that we became like fish in water where what surrounds us, abundant energy in our case, is so ubiquitous that it is the last thing most of us notice.[1] This point is particularly true of the growing urban populations that have become dependent on rapidly growing inputs of both renewable and nonrenewable resources but rarely if ever see the means of primary production (unless you are a character in Alice Munro's short story). The renewable resources are produced from functioning natural systems including farms, wetlands, forests, prairies, and waters. These natural resource flows include both materials that are consumed directly, such as wood and food, and ecosystem services without which the economy would not function. These services include the hydrologic cycle that provides water for both direct consumption and indirectly as rainfall on crops (weather forecasters may apologize to New Yorkers for raining on their parade, but this is unlikely to occur in agricultural areas during the growing season) and water flow that supports hydropower. But high human population pressure on the biosphere is degrading natural renewable resources.[2] Increasing pressure on natural resources and the generation of wastes produced at an industrial scale are degrading the renewable resource production potential of natural ecosystems. For example, nitrogen pollution in the runoff from large commercial industrial agriculture impairs water quality and associated natural resources the world over. Climate change is perhaps the most high profile example of the degenerative impacts of the consumer-based economy in the petroleum age. We will address each of these issues in more detail in later chapters.

The Myth of the Self Sufficient City

In spite of their enormous requirements for materials and energy, and their enormous generation of wastes, many see urban living as the sustainable future for most of humankind in the twenty-first century.[3] But there are

---

[1] Joseph Tainter and Tadeusz Patzek. 2012. *Drilling Down – The Gulf Oil Debacle and Our Energy Dilemma*. Springer, New York. 242 p.

[2] Vitousek PM, Mooney HA, Lubchenco J, Melillo JM. 1997. Human domination of earth's ecosystems. Science, 277, 494–499.; Smil, V. 2013. *Harvesting the Biosphere: How Much We Have Taken from Nature*. The MIT Press, Cambridge, MA, 312 pp. This is discussed in more detail in Chap. 5.

[3] e.g., Better, Smarter Cities. We have seen a brighter future, and it is Urban. Scientific American Special Issue, September 2011.; Robert Kunzig. 2011. The City Solution; Why cities are the best cure for our planet's growing pains. *National Geographic*, December.

serious issues for urban areas, especially very large ones in both the developed and developing world, given the interrelated problems of climate change, and energy and resource scarcity, and the importance of natural systems for society. These interrelated problems will pose constraints for all of society, but they will be much more challenging and difficult to solve for very large urban areas. Let's look at this in more detail.

Cities are touted as efficient living areas compared with suburban and rural areas. Why are cities considered as the solution to many of the problems that we face in this century? The efficiency of mass transit, bicycles, and walking, shorter commutes, and smaller more efficient living spaces are reasons often cited. David Owen, in a well-quoted article in *The New Yorker*, wrote that "the average Manhattanite consumes petrol at a rate the country as a whole hasn't matched since the 1920s."[4] Owen goes on to state that "eighty-two percent of Manhattan residents travel to work by public transit, by bicycle or on foot" or "ten times the rate for Americans in general." But in the greater New York City area, 56 % of residents drive to work.[5] According to this same study, New York City has the longest commute time at 35 min and the fifth highest congestion score of the ten cities in the U.S. with the worst traffic. Elizabeth Farrelly writes in *Blubberland* that density is the key to sustainability. "If we were to design a green settlement-pattern from scratch, it would not be suburbia, or urban villages, or Greek fishing towns, or even say, Barcelona. It would be Manhattan. Manhattan – or something like it – is the greenest city on earth."[6] Is Manhattan really that green? We will come back to the concept of green later, but one point is clear. There will need to be fishing towns in Greece and many other places if the millions inhabiting Manhattan want to continue to eat fish!

In *Triumph of the City*, economist Edward Glaeser states that "New York State's per capita energy consumption is next to last in the country, which largely reflects public transit use in New York City."[7] Glaeser goes on to state "traditional cities have fewer carbon emissions because they don't require vast amounts of driving. Fewer than a third of New Yorkers drive to work, while 86 percent of American commuters drive. Twenty-nine percent of all the public-transportation commuters in American live in New York's five boroughs." But is less driving the reason New York State

---

[4] David Owen, Green Manhattan. *The New Yorker*, October 18 2004.

[5] http://autos.yahoo.com/photos/the-10-u-s-cities-with-the-worst-traffic.

[6] Elizabeth Farrelly. 2008 Blubberland: *The Dangers of Happyness*. The MIT Press, Cambridge MA. 218.

[7] Edward Glaeser. 2011. *The Triumph of the City*. Penguin Books, London, 338 p.

has low per capita carbon dioxide emissions? We will come back to this question shortly.

In the spirit of the efficient and green city, *Scientific American* magazine devoted an entire issue to the idea that cities are the future for mankind.[8] It featured a two-page spread (pp. 74–75) entitled The Efficient City. The picture is of a thoroughly modern high-density city with gleaming multistory buildings interspersed with (relatively small) green spaces and clean streets with little traffic. This city apparently doesn't have poor people as there is nothing that looks like a low-rent district. Included in the example are a number of "creative solutions to reduce energy consumption, water use, waste and emissions." Some of the features are solar power (photovoltaic, hot water heaters, and solar films), high efficiency windows, carbon sequestering concrete, green roofs, vertical farms (no need to depend on ecosystem services for this agriculture; we will come back to the idea of urban farming in a later chapter on food), hybrid taxis, underground and smart parking, irrigation systems controlled by satellites, sewage sludge incineration, efficient appliances, and storm-surge gates to protect low-lying parts of the city (especially important for New York and New Orleans, two cities featured in the next chapter). Some of these solutions, such as solar power, high efficiency windows, and efficient appliances, could be used for suburbs and even rural areas, but high-density living likely does foster lower *direct* energy use for transit and heating and cooling. However, almost all gasoline consumed in the U.S. is consumed in metropolitan areas (in the broader sense) because this is where most people live (see map below). In addition, there are large indirect energy and material demands that are not often accounted for in calculations of urban energy use. An example is all the energy and materials it takes to get gasoline to an urban consumer, whether for a car or bus, (searching for and drilling wells, pumping the oil to a refinery that converts the oil to gasoline, and transporting the gas to a network of gas stations) and, in fact, almost everything from building materials to artificial lighting to clothing to food that is used in a city is produced elsewhere, often at high energy costs. The same issues are at play on a global scale. When economists claim that the US economy as a whole has become less energy intense (defined as energy used per unit of gross domestic product or GDP), part of what they are really saying is that energy-intensive and highly polluting industries such as aluminum- and steel-making have been outsourced to regions and countries where labor is cheaper and environmental regulations are less stringent. However, our

---

[8] *Scientific American* 2011, ibid.

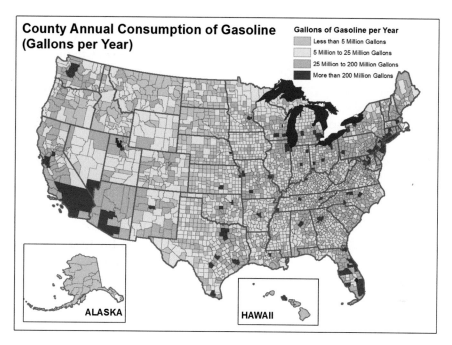

Fig. 3.1 Total gasoline use by county in the United States. (Source: National Resources Defense Counsel (NRDC). This map originally appeared on NRDC's Switchboard blog: http://switchboard.nrdc.org/blogs/dlovaas/CNTYgas.consumption.jpg)

need to consume these imported products, often shipped over long distances, only increases, and particularly in densely populated urban areas with the sophisticated infrastructure of "The Efficient City."

The red and orange areas on the map (Fig. 3.1) below showing high gasoline use would overlay well on the map of the population megaregions discussed in Chap. 2, showing that heavily populated areas burn more fuel. Any proposals to reduce gasoline use need to focus on metropolitan areas. Manhattan may have a high use of public transit, but what about the whole 21 million strong metro area in New York and northern New Jersey?

Let's come back to greenhouse gas emissions. The U.S. Energy Information Agency produced an inventory of direct carbon dioxide emissions by state from 2000 to 2010.[9] The information in this report is illuminating. If we look at energy-related per capita carbon dioxide emissions by state in 2010 there are some striking differences. At one end of the spectrum

[9] U.S. Energy Information Agency. *State-Level Energy-Related Carbon Dioxide Emissions, 2000–2011*. U.S. Department of Energy, Washington. August 2014. 16 p. http://www.eia. gov/environment/emissions/state/analysis/

are states with high CO_2 emissions per person. Kentucky, Montana, Louisiana, West Virginia, Alaska, and North Dakota all have emissions of more than 35 metric tons of CO_2 per person with Wyoming being the highest at 118 tons per person. At the other end are states with less than 15 tons per person. These include almost all of the northeast states as well as California, Idaho, Oregon, Washington, and Florida. New York is last among the states at 8.8 tons per person. Only the District of Columbia is lower at 5.4 tons per person.

If we look at how many tons of CO_2 are generated per million dollars of GDP, there are also some striking differences among states. Most of the states in the Northeast plus Washington, Oregon, and California produce less than 350 tons of CO_2 per million dollars of GDP. By contrast, Alaska, Kentucky, Louisiana, Montana, North Dakota, West Virginia, and Wyoming all produce more than one thousand tons CO_2 per million dollars GDP. Thus, the states that have low per capita direct emissions also produce less direct emissions per unit of GDP. And the states that have high per capita direct emissions also have high direct emissions for each million dollars of GDP.

Are there lessons that the folks in states like Louisiana, North Dakota, and Kentucky can learn from people in New York, Rhode Island, and Vermont about lowering their emissions and being more efficient? Yes and no.

Lower emissions are not necessarily due to the fact that people in New York State, Vermont, or Massachusetts drive less or have smaller houses that they have low per capita total emissions. And it's not necessarily because the average West Virginian, Louisianan, or Montanan personally uses huge quantities of energy or drives long distances that they have such high per capita total emissions. At least partially, it has to do with the nature of the economy in the different states and the particular area (central city versus suburbs) that is being considered.

Let's compare two states at different ends of the emissions spectrum, New York and Louisiana. On average, New York annually produces about 10 metric tons of CO_2 per capita compared to about 50 tons per year for Louisiana. But here's the most important difference in energy use between the two states: the industrial sector. Each year, the industrial sector in Louisiana produces 128 million metric tons of CO_2, compared to 9.1 million tons in New York. Texas is even higher with 211 million tons. This is because Texas and Louisiana have the highest concentration of refining and petrochemical industries in the nation.

But here's the key point. Very little of the energy and industrial products produced in Louisiana, Texas, and North Dakota and similar states are used personally by the citizens of these states. In addition to refining and petrochemical industries, Texas and Louisiana are major producers of oil and natural gas. Montana, Kentucky, and West Virginia produce a lot of coal. And Alaska and North Dakota produce large quantities of petroleum. All of these energy-intensive and polluting industries exist to meet demand for their products all over the country, but particularly in urban population centers since this is where most people live. This demand involves both direct and indirect uses, including fertilizers on the farm for high-yield agriculture that is needed to feed these urban populations. So, to be fair, most of the industrial CO_2 emissions should be removed from the balance sheets of Texas and Louisiana and other "energy" states (where citizens use relatively little of what is produced) and added to the emissions of the states that are end users, directly or indirectly, of the products of energy production and industrial activity.

In answer to the question, "are there actions that the states with high per capita emissions of CO_2 could take to lower them?" One option is to outsource energy intensive activities to other states or even to foreign countries and develop activities such as finance, tourism, or high tech. But this is not likely to happen. Oil, natural gas, and coal are mined where they exist. And the petrochemical and refining complexes in Louisiana and Texas are strategically located near production areas and international trade routes, and are aided and abetted by state regulatory agencies that are not as stringent as in other states. As long as we live in a modern industrial society, there will be a need for energy production, refining, and petrochemical industries somewhere that, by their very nature, generate high emissions to produce products consumed elsewhere. You can get rid of your emissions by foisting off the high emissions part of your life and economic support systems to other geographic areas, but is this green or efficient or ethical? As we noted, the industrialized first world has done this by exporting a lot of industrial activity to China and other developing countries with the result that China is now the largest greenhouse gas emitter in the world. But one can also make direct choices—such as driving less, eating less meat, and living in a smaller energy-efficient home—that do have a considerable impact on overall emissions. So to understand what activities and regions are producing more or less, one must consider both direct and indirect emissions. The result is that cities are the proverbial double-edged sword, but in the end high population density puts great pressure on all ecological support systems.

Christopher Jones and Daniel Kammen of the University of California at Berkeley carried out studies of carbon footprint and greenhouse gas emissions in urban areas across the country.[10] They analyzed, in detail, energy use for different U.S. zip codes, cities, counties, and metropolitan areas.

---

[10] Christopher Jones and Daniel Kammen. 2011. Quantifying carbon footprint reduction opportunities for U.S. households and communities. *Environmental Science and Technology*, 45, 4088–4095. Jones and Kammen. 2013. Spatial distribution of U.S. household carbon footprints reveals suburbanization undermines greenhouse gas benefits of urban population density. *Environmental Science & Technology*. doi:org/10.1021/es4034364.

They considered demand for energy, transportation, food, goods, and services. They explicitly considered both direct and indirect costs. So, as we discussed earlier, oil produced in Texas or corn grown in Iowa was apportioned to the final user. They found that direct emissions account for only 23 % of total emissions while indirect emissions account for 77 %. Therefore, most of those high industrial emissions in Louisiana, Texas, and similar states should be put in the balance sheet of people living in other states, like New York and New Jersey, who indirectly use them.

Jones and Kammen found that the factors explaining most of the differences in emissions among the different areas included number of vehicles, income, carbon intensity of electricity generation (coal vs hydropower, for example, but local people can't do anything about this), home size, number of people in the home, and population density. Income is the single most important contributing factor to household carbon footprint (poor people use less energy—direct plus indirect). The factors that would be most effective in lowering emissions are smaller homes, shorter driving distances, and most importantly lower incomes. They found striking differences in emissions, even on small spatial scales. In the Northeast megalopolis, dense urban cores had significantly less emissions than the sprawling suburbs around them. This was due to smaller homes and less driving, but also to more people with low incomes. They concluded: "As a policy measure to reduce GHG emissions, increasing density appears to have severe limitations and unexpected tradeoffs...Generally, we find no evidence for net GHG benefits of population density in urban cores or suburbs when considering effects on entire metropolitan areas." But as stated earlier, there are things that individuals can do by making personal choices in the way we live.

The maintenance of large urban megaregions requires enormous and continuous inputs of energy and materials most often transported over long distances. William Rees of the University of British Columbia argues that modern industrial society and modern cities are inherently unsustainable.[11] From an energy standpoint, he states "cities are self-organizing far-from-equilibrium dissipative structures whose self-organization is utterly dependent on access to abundant energy and material resources." What does this statement mean? It reflects the laws of thermodynamics, the fundamental energy laws that govern the universe. The first law of thermodynamics is the conservation of matter and energy. It states that energy and materials cannot be created or destroyed, only changed. Einstein's famous equation, $E = mc^2$, is a manifestation of this law, showing how energy and

---

[11] Rees, W. E. 2012. Cities as dissipative structures: Global change and the vulnerability of urban civilization. In: *Sustainability Science*, pp. 247–273. Springer New York.

matter are interchangeable. The second law of thermodynamics states that no process can be 100 % energy efficient. When energy is used to do work, part of the original energy must be lost as low grade heat that is not available to do further useful work. For example, when an air conditioner cools down a room, it must heat up the outside more than the inside it cools. This is the famous entropy law, and it is what physicists mean when they say that for any useful work to be done, there must be entropy generation. These basic laws constrain the functioning of cities (we will discuss these laws in more detail in Chap. 7 on energy). Portland, Oregon is an example of how high energy and material inputs conflict with the idea of "green". Portland is considered one of the "greenest" cities in the U.S., but has enormous inputs of energy and materials that generate large quantities of wastes.[12] Later, we will take a closer look at Portland and the idea of green.

Thus, cities are open systems that are dependent on the materially-closed biosphere. By this, we mean that enormous quantities of materials and energy flow into cities, while large amounts of waste flow out. But the biosphere—the earth in other words—stays the same size. It can only produce so much for cities, and humankind in general, and can only accept so much of its waste without becoming over taxed. Modern first world cities especially, like Portland and New York, are concentrated areas of material and energy consumption and waste production that are dependent on large areas of productive ecosystems and waste sinks most often located far from cities. In the chapter on ecosystem services, we will show that the "ecological footprint" of humanity now exceeds the ability of the biosphere to support it sustainably.

In a review of energy and material flows through the world's 25 largest urban areas, Ethan Decker and colleagues conclude that large urban areas are only weakly dependent on their local environment for energy and material, including food, inputs, but are generally constrained by their local areas for supplying water and absorbing wastes.[13] Los Angeles is an exception since it gets almost all of its water from hundreds of kilometers away, making it even more unsustainable than urban areas that obtain water locally. New York City produced 11,000 tons of solid waste per day in 2002, enough to fill 550 20-ton tractor-trailers stretching 9 miles end to end.[14] Rees contends that if cities are to be sustainable in the future, they must rebalance

---

[12] Burger, J. R., et al. 2012. The macroecology of sustainability. *PLoS Biol*, 10(6), e1001345.

[13] Decker, E., et al. 2000. Energy and material flow through the urban ecosystem. *Annual Review of Energy and the Environment*, 25, 685–740.

[14] Brown L. 2002. *New York: Garbage capital of the world*. Earth Policy Institute. http://www.earth-policy.org/plan_b_updates/2002/update10.

production and consumption, abandon growth, and re-localize. The trajectory of megatrends of the twenty-first century will make this difficult for all
large urban regions in the U.S. It will be impossible for some, in part
because of low-quality ecological services in the surrounding area. Open,
far-from-equilibrium systems have the capacity to radically re-organize in
unpredictable ways following disturbance. This implies that the future for
complex systems, such as large cities, is inherently unknowable in many
ways and that surprise is inevitable. People have choices: they can move
and/or reorganize their activities, but it is difficult to clearly see from the
present what the overall results will be. But in the end, large urban areas,
and the global economic system in general, require enormous inputs and
outputs to continue to exist as they are.

Some Conclusions and What's Next

Industrial age cities and the 11 major megaregions of the U.S. were able to
develop due to the availability of large concentrations of cheap concentrated energy in the form of fossil fuels, electricity, and abundant natural
resources. In 2009, about 85 % of global primary energy production was
based on fossil fuel combustion. Civilization crossed the threshold of
greater than 50 % reliance on fossil fuels for energy production at some
point in the late 1890s.[15] Modern large urban areas could not have developed without abundant and cheap supplies of fossil energies.

All major urban areas generally have similar patterns in the consumption of energy and materials at a large scale, but as Jones and Kammen
found, there can be striking differences even within individual urban areas,
depending on individual behavior and income level. Prior to the industrial
revolution, most cities were largely rooted in local wealth and natural
resources that helped define local culture, as was the case for colonial cities
in America. Cities functioned as nodes of regional commerce where merchants and traders could meet and trade goods that were produced mainly
in the local region. The advent of modern energy-intensive transportation
allowed the spread of trade outside regional areas, becoming truly global.
Climate controlled storage combined with cheap delivery of goods to far
flung markets allowed competitors to establish businesses far away from
what were previously regional markets defined by production limits of the

---

[15] Smil, V. 2011. Harvesting the biosphere: the human impact. *Population and Development Review*, 613–636.

local labor pool, climate, and availability of resources. For example, without cheap fossil fuels that subsidize modern industrial agriculture and allow for cheap air, rail, and truck transportation, it would not be possible for an urban dweller in the northeast U.S. to eat fresh produce from South America or even from other states in the U.S. These factors have further defined the lifestyles of many urban dwellers throughout the world, and especially in the first world.

The importance of understanding the energy and resource requirements of the urban environment in relation to the outlying rural environment cannot be overemphasized. Before the industrial revolution, for most cities, the region was the city's territory and the city was the region's center.[16] Products increased in value as they flowed to the city, but some of the wastes of the city were recycled to the surrounding rural area, and a relative balance was achieved. An example of this is the grazing system that existed in the heather lands of the central Netherlands in the Middle Ages.[17] Sheep grazed on heather growing on sandy soils. At night, shepherds herded the sheep into pens on the city fringe. Farmers collected sheep droppings and used them to fertilize farm plots surrounding the city. This produced an abundance of food that helped the city to prosper. The system was sustainable until population growth led to overgrazing and deterioration of the heather lands.

Upon closer analysis, the argument that large cities are more sustainable and less energy intensive does not necessarily hold when one broadens the boundaries of the analysis to look beyond the limits of the individual or city to incorporate the materials and energy input demands and outputs of people living across the landscape. Some urban dwellers do consume less because they live in smaller houses or drive less, and others because they have low incomes. But in aggregate, large cities consume high levels of energy and materials. These needs are mostly met by outsourcing energy and materials production demands to other places. Integrated transportation networks allow access to resources produced outside of the city. The population of New York City probably could not feed itself for a single day based on the amount of "green" land within its boundaries. Inputs must come from far outside the city boundaries. Household income strongly correlates with embodied energy and material use.[18] The goods and services

---

[16] Howard T. Odum. 2007. *Environment, power, and society for the twenty-first century: the hierarchy of energy*. Columbia University Press.

[17] Webb, N.R. 1998. The traditional management of European heathlands. *Journal of Applied Ecology*, 35, 987–990.

[18] Weisz, H., & Steinberger, J. K. 2010. Reducing energy and materials flows in cities. *Current Opinion in Environmental Sustainability*, 2, 185–192.

people consume, directly and indirectly, are just as important for the wealthy living in the urban or rural environment when it comes to carbon emissions.[19] Financially poor people consume less than the financially well off, and as a result, have total smaller energy demands and CO_2 emissions. According to the *New York Class Index*,[20] the per capita income in Manhattan is $93,377 while that of the Bronx is $23,513. So greenhouse gas emissions of the average inhabitant of Manhattan is nearly four times that of the average Bronx resident.

In the next chapter, we consider in more detail 12 characteristic cities in 10 regions of the U.S. to develop a more detailed understanding of what fueled manifest destiny and what the future portends for different areas of the country.

---

[19] Heinonen, J., & Junnila, S. 2011. Implications of urban structure on carbon consumption in metropolitan areas. *Environmental Research Letters*, 6(1), 014018.

[20] http://www.thelmagazine.com/2008/03/new-york-class-index/.

Chapter 4

A Tale of Twelve Cities and Ten Regions

In the preceding chapter, we discussed the growth of the United States from 1790 to 2010 as the population expanded and spread across the continent. People did not diffuse evenly across the land. Population centers developed in resource-rich areas near the coast and along waterways where farming first dominated the landscape and along trade routes such as the Mississippi River. Changes in both rural and urban economic opportunities resulted in the movement of people from rural to urban areas. "The flow of labor off U.S. farms during 1940–85 was driven by an increase in economic opportunity in cities relative to what existed on farms, indicating the dominant role of economic incentives."[1] At the same time, mechanization of farming reduced the number of jobs needed on farms.

As you read on, one image to keep in mind was provided by Howard Odum in his book *Environment, Power, and Society*.[2] Odum showed daily inputs of solar energy, what scientists call "flows," resulting in a field of vegetables, a familiar scene whether the crop is wheat, corn, or cabbages. In contrast, a tanker at a terminal inputs oil, really an ancient storage of solar energy, which is used to power vehicles, heat buildings and grow roads and skyscrapers instead of cabbages and tomatoes, and which—unlike solar energy—once used is not replenished. Of course, building materials and cranes are only the beginning, as people have to be moved large vertical

---

[1] Barkley, A. P. 1990. The determinants of the migration of labor out of agriculture in the United States, 1940–85. *American Journal of Agricultural Economics*, 72(3), 567–573.

[2] Howard T. Odum, 1971. *Environment, Power, and Society*. Wiley-Interscience, John Wiley and Sons, New York.

© Springer Science+Business Media New York 2016

J.W. Day, C. Hall, *America's Most Sustainable Cities and Regions*,

DOI 10.1007/978-1-4939-3243-6_4

distances every day and artificial light must illuminate this newly available space. Daily flows of solar energy could make the lights shine, but oil is the main form of energy with the necessary properties to fuel most of these now familiar activities. In fact, the growth of the economies of the US and indeed all the world over the last two centuries is essentially the same exponential growth pattern as the use of fossil fuels—primarily oil but also natural gas and coal.

As Odum knew and we will see later, oil and gas, in the form of fertilizers and pesticides, is also needed to grow enough wheat, corn, and cabbages to feed the concentrated urban population living and working in the high-rise buildings. In this chapter we visit 12 cities in 10 regions to look more carefully at the interdependencies of the social and physical phenomena of urbanization (Fig. 4.1). Later in the book, we will come back to these regions and dependencies to discuss how each area may fare during the rest of the twenty-first century. These particular cities and regions were chosen to capture the broad range of urban and regional conditions that

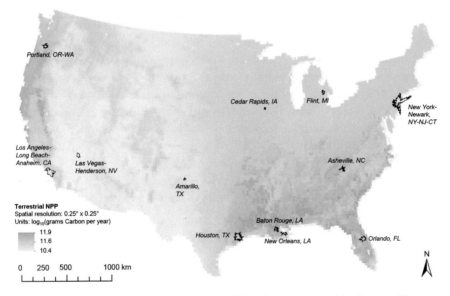

Fig. 4.1 The 12 cities and 10 regions selected for discussion in this chapter. The outlines for each city show the urban boundaries. *Color* indicates terrestrial net primary production (*NPP*), a measure of how naturally productive different regions are (*darker green* = more productive). This is examined in more detail in Chap. 5 on the value of natural ecosystems (Data distributed by the Socioeconomic Data and Applications Center (SEDAC). http://sedac.ciesin.columbia.edu/es/hanpp.html.; Imhoff ML and Bounoua L. 2006. Exploring global patterns of net primary production carbon supply and demand using satellite observations and statistical data. *J Geophys Res* **111**: D22S12)

exist across the U.S. Many of them are immediately recognizable as characteristic urban icons of the American landscape (e.g., New York City and Los Angeles). Others are less well known (e.g., Cedar Rapids, Iowa and Amarillo, Texas) but represent characteristic city and regional types. What we want to discuss in this and later chapters is not so much the iconic images, but rather how the cities and regions got started and developed, and what it takes, and will take, to sustain these areas. In the end, the availability of energy and other resource constrains what feeds the urban metabolism. In this context, we will consider the limits of urban efficiency and self sufficiency, at least on the scale of the modern metropolis.

Cities are found throughout the world in many different configurations and for many different reasons. In general, cities are found where there is an economic reason for their existence. Many of these reasons are energy related (i.e. fertility, cheap transportation, resources). These economic reasons involve many factors: availability of local resources, cheap transportation, concentration of various skills and so on. In general we may consider cities as locations of intense consumption of food, energy, resources surrounded by large areas of production. Cities on a landscape show patterns similar to natural ecosystems that correspondingly tend to have areas of intense metabolism surrounded by large areas of lower-level production.[3] Thus cities tend to be spread out at intervals within a matrix of forests, agriculture, and other natural systems. We will come back to these ideas in a discussion of one more city, Syracuse New York, in Chap. 7.

New York City, New York: The Capital City of the World

New York City is perhaps the most widely recognized city in the U.S. and could be considered, in many senses, as the capital city of the Western world. Many iconic images of the U.S. depict New York City, including its skyline, the Statue of Liberty, Wall Street, and other similar attractions. It is at the center, figuratively and literally, of the East Coast megalopolis that contains almost 60 million people, or nearly a sixth of the U.S. population. This region exerts enormous influence on the U.S. and consumes huge amounts of energy and resources. For many people, New York City is a monument to human creativity, with its immense wealth, massive skyscrapers, Broadway hits, famous museums, and glitzy, trendsetting lifestyles. But what other factors brought the city to its current exalted status, and what will it take to maintain its status now and in the future? (Fig. 4.2).

---

[3] Odum, 1971, ibid.

Fig. 4.2 A night view of Manhattan with the Empire State Building at center (Photograph by: Javiar Gil, http://commons.wikimedia.org/wiki/File:Vista_aérea_de_Times_Square_desde_el_Empire_State_Building.jpg)

Population

According to the U.S. Census Bureau, the five boroughs of New York City had a population of 8,175,133 in 2010. The City Planning Department estimates that by 2030, the population will increase by 13.9 % to just over 9.1 million. The Big Apple is the most populous U.S. city with approximately 4.3 million more people than Los Angeles and 5.4 million more than Chicago (U.S. Census Bureau 2000). At the same time, the population of the New York–Northern New Jersey–Long Island Metropolitan Statistical Area (MSA) was estimated at over 18.9 million people. The so-called Combined Statistical Area (CSA)[4] around New York City included 22.2 million people in 2011 according to the U.S. Census Bureau. New York City is part of the Northeast megaregion or megalopolis, stretching from north of Boston to

---

[4] The Metropolitan Statistical Area (MSA) includes New York-Northern New Jersey-Long Island while the Combined Statistical Area contains almost two times the land area and is known as the New York-Newark-Bridgeport CSA.

south of Washington D.C., which included approximately 52.3 million people in 2000 (projected to approach 60 million by 2025), and produced 20 % of the nation's GDP on 2 % of the nation's land area.

Population growth has been a major factor in the expansion of New York City since 1700, when there were approximately 5000 inhabitants. It took until the middle of the eighteenth century for the population to double to 10,000, but only 20 years for the next doubling to 20,000. By 1800, the population reached over 60,000, and by 1850 the five boroughs comprising New York City had a population of well over a half million people. Large-scale immigration from Europe was largely the cause of the exponential increase in population throughout the nineteenth century as the population swelled to 3.4 million people by 1900. The population had doubled again by the time the stock market crashed in 1929. The growth slowed throughout the latter part of the century, adding only approximately 1.25 million more residents to the 8,175,133 five-borough total at the time of the U.S. Census in 2010.

History

Initially the location of NYC was a consequence of its abundant resources (clean flowing rivers, beavers, fish and shellfish, including 350 square miles of oyster reefs and cheap water-borne transportation), but over time it evolved into one of the world's largest financial centers. The growth and development of NYC is a history of the economic evolution of capital, centered around trade, commerce, and the exploitation of the resources of the new world. It is also one of the cultural capitals of the world. It is widely believed that the first European to explore the region was Estevao Gomez in the spring of 1524. Gomez was a Portuguese cartographer and explorer working for the Spanish crown who had previously been a part of Ferdinand Magellan's fleet. Gomez deserted Magellan and was incarcerated for his actions upon his return to Spain. He was subsequently released and then set out in search of the fabled Northwest Passage. That expedition, which included a stint from Nova Scotia and south, produced the most detailed map of the east coast of North America at the time. During the expedition, he likely entered New York Harbor and spied the Hudson River. However, it was not until Henry Hudson established the first permanent European settlement in 1609 that New York City began to grow into a major Dutch trading center known as New Amsterdam. New Amsterdam was the capital of the Dutch-ruled New Netherlands that included Fort Amsterdam, a fortified area located strategically on the southern tip of lower Manhattan. The abundant beaver population, noted by Hudson in his initial journey, formed the foundation of an economy, protected by Fort Amsterdam, that would grow to be the largest city economy in the U.S., and second largest in the world behind Tokyo. New York is also surrounded by a rich coastal ecosystem with large areas of oyster reefs, abundant shad, sturgeon, striped bass and many other fish species.

Approximately 5000 Lenape natives living in 80 settlements inhabited the NYC area when the first European settlers arrived. The Lenape were a Native American culture that settled mostly along the northeastern seaboard from the Hudson River down to the Delaware River. They had inhabited the region for approximately 11,000 years by the time the Europeans arrived. These native settlers survived largely off of the abundant resources provided by the lush river valleys and coastal systems of the region. To survive, the Lenape hunted wild game, gathered roots and vegetables, fished, and employed early agricultural techniques. They used slash and burn agriculture to grow maize as a primary food source. The population migrated seasonally along the northeastern seaboard throughout the year. In addition to fishing and hunting for food, the Lenape were skilled at trapping beavers for clothing and trade. The high density of waterways in the region provided prime habitat for beavers. The Dutch East India Company explorers duly noted this fact after hearing directly from Henry Hudson that a supply to meet the growing demand for fur in Europe was available in the New World. Fort Amsterdam and the area now known as Lower Manhattan are located on the southern tip of an island that served as a natural harbor at the confluence of three major waterways: the Hudson River, New York Bay, and the East River. The Lenape people quickly developed trade relationships with the Dutch settlers whereby they traded beaver pelts for European made goods, such as farming implements. However, supplying the insatiable demand of the European fur markets led to overharvesting and a drastic decline in beaver pelt exports. Soon after, another fashion accessory known as wampum became the major trade export for the Lenape people living around Manhattan Island. Throughout this time more European colonists moved into the region, exposing the Lenape peoples to a host of new diseases, including smallpox and measles, for which their native immune systems were ill adapted. Population declines occurred as a result, and the colonists began to take over traditional tribal lands. Thus after living in the area for over ten millennia, the Lenape faded away in less than century.

The Dutch handed over New Amsterdam to the British without a fight and the British developed it into a major port city throughout the 1700s. The slave trade was a major part of that growth. Slaves worked in the ports and shipping industry, and nearly half of the residents in the city had slaves working in their homes. New York City had the second largest slave population of any city in the nation by 1730; second only to Charleston. Throughout the eighteenth century, New York City was a hub of activity that included seven major yellow fever epidemics, the largest battle of the American Revolutionary War and a Great Fire. It became the capital of the newly formed government of the United States, just as its population grew to become the largest city in the U.S., surpassing Philadelphia in the last decade of the century. It has remained the largest city up until the present.

Commercial trade in the port of New York ramped up in the 1800s as the Erie Canal made the interior of the United States accessible to Atlantic-bound ships. Surplus agricultural products could be shipped for little additional energy from the Great Lakes region through New York City via the Hudson River and onward to global markets. In return, the Midwestern markets opened to the machinery and manufactured goods produced in the east. The Erie Canal supplanted the draft animal drawn carriages as the main mode of trade transport for the region and greatly decreased transport costs, and allowed much faster trade with interior western states. The Canal also functioned as a corridor for immigrants traveling west from New York City after arriving from overseas.

By mid-century, the immigrant population in the city had increased dramatically as a result of the Irish potato famine and an influx of German immigrants. During the second half of the nineteenth century, New York City experienced the largest civil insurrection in American history when working class men refused to be drafted to fight in the ongoing Civil War. Many were longshoremen, who were paranoid that newly freed slaves would compete for their jobs in the port. They soon turned their initial anger against the federal government toward the local free black population. The New York City draft riots, as they are commonly known, were depicted in a 1928 book entitled *Gangs of New York*, and later made into a popular motion picture in 2002. The riots forever changed the racial demographics of the city as many of the black population were displaced from housing in Manhattan and moved into Brooklyn and New Jersey.

The political demographics of the city also changed as a result of the Civil War. Prior to the war, a formidable Democratic Party controlled the city. It was aligned with the commercial interests of the Confederacy. By 1820, cotton produced in the South composed half of the exports from the New York port. At the outset of the Civil War in 1861, the mayor of New York City, Fernando Wood, announced his desire for secession along with the southern states. Much of the mayor's rationale was based on economic interests, whereby southern cotton growers were financed by New York banks, protected from loss by New York insurers, and their harvest transported in New York ships. However, the political tides would turn as finance and manufacturing industries eventually backed the Union cause. The growing banking sector financed the Union army and the industrial sector produced the military supplies necessary to defeat the Confederates. At the time, the industrial capacity of New York State was greater than that of all of the southern states combined.[5]

By 1898, New York City was officially consolidated into the five boroughs of The Bronx, Brooklyn, Manhattan, Queens, and Staten Island. The first New York subway opened in 1904, carrying passengers across the city at 40 mph, or seven times faster than a horse. Grand Central Station and Penn Station allowed for further public transport options for workers, resulting in the expansion of the metro area into bedroom communities along the outskirts of the city and surrounding region. Transport options grew further as the personal automobile arrived on the roadways and bridges of the city. Automobiles became a centerpiece of New York City urban planning and growth under the direction of Robert Moses, the "master builder" of the city, who developed many bridges, parkways, and expressways during and after the Great Depression in the 1930s. World War II weighed heavy on the population and the economics of trade in New York City as German U-boats attacked ships leaving New York Harbor. City leaders decided to dim the lights at night throughout the five boroughs for 18 months, so as to prevent attack from the air and surrounding waterways. However, the lights shone bright again across the city as New York emerged from WWII as one of the leading cities of the world, as evidenced by trade on Wall Street and the establishment of the United Nations on Manhattan's East Side.

The nature of the New York City economy changed after the WW II as the industrial-based economy of large shipbuilding and garment manufacturing declined and the service sectors of finance, education, medicine, tourism, communications, and law increasingly dominated the economy. NYC of the 1960s was the cultural hot bed of the east coast, as beats blended with folkies in the coffee shops of Greenwich Village and the 1964 World's Fair brought attention to NYC as a major international city. However, the municipal spending spree that occurred in the 1960s and through the early 1970s led to a severe economic crisis for the City in 1975, as the banks and bond markets that lent the City the money previously cut off the flow. This economic crisis was related to the first oil crisis of 1973. The City went through at least 3 years of economic woes that were multiplied by white flight to the suburbs and the movement of several major economic players

---

[5] Sam Roberts. "New York doesn't care to remember the Civil War" *New York Times*. December 26, 2010.

outside the City limits. Municipal service bills went unpaid, construction projects were abandoned, and City streets and sidewalks fell into disrepair as the City faced bankruptcy. The grittiness of NYC—with people living in a high density, low wage environment offering little material support or disposable income—began to show. Once 65,000 City workers were laid off, services had been cut, and taxes raised, the City was finally bailed out by the federal government. The socio-economic woes reached their nadir during the Blackout of 1977.

The New York City area has experienced several disturbances over the course of the last half-century. Besides the terrorist attacks of 9/11, large-scale disturbances generally occurred in the form of electrical grid failures lasting less than 24 h. The incidents created considerable disruption in the normal pattern of life in the city that never sleeps, and it is difficult to imagine what might ensue if such a disruption were more prolonged. New York City experienced electricity blackouts in 1965, 1977, and 2003. The blackouts in 1965 and 2003 were northeast regional blackouts that affected a wider swath of the region while some neighborhoods of the City remained illuminated. The Blackout of 1977 was more localized and led to a great deal of looting, vandalism, and arson throughout the neighborhoods of the City. Things came under control again once over 1000 fires were extinguished, 550 police officers were injured, and well over 3500 looters were arrested. The incident is depicted in a BBC documentary from 2007 entitled "Nightmare in the City that Never Sleeps". Not only NYC but also the nation as a whole experienced an economic downturn during this time.

The hangover from the 1970s and subsequent recovery stretched well into the 1980s as unemployment and crime rose. The stock market took a dive in 1987, and the loss of tens of thousands of Wall Street jobs contributed to the general employment woes of the City. In the 1990s, NYC experienced a rebirth as many of the former industrial neighborhoods turned to high end residential in a process called "gentrification," as young professionals moved into the urban core again. The real estate market grew, and Wall St. experienced a resurgence, especially as a result of the dot com bubble. Since the terrorist attacks of 9/11, large-scale redevelopment of sections of lower Manhattan has occurred, while the financial markets have withstood, to some degree, another major shock as a result of the 2008 housing mortgage crisis. The U.S. economy has not fully recovered by a number of measures, but Wall St. and the world's largest stock exchange (NYSE) as measured by market capitalization, still turns out millionaires and billionaires while the New York City economy remains buoyed by finance, insurance, health care, and real estate. New York also serves as the largest mass media center in the nation. The financial services sector accounts for at least 35 % of the City's employment income.

In addition to the financial services sector, NYC continues to grow as an international commercial trade hub. China remains the major growing export market for trade. Foreign capital shows a significant presence in the economy. A growing number of foreign-owned businesses make their headquarters in NYC, and many have been investing heavily in the City's real estate market. However, Hurricane Sandy in 2012 starkly showed the city's vulnerability to nature, as Katrina did to New Orleans. Was Sandy an extremely rare event or a harbinger of the future? Was the Great Recession linked to the growing scarcity of cheap oil and also a sign of things to come?[6] We will look into these questions in subsequent chapters.

New York's success depends on the ability of people and natural systems the world over to continue shipping their raw material, energy, and wealth to the city in exchange for the services and intellectual capital that the city provides. When we revisit New York, we will consider how the city's economy might respond to changing economic conditions, and what difficulties may be encountered in maintaining the imports of energy, materials, and money that sustain the city. We will also consider if climate change is likely to lead to more events like Sandy (Fig. 4.3).

Flint, Michigan: Rust Belt Return to Nature

Flint, Michigan, is an example of a post-industrial, depopulating city of a type common across the Northeast and Midwest "rust belt". The population of Flint since 1860 illustrates this pattern of boom and bust. In 1860, the population stood at 2950 and grew to just over 13,000 by 1900. Between 1900 and 1930, the number of people living in Flint grew to over 156,000, increasing by 194 % in the first decade of the twentieth century, and by 137 % from 1910 to 1920. There was a slight population decrease in the 1930s, likely associated with the Great Depression. Growth resumed in the post war years, reaching nearly 197,000 in 1960 when Flint was the second largest city in Michigan. The 1950s and 1960s were the city's most prosperous years. After 1960, the population began to decrease, reaching about 102,000 by 2010, with decreases of nearly 20 % in some decades. In 2011, the population stood at just over half of what it was in 1960. Other "legacy" rust

---

[6] James D. Hamilton, professor of economics at University of California San Diego, has written extensively on the influence of oil shocks on the economy. For a list of his publications see the following link: http://econweb.ucsd.edu/~jhamilton/#publications.

Fig. 4.3 New York produces about 11,000 t of solid waste per day (http://www.dailymail.co.uk/news/article-2003251/Top-10-Americas-dirtiest-cities-New-York-Miami-Las-Vegas.html; Image Source: "42nd Street" by Ralph Hockens, [CC by 2.0] https://creativecommons.org/licenses/by/2.0/legalcode)

belt cities in the region suffered similar fates.[7] For example, the population of Detroit fell by more than half, from about 1,670,000 to 714,777, between 1960 and 2010 (Figs. 4.4 and 4.5).

History

Flint was founded by Jacob Smith in 1819 as an outpost for fur trading with the local Ojibwa tribes. The town was also a stopover on the land route between Detroit and Saginaw. The village was incorporated in 1855. Flint and other towns in southern Michigan, such as Detroit, prospered as agricultural communities based on the rich soils of the area. In the second half of the nineteenth century, Flint was an important town for the lumber

---

[7] Jordan Rappaport. *U.S. Urban Decline and Growth, 1950 to 2000.* Federal Reserve Bank of Kansas City. 44 p. www.kc.frb.org.

Fig. 4.4 Flint River in Flint, MI. http://commons.wikimedia.org/wiki/File:Flint_ River_in_Flint_MIchigan.jpg (By U.S. Army Corps of Engineers, [Public domain], via Wikimedia Commons)

Fig. 4.5 Flint's 235-acre complex known as Buick City was in operation from 1904 until 1999. The facility was recently demolished. According to Wikipedia (http:// commons.wikimedia.org/wiki/File:Buickcityflint)   (Photograph   by   user Blueskiesfalling (Own work) [Public domain], via Wikimedia Commons)

industry based on exploitation of extensive old growth forests of the region. Growth of the lumber commerce resulted in the development of a carriage-making industry. As horse-drawn transport gave way to vehicles powered by the internal combustion engine, Flint developed as one of the most important locations for the growing auto industry. It was the birthplace of General Motors as both Buick and Chevrolet established plants in Flint. The AC Spark Plug Company also originated in the city. The founding of the United Auto Workers union grew out of the Flint Sit-Down strike in 1936. During the Second World War, the city prospered as it manufactured tanks and other war machines. Up until the mid twentieth century, Flint's prosperity paralleled the explosive growth of the auto industry as the U.S. transformed into an automobile-dominated society. Flint benefited from a regional supply base that included steel mills in Pennsylvania, coal mines in Kentucky and West Virginia, and iron ore from Minnesota.

Flint's fortunes reversed after the 1960s as a result of a number of major social and economic trends. Central to these was increasing prices of oil and the development of more fuel-efficient vehicles by the Japanese. The collapse of the US auto industry and the general deindustrialization of the northeast and Midwest led to the dramatic loss of population and urban decay of the city. Employment by General Motors fell from about 80,000 in the late 1970s to less than 8000 by 2010. The plight of Flint was vividly described in Michael Moore's documentary film *Roger & Me*.

The economic crisis led to a series of financial emergencies and political upheavals as the city continued to decline. This resulted in the state of Michigan taking over much of the city government operations. Flint now stands as a city fundamentally changed from its heyday of the 1960s. The population has been reduced by half and it is unlikely that the city will ever regain its industrial prowess. Much of the city and residential neighborhoods have been abandoned, and blighted housing is widespread. In 2002, the county treasurer Dan Kildee founded the Genesee County Land Bank. The purpose of the land bank was to take charge of dilapidated and abandoned properties, making the properties available for resale or demolition. In 2009, the land bank owned 14 % of property parcels in the city, much of which is open space. Detroit has suffered a similar fate. In 2009, of Detroit's 139 mile^2, 40 mile^2 were vacant. This could rise to 80 mile^2 if population stabilizes at about 600,000 as projected. There are active plans in both Flint and Detroit to develop urban agriculture on vacant land. "Urban farms" from a few acres to several hundred acres have sprung up in both cities with vegetables, fruit trees, chickens and eggs (Fig. 4.6). Detroit is trying to get a food processing facility established in the city. Thus, in the face of pervasive urban decay and collapse, these cities may be able to produce a significant amount of food. This is potentially important if industrial agriculture begins to falter because of high energy prices and climate change.

Fig. 4.6 Photo of a typical urban garden (http://commons.wikimedia.org/wiki/ File:North_view_of_a_Chicago_urban_garden.jpg, By Linda from Chicago, USA [CC BY 2.0] (http://creativecommons.org/licenses/by/2.0)], via Wikimedia Commons)

Flint, Detroit, and other similar cities serve as examples of both the perils and possibilities of post-industrial cities in the Midwest and elsewhere. In many ways, Flint and similar cities are on the leading edge of change that will come to most urban areas as the mega trends of the twenty-first century sweep across the landscape. We will revisit this later in the book.

Asheville, North Carolina: Blue Ridge Chic

Asheville is a city nestled in the Blue Ridge Mountains of western North Carolina at the confluence of the French Broad and Swannanoa Rivers, waterways that provided avenues for trade (Fig. 4.7). The population was estimated to be 84,458 in 2011. The humid sub-temperate climate it shares with much of the rest of the southeastern region is tempered by its higher elevation. The city has predominantly served as a resort town with a tourism-based economy for most of its modern history. Much of the present

Fig. 4.7 Biltmore Estate is a popular tourist destination in Asheville and is a reflection of the attractiveness of the city as a place to live (http://commons.wikimedia.org/wiki/File:Biltmore_Estate-27527-2.jpg, By Ken Thomas (KenThomas.us (personal website of photographer)) [Public domain], via Wikimedia Commons)

downtown area consists of the sorts of places where discretionary income is spent: trendy bars and restaurants, antique and vintage stores, and other attractions for younger and older folks alike with surplus income who frequent such establishments. Bluegrass music, biscuits and gravy, and microbrews are common fare on the menu. Asheville is also an attractive area for retirees seeking moderate temperatures throughout the year, a beautiful natural environment, a location outside of the coastal zone where real estate investments are less likely to be devastated by hurricane activity, and a progressive outlook. However, there are times when the river rises, generally after a large storm makes its way inland, and flooding occurs through the river valleys and across the adjacent floodplain properties.

Population

The population of Asheville has grown slowly since Hernando DeSoto became the first European to explore the region in 1540. Approximately 1000 non-native American settlers were living in the Asheville region in 1790, and there were 1400 people living in the city at the time of the 1870 Census. The population grew to over 14,000 by the turn of the twentieth century and doubled to over 28,000 by 1920. The decade of the 1920s saw the largest increase in total population for any decade since Asheville became a city. More than 50,000 people were reported to be living in Asheville by 1930. The population increased to 60,192 by 1960, before declining to 54,022 in 1980. Over the next 30 years, the population climbed to an estimated 83,393.

History

The native inhabitants of the area were Cherokee Indians whose trade routes crossed at the site of present day Asheville, likely because of the confluence of the two rivers. The Cherokee nation remained largely intact in the region until late in the eighteenth century, as British colonists had not expanded their territories into the Blue Ridge Mountains by the time of the Revolutionary War in 1776. The natural geographic boundary of the mountains allowed the Cherokees to maintain their culture and traditional lifestyles for longer than many tribes in the east that were devastated and relocated by colonists much earlier on. However, the relative peace and separation between the colonists and natives did not last long. Battles between the two sides raged until the majority of the native population were killed off by disease or forcibly relocated to Oklahoma during the Trail of Tears in 1830.

Many of the early European settlers of the region were Scotch–Irish immigrants skilled in the wool and textile trades who had fled Northern Ireland due to the high tariffs placed upon their goods by the British Crown. Major William Davidson and Col. Samuel Vance established the first permanent settlement in the newly named Buncombe County in 1792. The settlement occupied a well-drained plateau at the crossing of the two old Native American trails. The original city plan of 1793 consisted of 42 half-acre lots that hit the market for a whopping $2.50 apiece. Initially, the Buncombe county seat was called "Morristown", until a name change occurred in honor of Governor Samuel Ashe in 1797.

The early Native American trading routes became the first roads used by settlers in the region. Travel was difficult in the area due to the surrounding rivers and the mountainous terrain. River crossings were difficult during high water events, so a ferry system was employed using canoes and flat boats to shuttle travelers and their belongings across. Asheville remained a secluded mountain settlement until the Buncombe Turnpike, a road that paralleled the French Broad River running west into eastern Tennessee, was established and the flow of goods and people from the west increased considerably. Farmers traveled along the road with their livestock on their way to market. Hogs were a major source of revenue for local growers, and Buncombe County's early commercial success was largely a result of the hog industry. Hogs were raised on the corn grown in the region, and they were much easier to transport along the road than sacks of corn since they had their own four legs. It is estimated that between 150,000 and 175,000 hogs passed through Asheville each year during the last 3 months of the year. Many of the first tourists to come to the region arrived via stagecoach on the Turnpike. They would often stop over at one of the many hotels that grew up along the commercial corridor in Asheville. From there, they could travel west to Nashville, Tennessee and thence to cities throughout the region.

The Civil War directly affected Asheville and its commerce. The city was the scene of a Confederate battle victory, but the negative impacts of the larger war on the surrounding agricultural communities led to the disappearance of much of the commercial traffic that had allowed the city to grow during the first half of the nineteenth century. The city and local economy were mired in an economic depression for many years after the Civil War. The Turnpike was turned over to the North Carolina Railroad Company and many of the agricultural commodities that once walked the roads were increasingly being shipped on train cars since it was faster and the animals lost less weight on the journey. The railroad brought many new settlers to Western North Carolina from the 1880s onward, and the area became known popularly as the "Land of the Sky".

It was during the last two decades of the nineteenth century that George Vanderbilt, a young aristocrat whose family had made their fortune in shipping and railroads, moved to Asheville and constructed the largest home in the United States, the Biltmore House (Fig. 4.8). The 250 room house remains the largest privately held home in the U.S. and the land his widow sold to the government became Pisgah National Forest. The extensive park-like areas around the estate were designed by Frederick Law Olmsted, and the forests were managed by Gifford Pinchot, who would go on to become the first chief of the U.S. Forest Service. The Biltmore Estate endures as one of Asheville's major tourist attractions and employers with over a million visitors a year.

The Asheville economy improved in the early 1900s as city business leaders launched a national advertising campaign claiming it to be one of the "leading convention cities in the country". The city built a convention auditorium and a grand opera house that attracted visitors from across the nation. Several luxury resorts were also constructed in the area around the same time, and downtown architecture benefited from architects working in the Art Deco style. Many of the municipal buildings in use today were constructed during the 1920s. The climate and natural beauty of the region

Fig. 4.8 Asheville in 1854 (*Source*: F.A. Sondley, *Asheville and Buncombe County* (Asheville: The Citizen Company, 1922), p. 121. Author: C. H. G. F. Loehr (drawing) http://commons.wikimedia.org/wiki/File:Asheville-loehr-1854-nc1.jpg   [Public domain] Via Wikimedia Commons)

also attracted visitors seeking remedies to respiratory ailments including tuberculosis. It was not uncommon for doctors to recommend the mountain city to their sick patients as a refuge for health and healing. In addition to the resorts, there were numerous rooming houses that catered to the less well affluent who came in the summer to escape the lowland heat. The mother of Thomas Wolfe, the noted American novelist, operated a rooming house. Wolfe's semi-autobiographical novel, *Look Homeward Angel*, chronicles life in Asheville in the early decades of the twentieth century. Literary buffs can still visit Mrs. Wolfe's house. In the summer of 1916, two hurricanes, one moving northward from the Gulf Coast, and the other having come ashore at Charleston, SC, combined with a series of heavy rains to inundate parts of the city, killing at least 80 people and causing approximately $22 million in damage.

When the stock market crashed, Asheville had the highest per capita debt of any city in the country. The Great Depression hit the region hard. It was not until 1976 that the city was able to pay back the outstanding debts incurred as a result of the 1920s spending spree and subsequent market crash. The architectural styles on display throughout Asheville today have survived largely as a result of the prolonged debt repayment period and the lack of "urban renewal" that many similar sized cities experienced after WWII. During the Depression, many local laborers joined the Civilian Conservation Corps for employment. They went on to build infrastructure in Great Smokey Mountains National Park and constructed the Blue Ridge Parkway. These recreation hot spots remain a hub of activity throughout the year, and many of the towns in the vicinity are heavily dependent on the surge in seasonal tourist traffic they experience during the warmer half of the year. Asheville has seen a resurgence since the 1990s, as the popular appeal of a quaint downtown dominated by independent businesses located in a unique mountain environment resonates with the young and old alike. The city is home to the University of North Carolina–Asheville. Several small private colleges are also located nearby, adding to the pool of young energy that helps to fuel the local economy. The National Climatic Data Center is located in Asheville, which is the world's largest active archive of weather data. The largest employer in the city is the hospital, followed by the local county school system, and the local grocery store chain. Asheville experienced more dramatic flooding in 2004, as the remnants of Hurricane Frances filled the local rivers, inundating adjacent properties that had developed in the floodplain, since the last major flood event in 1916.

Asheville's location in the foothills of the Smokies will buffer some of the extreme heat projected from climate change in this century. But decreasing disposable income may force it to move away from an economy so heavily based on education and tourism. Its relatively low population will help make it more sustainable than larger urban areas.

Orlando, Florida: Rollercoaster to an Uncertain Future

Orlando, previously known as Jernigan before it gained its nicknames as "The City Beautiful" and the "Theme Park Capital of the World", is a sunbelt city located in central Florida. The climate in Orlando is humid and subtropical, with just over 50 in. of rain per year.

Population

Orlando had a population of 2481 at the turn of the century in 1900. The population grew dramatically during the first half of the twentieth century, reaching 52,367 in 1950 before quadrupling to over 238,000 by 2010. The dramatic rise in population after WWII can be largely attributed to

Fig. 4.9 The entrance to Walt Disney World in Orlando (http://commons.wikimedia. org/wiki/File:Walt_Disney_World_Resort_entrance.jpg, By Jrobertiko (Denis Adriana Macias) [CC BY-SA 3.0] (http://creativecommons.org/licenses/by-sa/3.0) (http://creativecommons.org/licenses/by-sa/3.0) or GFDL (http://www.gnu.org/ copyleft/fdl.html)], via Wikimedia Commons)

development of the city as a modern international tourism destination. Like Asheville, the fossil fueled growth in affluence has allowed for much more tourism (Fig. 4.9).

History

The principal inhabitants of the area prior to European contact in 1536 were the Timucua tribe of Native Americans. The area remained sparsely settled by Europeans until the 1850s. In 1838, during the Second Seminole War, the U.S. Army established Fort Gatlin. The Seminoles were descendants of several tribes of southeastern Native Americans. Fort Gatlin was the epicenter of European settlement, offering protection to pioneering homesteaders such as Aaron Jernigan, one of the first settlers in the area, after whom the town was named. In 1842, after a treaty was signed with the remaining natives, the Armed Occupation Act offered 160 acres to any homesteading pioneer who would settle in the area for 5 years. Settlers' diets were subsidized by the natural ecosystem. Deer and wild turkey were abundant throughout the swamps and pine forests, while large amounts of fish could be harvested from the nearby pristine lakes that also provided clean drinking water for herds of livestock. These still, warm waters of the central Florida wetlands also provided breeding habitat for mosquitoes. Jernigan was located in Mosquito County until the name was changed to the more appealing Orange County in 1845, the same year that Florida became the 27th state in the Union. A post office was established in Jernigan in 1850. After the Third Seminole War, the name was changed from Jernigan to Orlando and the town became the official county seat. Several stories exist regarding how Orlando got its name, but one of the most common is that it is named after a night watchman named Orlando Reeves who was killed by Native Americans in the area during the final time of native unrest.

Orlando remained a rural backwater until after the Civil War when a sizeable population moved into the area during the Reconstruction Era. During this period, an increasing number of cattle ranchers began settling in the area. Cattle ranching remains a dominant land use today, alongside citrus farming, which is still common on much of central Florida's sandy soils. Prior to the advent of the citrus industry, cotton was the most popular crop grown in the region. However, cotton declined during and after the war with the loss of the cheap labor pool of slaves that existed prior to the war, a workforce that helped to maintain plantation profitability. Citrus farming in the area began in the middle part of the century, and many of the former cotton fields were planted in citrus in the 1870s. The industry grew with the establishment of a rail line in 1881, affording growers access to a much larger market for their produce. This market expansion led to much of the early Reconstruction Era development in Orlando. By 1884, there were 1666 residents in Orlando, and central Florida exported 600,000 boxes of oranges out of the state, the majority from the Orlando area (Fig. 4.10).

The town was incorporated as a city by 1885. Churches, schools, and hotels were built with the profits made from agricultural goods produced in the pastures and orange groves. A bank, newspaper, and 50 other businesses allowed the Orlando economy to flourish. During this time, the city had also become a popular resort town, north of the more tropical paradise of Miami. However, mother nature slowed the rapid growth of this agriculture based economy in 1894, when a 3 day freeze, destroyed an estimated $100 million worth of citrus and the city of Orlando was set back 15 years.

At the turn of the century, Orlando had 2481 residents and grew to 4000 by 1910, despite flooding in 1904 and drought in 1905. The population doubled between 1910 and 1920. During the 1920s, Florida experienced its first housing bubble commonly known as the "Florida Land Boom". It was the roaring 1920s, and a combination of events—including the growth in popularity of the American automobile, rising discretionary incomes, and

Fig. 4.10 "View of orange groves from summit of Bok Tower in Lake Wales, Florida" near Orlando in the first half of the twentieth century (http://commons.wikimedia. org/wiki/File:Bok_Tower_view.JPG, By Averette (Own work) or [CC BY 3.0] (http:// creativecommons.org/licenses/by/3.0)], via Wikimedia Commons)

available credit—led to more middle aged, middle class families visiting Florida in search of the warm climate, sunshine, and investment opportunities. Prior to this time, visitors to the region were mostly rich, elderly, ill, or a combination, who bedded down in hundred-dollar-a-night railroad hotels. Northern newspapers popularized the land boom, and the most popular mode of acquisition by speculators was via the mail. About two-thirds of the land purchased in Florida during the Land Boom was bought site unseen. By 1925, sellers outnumbered buyers, and the land bubble burst. An article in the July 26, 1926 edition of the *Nation* magazine succinctly summed up the housing bust as "The world's greatest poker game, played with lots instead of chips, is over. And the players are now . . . paying up." Shortly thereafter, as if to add insult to injury, the local economy received another blow as a Mediterranean fruit fly infestation wiped out the citrus harvest.

After World War II, many veterans, who were formerly stationed at nearby Pinecastle Army Air Field and Orlando Army Air Base, settled into civilian life in the Orlando area. These trained airmen provided skilled labor for the aircraft and aviation parts industry centered in what was then "Florida's Air Capital". The city established its second airport during WWII. In addition to these government sponsored defense outposts, the defense and aerospace company now known as Lockheed Martin moved to town and established a plant that provided jobs to thousands of Orlando residents. Lockheed Martin was the largest employer in Central Florida at the time. After WWII, the U.S. Army installations were turned into Air Force bases. They served that purpose until one became the international airport in the 1970s and the other a housing development in the 1990s. In 1950, the population of Orlando had reached 51,826 as an estimated 4.5 million tourists visited the state of Florida that year. The Orlando economy got another shot in the arm as the Cold War began and the U.S. space program established NASA headquarters 55 miles to the east at Cape Canaveral.

The 1960s marked a turning point in the development of Orlando as an international tourist destination. Walt Disney flew over the swamps around the outskirts of town and decided to invest in a tract of real estate that would forever change the landscape, economy, and culture of the area. Secretly, he purchased five 20,000 acre tracts of rural property in the area. On November 15, 1965, Walt Disney announced that a theme park "bigger and better than Disneyland" would be situated on his newly acquired real estate. Orlando offered Walt Disney World an inland location that was partially protected from hurricanes, unlike other potential coastal sites in Tampa and Miami. The development of Walt Disney World in 1971 at a total cost of $400 million led to a huge number of construction jobs, and the theme park building frenzy continued for several decades. Massive aquaria were constructed to house the aquatic megafauna of Sea World, which opened in 1973. A temple to the space age, future cities, and monorail transport was erected and opened as Epcot Center in 1982. In 1989, Disney-MGM Studios theme park brought the magic of the movies to real life in what had been backwater swamps of central Florida. Theme park construction led to a commercial real estate frenzy in the surrounding areas as hotels, restaurants, and shopping centers catering to visitors accompanied the development. Two million visitors passed through the gates of Disneyworld in its first 2 years of operation, and 13,000 Orlando residents had jobs as a result. Since then, a dizzying array of theme park-based construction and service industry employment jobs have sprung up to buoy the entertainment tourism-based economy.

During the next two decades, the Orlando economy also began to diversify, as many high tech industries such as digital media, software, and biotechnology based companies moved to the city. In addition, many military-industrial based companies continue to operate in Orlando, including Lockheed Martin, Raytheon, Boeing, Northrop Grumman, and General Dynamics. This combination of military-based companies, combined with the new high-tech economy, has led to the establishment of a large research park that is the hub of U.S. military simulation and training programs.

When the housing bubble burst in 2008, Orlando residential real estate values declined. The story is somewhat reminiscent of the historic speculatory fervor that accompanied the 1920s real estate boom. Throughout the first decade of the twenty-first century, there was a dramatic run up in housing prices in Orlando from a median value of $182,300 in 2004 to $264,436 in 2007. From there, prices in the market plummeted to a median value of $110,000. Prices have recovered little since then, as the median value for a home stood at $116,000 in the spring of 2012.

Orlando is mainly a tourism-based economy that relies heavily on theme park seekers and conventioneers as economic stimuli. The city is therefore heavily dependent on cheap fuel and the discretionary income of visitors. The city is a tourist Mecca that hosted 48 million visitors in 2004. A cursory review of the theme park listings suggests Orlando is home to at least 14 legitimate amusement parks.[8] Many of the theme park visitors stay in hotels, of which Orlando has the most in the country. Tourists can also shop in the nearly 4 million square feet of retail space included in four major malls. According to a 2009 Pew Research Center poll, Orlando ranked as the fourth most desirable metropolitan area to live in, according to the 46 % of Americans who have wanderlust.[9] From its humble military outpost and agricultural origins, Orlando has grown into an international tourism destination and a high tech hub of central Florida. Critical to its future success is the continued availability of high levels of discretionary income and cheap energy. We will come back to these issues later in the book.

The Lower Mississippi River and the Central Gulf Coast: Wetlands, Steamboats, and Oil

The lower Mississippi River and the central northern Gulf Coast that encompasses New Orleans, Baton Rouge, and Houston includes some of the most important petrochemical complexes, refineries, and oil and gas fields in the U.S. The port facilities from Baton Rouge to New Orleans

[8] Theme parks and amusement parks are synonymous in this context according to the literature. The first theme park established in Orlando was "Gator Land" (1949) which is still in operation today, and the latest one is the "Holy Land Experience" (2002), a registered non-profit, non-denominational Christian theme park and church that includes an exhibit where guests may partake of the last supper with Jesus and his disciples.
[9] Pew Research Center Publications "*For Nearly Half of America, the Grass is Greener Somewhere Else*" January 29, 2009 http://pewresearch.org/pubs/1096/community-satisfaction-top-cities.

and further downriver ship about 450 million tons of cargo, more by tonnage than any other port facility in the world, including agricultural products, coal, petrochemicals, and manufactured goods. Nearly 7500 ocean going vessels leave the mouth of the Mississippi each year. Many of them transport grain from the Midwest and coal from the plains so that a significant portion of the economy of the central U.S. is dependent on a functioning lower Mississippi River. The port of Houston handles more international cargo than any other port in the country and ranks second in total tonnage. Thus, a large part of the energy used in the U.S. is pro-duced in—or flows through—this region. The lower River flows through and discharges into one of the most ecologically and economically impor-tant ecosystems in America, the 25,000 km² Mississippi delta. The area around the delta has been called the fertile fisheries crescent because of the enormous fishery. Unfortunately, the delta is collapsing, primarily due to human causes. Finally, south Louisiana is home to one of the most unique cultural complexes of the country; the French Creole-Cajun-Anglo-African American mix of music, food, festivals, and folkways. New Orleans especially is known for its food, music, and fun. The culture was and is hugely influenced by the rich ecosystems of the region. This rich ecological and economic zone stretches from Galveston Bay and Houston in the west to Mobile, Alabama in the east, also an important port. The people of this region, in the words of the French Cajuns, "laissez les bon temps rouler". They let the good times roll, both economically and cultur-ally (Fig. 4.11).

New Orleans, Louisiana: The Big Easy in the Big Bowl

New Orleans is one of the oldest and most interesting cities in America. It is often referred to as the most European of American cities due to its Spanish architecture, Creole French cuisine, and the laissez-faire lifestyle adopted by many of the city's residents. The city is the southernmost port city established in the early eighteenth century along the Mississippi River in a flood prone, mosquito-ridden site.[10] In the words of the geographer Pierce Lewis, "it was the impossible but inevitable city."[11]

---

[10] Lawrence Powell. 2012. *The Accidental City: Improvising New Orleans*. Harvard University Press.

[11] Pierce Lewis. 2003. *New Orleans, the Making of an Urban Landscape*.

Fig. 4.11 Mississippi River plume. An aerial view of the Mississippi Delta during the Great Flood of 2011 shows a plume of sediments jutting into the Gulf of Mexico. The Mississippi River coursing through south Louisiana is the lifeblood of the region (NASA Images. Visible Earth Collection. https://archive.org/details/visibleearthcoll ection?and%5B%5D=subject%3A%22Where%20--%20Mississippi%22)

Population

The first official U.S. census in 1810 (after the Louisiana Purchase in 1803, almost a century after the city's founding) counted 17,242 people. The population rose to 116,375 by 1850 and grew further to 287,104 in 1900. It reached an all time high of more than 620,000 residents in 1960 before contracting to a bit more than half that number in 2010. The changing job

Fig. 4.12 Jackson Square (the Plaza d'Arms) and St. Louis Cathedral in the center of the French Quarter in New Orleans. This is where New Orleans was founded and is one of the most recognizable views in America (http://commons.wikimedia.org/wiki/File:Jackson_Square.jpg, By Daniel Schwen (Own work) [CC BY-SA 4.0 (http://creativecommons.org/licenses/by-sa/4.0)], via Wikimedia Commons)

market with respect to fewer oil and gas related jobs in the city after the 1970s, as well as the devastating impacts of Hurricane Katrina, led to population declines over the last several decades. The ecosystem that has supported the unique, vibrant culture of the city is rapidly eroding into the sea as the impacts of sea level rise, levees, and oil industry canals exacerbate the natural land loss rates of the subsiding deltaic wetland environment that surrounds the city. The history of New Orleans is a portrait of dynamic change through time from the pre-industrial era through the petroleum age and the rise of an international tourism-based economy (Fig. 4.12).

History

Amidst vast cypress swamps and frequently flooded delta lowlands, the site of New Orleans had a natural advantage as the first high ground located roughly 100 miles upriver from the mouth of the Mississippi.

By high ground, we mean 5–10 feet above sea level rather than less than half that south of the city. Native American tribes inhabited the area, establishing a canoe portage from the Mississippi River to the coastal waters of Lake Pontchartrain north of the city. New Orleans has been referred to as "the impossible but inevitable city" due to its vulnerability to flooding from both the river and hurricanes, large populations of snakes and mosquitoes, and the pestilence-prone moist, humid subtropical climate. However, the strategic location of the site near the mouth of the continent's largest river made the settlement of New Orleans inevitable.

It is important to appreciate the geography of the Mississippi delta to understand human settlement patterns in the region. The Mississippi delta formed over the past 6 to 7000 years, as the lower Mississippi River occupied seven different major and many minor courses on its way to the sea. These are called distributaries because they distribute water and sediment over the landscape as opposed to upstream tributaries above the delta that come together to feed the growing river. As the river switched back and forth across the landscape, it formed the 10,000 mile^2 delta that stretches more than 300 miles wide. When the river overflows its channel within the delta, it deposits heavier sediments near the channel. Thus the highest ground is adjacent to the river and land elevation slopes down away from the river edge towards sea level. It was on this higher ground that New Orleans was first settled. All early settlement in south Louisiana occurred on these elevated natural levees with names such as Bayou Lafourche, Bayou Teche, Bayou la Loutre, and Bayou Terrebonne.

The first Europeans to view the site of New Orleans were Luis Moscoso and the survivors of Hernando De Soto's expedition who sailed down the Mississippi River in 1543. They were seeking an outlet to civilization after their captain died along the banks of the River further upstream.[12] In 1682, Sieur de la Salle visited the Native American village of Maheoula during his expedition south from the Great Lakes, close to the present day location of New Orleans. De la Salle placed a cross in the ground just south along the River, thereby claiming the territory for King Louis XIV of France. Shortly thereafter in 1699, in a French party led by Pierre Le Moyne, Sieur d'Iberville, was met by Bayougoula and Mongoulacha Indians along the River, just prior to erecting a cross and marking some trees for the French, at the site of the French Quarter in modern day New Orleans.

In mid-September 1700, British forces were turned away on a bend in the lower Mississippi River just a few miles below New Orleans. The 19 year-old French navy Lieutenant Jean-Baptiste LeMoyne, Sieur de Bienville,

---

[12] The Federal Writers' Project. 1938. The WPA Guide to New Orleans: Guide to 1930s New Orleans. Pantheon Books. New York City.

(Iberville's brother) bluffed a northbound English sea captain and his better-armed crew into turning south and heading back downriver, thereby allowing Bienville to preserve New Orleans as a French territory.[13] For this reason, the site is today known as English Turn. It was not until 1718 that Bienville's engineers plotted out the original street map for the city that still exists today as the French Quarter. New Orleans superseded Biloxi to become the capital of the French colonial empire of Louisiana in 1723. New Orleans flooded the year after it was founded and the era of levee building began.[14] Eighteen miles of levees were constructed along the river to help prevent flooding. This is a battle that still continues.

During the early years of the city's life, manufacturing in and around the city had only to do with supplying the needs of the colony. The surrounding cypress swamps provided a high quality stock of lumber to city saw-mills. Sawmills were in operation soon after the town was founded and by 1729, brick, pottery, and tiling were being sold.[15] The port of New Orleans did not begin to realize its potential until the second half of the eighteenth century, due to trade restrictions imposed by French authorities against England, Spain, Mexico, Florida, and the West Indies. This led to a considerable market for smugglers and pirates, the lore of which is present throughout the city even today. Slaves were an important component of the city's economy up to the Civil War, consisting of black slaves from Africa and the West Indies, Native American prisoners of war, and impoverished Europeans sold to planters for 3 years' servitude by ship captains in return for their ocean passage. Early on, slaves and European merchandise were received in exchange for lumber, pitch, tar, wax from the wax myrtle, brick, rice, indigo, sugar cane, cotton, sassafras, and fur pelts. It was not until the Midwest became populated that agricultural products from the heartland began making their way through the city's port.

New Orleans came under the control of Spain through treaties in the 1760s. The Spanish supported the American forces during the Revolutionary War by allowing the establishment of American bases in the city that supplied the Atlantic colonies with munitions and supplies. The Spanish then went on to rout the British during a series of raids in 1779 at Manchac, Baton Rouge, Natchez, Mobile, and Pensacola. The end of the eighteenth century was an eventful time in New Orleans, as The Great Fire of 1788 destroyed four-fifths of the city, a sugar cane planter discovered the process for granulating sugar, a transport thoroughfare in the form of a canal

---

[13] Powel, 2012, ibid.

[14] Lewis, 2003, ibid.

[15] Federal Writers' Project, 1938, ibid, pg. 12.

connected the city to Lake Pontchartrain, and another less severe fire destroyed parts of the city in 1794.

In 1803, the city was transferred back to French control prior to the Louisiana Purchase, the monumental land deal that brought the city and the enormous Louisiana territory under control of the United States. At the time of the sale, the French-Creole population in the city outnumbered the American contingent 12–1. The city consisted of 12 to 1400 buildings and a population of 10,000, including 4000 whites, 2500 free blacks, and 3500 slaves. The main industries in the city consisted of two cotton mills and a crude sugar refinery. Street flooding was a persistent problem that impeded travel along the muddy streets of the city. Social and cultural gatherings that included dancing, bull and bear baiting, voodoo rituals, and French theater provided a wide variety of entertainment options to residents and visitors of the city. The arrival of 6000 refugees from Santo Domingo in 1809 added to the city's population, which reached 24,552 in 1810.

During the late eighteenth and early nineteenth centuries, flatboats began to bring agricultural goods and other products of the Ohio and upper Mississippi valley down river to New Orleans. It was a one-way trip, as the boats had no way of going back up stream. They were most often broken up and the timbers used in construction. The river men on these boats were known as Kaintucks and were a rough and ready bunch. They either made their way back north on the famed Natchez Trace or settled in the city. Those that stayed settled upriver from the French Quarter in the "American" zone. There was a clash between the cultured, urbane creoles and the rustic Kaintucks. The two groups did not get along and it was dangerous to go into each other's neighborhoods. Canal Street was the dividing line and a neutral zone between the American and French sectors. To this day, the green median in south Louisiana boulevards is known as the neutral ground.

The first steamboat arrived in New Orleans in 1812, the same year that Louisiana entered the Union, putting an end to the flatboats. The steam engine solved the problem of upstream navigation along the Mississippi River, and this energy-based technology has been attributed as the primary factor in the rapid growth of New Orleans as a major North American port. In 1815, Andrew Jackson and his motley crew of 5000 troops successfully turned back a British advance at the Battle of New Orleans. That year the value of exports passing through the port was almost $10,000,000. The period from 1815 until the Civil War was a period of tremendous growth for the city and traffic along the river. The number of steamboats operating on the river increased from 21 in 1814 to 989 in 1830. Exports were dominated by cotton, tobacco, grain, and meat while large quantities of sugar, coffee, and European manufactured goods made up the majority of imports.

Fig. 4.13 View looking downriver from the foot of Canal Street, New Orleans. "The Levee at Canal Street, New Orleans, no. 53538." shows steamboats on the Mississippi River, goods in sacks and barrels stacked on the levee, groups of stevedores, horse carts (A photochrom postcard published by the Detroit Photographic Company in 1900, http://commons.wikimedia.org/wiki/File:Detroit_Photographic_Company_ (0352).jpg, [Public domain], via Wikipedia Commons)

The first railroad in the city, a four and a half mile stretch from the city to Lake Pontchartrain, was built in 1831 (Fig. 4.13).

A total of 10,000 people in the city died from an epidemic of yellow fever and Asiatic cholera during 1832 and 1833. In the late 1830s, the city divided into three different municipalities after incidents involving the aforementioned Creole and American animosities. By 1840, the city was the fourth largest in the country with a population of 102,192 and the second largest port in the nation after New York. However, the trade in imports as a percentage of total trade was on the decline since the construction of the Erie Canal and the building of the railroads from the Atlantic seaboard to the Midwest. Commercial trade in the city was becoming increasingly reliant on cotton by 1850. The cotton trade went hand in hand with the slave trade, and New Orleans was one of the largest slave markets in the country.

Another yellow fever epidemic swept across the city in 1852 and 1853 killing 7189 people. Mosquitos, a constant presence in the humid region, served as the vector for yellow fever transmission. The streets and outlying areas of the city offered prime mosquito breeding habitat due to frequent flooding and stagnant pools of water that accumulated in the gutters and after rainfall events. The city did not have an underground sewerage system at the time. Another yellow fever epidemic visited the city in 1878 killing 3800 people.

At the onset of the Civil War in 1861, as a result of the economic reliance of the city on cotton and the slave trade, the city took on a secessionist, confederate bent until Union troops steamed upriver and took control of the city in 1862. Dark days fell upon New Orleans in the years following the Civil War. Violence and robbery were commonplace. Societal unrest persisted for more than a decade as the traditional roles of power and race were upended by the war's outcome. The commercial operations at the port recovered, but not to the level of trade seen prior to the war.

In the years after the Civil War, railroads expanded after Capt. James B. Eads succeeded in using jetties as a way of flushing sediment from the mouth of Mississippi River and deepening the channel, allowing increased commercial trade. Prior to this, the river was often impassable at low water and it was said that one could walk across the mouth of the river at low discharge. Legislative franchises were granted to enhance the city's access to rail and the cost of rail shipment proved more favorable than steamboats, thereby shifting the transports mode of many goods from water to land. Railroad expansion continued in the late nineteenth century as five large trunk lines entered the city and the volume of railroad business increased from 937,634 tons in 1880 to 5,500,000 tons in 1899. The first electrified street-car began operation in the city in 1892. The population of New Orleans in 1900 was 287,104, and the total value of commerce was estimated to be $430,724,621.

In 1921, the New Orleans Inner-Harbor navigation canal that connected Lake Pontchartrain with the Mississippi River was completed at a cost of $20,000,000. The 1920s New Orleans economy boomed. Products from the American Midwest and Latin America flowed though the recently renovated port. The city's footprint expanded into the surrounding wetland environment as the powerful new screw pump, developed by Albert Baldwin Wood, allowed wetlands to be drained and houses to be built atop the drained soils. The dried organic soils rapidly oxidized and much of the city sank below sea level.

By the middle of the twentieth century, Dallas, Houston, and Atlanta surpassed New Orleans as the economic centers of the south. The banking sector in New Orleans decreased in importance as the economies of other southern cities grew. The establishment of railroad and highway networks allowed for more diverse modes of freight transport that led to decreased reliance on river transport. The increased mobility offered to travelers by the personal automobile and interstate highway system, as well as the commercial airline industry, allowed New Orleans to rely more heavily on tourism as the major economic engine. By 1960, the population of New Orleans reached a historic high of 627,525.

As oil and gas extraction ramped up in Louisiana to an all time production high in 1970, many energy companies including Shell, Chevron, and Eni based their regional headquarters in New Orleans. After the peak in state and national crude oil production that year, many energy industry jobs would eventually move westward along I-10 to Lafayette, LA and mainly to Houston, TX.

Long established as a top tourist destination in the southern U.S., New Orleans emerged as one of the country's largest tourism markets in the last quarter of the twentieth century. The storied history and multi-cultural heritage of the city attracts tourists worldwide who visit and indulge in the food, music, culture, and laid back lifestyle of the Big Easy. The construction of the superdome and a large convention center, along with the rising popularity of Mardi Gras and other festivals brought New Orleans to the forefront of the modern tourism entertainment industry.

Throughout its history, New Orleans has been threatened with flooding, both from the river and from the Gulf of Mexico during hurricanes, and this has strongly influenced its development. River flooding caused frequent inundation of the city early in its history. The 1927 Mississippi River flood was one of the most deadly natural disasters in the U.S. during the twentieth century. The Mississippi reached flood stage on January 1, 1927 at Cairo, Illinois and stayed above flood stage well into June. Much of the upper Mississippi valley flooded, scores of people died and thousands were left homeless, and floodwaters entered the homes of nearly 1 % of the U.S population. The flood led to the intentional dynamiting of levees downstream from the city and flooding of most of rural St. Bernard Parish.[16] Floodwaters never entered the city of New Orleans, and indeed the breach in St. Bernard was not necessary, but $35 million (or $458 million in 2012 dollars) in economic losses occurred in St. Bernard parish where the levee demolition occurred.[17] Parish residents still talk emotionally about the 1927 flooding. *Rising Tide* by John Barry is a riveting account of the 1927 flood.

Hurricanes have played a pivotal role in the development of New Orleans. Approximately 180 hurricanes have raked across coastal Louisiana

---

[16] Barry, John M. (1997). Rising Tide: The Great Mississippi Flood of 1927 and How It Changed America. New York: Simon & Schuster.

[17] Gomez G.M. 2000. In *Transforming New Orleans and Its Environs*. C.E. Colton (Ed). University of Pittsburgh Press, pg. 120.

since Spanish Conquistadors documented one in 1559. Particularly destructive storms occurred in 1893, 1915, and 1940. The 1915 storm, with winds of 80–110 miles h^{-1}, killed 21 people in the city and caused $13 million dollars worth of damage as water from Lake Pontchartrain overflowed into the city, causing extensive flooding of low-lying areas. This initiated a period of more intense levee building. Then in 1965, Hurricane Betsy walloped the city, flooding a total of 164,000 homes in the city and surrounding area.

Katrina was the most disastrous hurricane in the history of New Orleans. The extreme destruction caused by Katrina was the combination of several human and natural factors. In 1965, the Mississippi River Gulf Outlet (or MRGO, pronounced Mister Go by locals) navigation canal was constructed. It connected the Gulf Intracoastal Waterway, which runs east–west through New Orleans, to the Gulf of Mexico. It was intended to be a quicker route to the Gulf from New Orleans because at about 70 miles, it is approximately 20 miles shorter than the twisting river. Under-utilized since its inception mainly because severe bank erosion forced ships to go slow, the MRGO was an economic disaster. One of the unintended consequences of its construction was the destruction of large areas of wetlands, including nearly 20,000 acres of cypress swamps that helped protect the city against storm surge. The MRGO and the loss of the wetlands were implicated in the inundation of the city 40 years later during Hurricane Katrina as the hurricane surge moved rapidly up the deep channel undeterred by wetlands. Thus, the MRGO was an economic, ecological, and flooding disaster.[18]

The disaster of Katrina is set in the context of the ongoing collapse of the wetlands of the Mississippi delta. The levees built along the river prevent the flooding which carried sediments and nourished the delta for thousands of years. Thousands of miles of canals, mostly for use in oil and gas exploration and production, also drastically changed natural water movement in delta wetlands. Thus the wetlands that for so long protected New Orleans from hurricanes have disappeared at an alarming rate. A former president of St. Bernard Parish, Junior Rodrique, tells the story of his experience as a teenager during the 1947 hurricane that hit the coast just east of the Mississippi River, taking a similar path to Katrina. He lived at Verrett next to the Mississippi River, just down stream of New Orleans. As hurricane winds rose, he asked an old man who had lived nearby for nearly 80 years if they should worry. "Don't worry," the old man said, "the swamps gonna stop that hurricane." In fact the swamps did, but by the advent of Kartina the swamps were mostly gone. The MRGO killed those swamps.

---

[18] Shaffer, G., J. Day, S. Mack, P. Kemp, I. van Heerden, M. Poirrier, K. Westphal, D. FitzGerald, A. Milanes, C. Morris, R. Bea, and S. Penland. 2009. The MRGO navigation project: A massive human-induced environmental, economic, and storm disaster. *Journal of Coastal Research*, SI 54, 206–224.

Hurricane Katrina's floodwaters changed the city landscape through the mass destruction of homes and property and the displacement of community residents. The city has rebuilt to some extent since 2005 but remains below its pre-Katrina population level.[19] Many of the more vulnerable, low lying neighborhoods in the metropolitan area that were developed during the post WWII housing boom, remain sparsely populated post-Katrina. Residential investors in the city have concentrated their energies on higher elevation neighborhoods such as the Bywater downriver from the French Quarter along the natural river levees that are not so flood prone. Large allocations of federal dollars—to the tune of about 15 billion dollars to improve hurricane protection and pumping capacity—have led to a heightened confidence in the city's capacity to weather storm events in the future. But there are few in the city that believe that such a disaster will never happen again. The population has decreased in size from 484,674 in 2000, to 343,829 in 2010. Long the most populated parish in Louisiana, Orleans Parish dropped to third place.

Another more recent disaster that has impacted the city came after the 2010 explosion of the Deepwater Horizon drilling rig and the subsequent oil leak from the Macondo well in the northern Gulf of Mexico. The tourism industry in the city and region was set back once again as oil washed ashore in the nearby estuaries where much of the seafood served in New Orleans restaurants is harvested. The ongoing litigation and payments to affected parties as a result of this man-made disaster will keep the city and its residents dependent, at least partially, on disaster funding into the future.

So at the end of the first decade of the twenty-first century, New Orleans is a city dependent largely on the navigation and tourism industries. It is chiefly below sea level, and the delta ecosystem that provides bounty and hurricane protection continues deteriorating while sea level increases and more strong storms can be expected. We will come back to the question of how much longer the Big Easy can laissez les bon temps rouler.

Baton Rouge, Louisiana: Red Stick Above the Waters

Baton Rouge is located about 130 river miles upriver of New Orleans at the first bluff on the Mississippi that is high enough to avoid flooding by the Mississippi. All points south of Baton Rouge along the river are located in the delta floodplain and would periodically flood were it not for levees.

---

[19] According to the U.S. Census Bureau, the population of New Orleans in 2013 was 76 % of its pre-Katrina level.

The population of Baton Rouge has grown steadily, but more slowly than New Orleans. In 1840, 2269 people were reported living in the city. By 1900, the population was 11,259, and a half-century later, there were 34,719 inhabitants. The population surged over the next 60 years as state government grew, there was extensive suburbanization, the regional petrochemical industry matured, and transplants from hurricane impacted areas of the Louisiana coast moved to higher ground. By 2010, Baton Rouge was home to more than 225,000 people and East Baton Rouge Parish with 440,171 inhabitants was the most populous in the state, having received a large contingent that fled New Orleans after Katrina. The port of Baton Rouge is the farthest inland port along the Mississippi River that can accommodate ocean-going tankers and cargo carriers. The port handled the 13th largest amount of tonnage in the nation in 2010. Over the years, many large industrial plants have located along the river corridor between Baton Rouge and New Orleans. Many of these industries are petrochemical facilities strategically located along the river, which serves as an abundant source of freshwater and a major transport corridor. By 2000, the population of the metropolitan area was nearly 603,000 (Fig. 4.14).

History

Baton Rouge was home to Native American tribes dating back 10,000 years. These tribes hunted, fished, grew crops and lived off of the abundant natural resources provided by the plants and wildlife that grew in the vast bottomland hardwood forests and adjacent uplands of the region. Members of these tribes constructed ceremonial mounds in the area, two of which are on the campus of Louisiana State University, with a third on the grounds of the State Capitol. These earthworks are arguably older than the Egyptian pyramids. The mounds at Poverty Point bordering the river's floodplain in northeast Louisiana were named a UNESCO world heritage site in 2014. The first documented European settlers to arrive in the area were French explorers in 1699. The French expedition led by Pierre Le Moyne Sieur d'Iberville was traveling up the Mississippi River when they encountered a red tinged cypress trunk (le baton rouge) on the first bluff, tethered with sacrificial fish heads and bear bones. The pole was the territorial boundary between tribal hunting grounds of the Bayagoula and Houma Indian tribes that inhabited the area known as Istrouma. The French renamed the place Baton Rouge, which translates to the English "red stick", due to this significant landmark that could easily be recognized by travelers along the River.

Fig. 4.14 Aerial view of the riverfront area of Baton Rouge USA (1990) with the Mississippi River in the foreground. The Exxon Mobile refinery and other industrial land uses can be seen in top left corner. The river between Baton Rouge and New Orleans has numerous oil refineries and petrochemical plants (http://commons.wiki-media.org/wiki/File:Baton_Rouge_Louisiana_waterfront_aerial_view.jpg, By Michael Maples, U.S. Army Corps of Engineers, [Public domain], via Wikimedia Commons)

1700–1810 was an eventful period in Baton Rouge, as political control shifted from France to England to Spain and finally to the United States, although throughout this time the settlement remained a small town along the River. By 1840, the population of Baton Rouge consisted of 2269 individuals. People were scarce compared to the 102,000 living downriver in New Orleans at the time. Fears of a concentration of political power in the state's largest city influenced the state legislature to relocate the state capitol to Baton Rouge in 1846. Shipping and river-associated transport dominated the local economy. When the Civil War broke out, the population of Baton Rouge was approximately 5500. Facing little resistance, Union troops occupied Baton Rouge in May 1862. After the war, many former slaves moved from rural areas into Baton Rouge in search of jobs and opportunity off the plantation. During this historic transition, African Americans went from less than one third of the population in 1860, to 60 % of the 7197 people counted in the census rolls in 1880. Baton Rouge remained a small town along the Mississippi River throughout the Reconstruction era and the latter part of the nineteenth century. Between 1888 and 1890, the Louisville, New Orleans and Texas Railways were constructed, connecting Memphis to New Orleans via Vicksburg and Baton Rouge. This transport line opened up an overland route to new markets for an economy that had been predominantly reliant upon river traffic for trade.

In the early twentieth century, significant oil discoveries were made in southern Louisiana. By 1909, the Standard Oil Company (now Exxon-Mobil), recognizing the strategic location of Baton Rouge on high ground along the river, constructed a large refinery and petrochemical complex that forever changed the economy of Baton Rouge and south Louisiana. The flood-proof site, on high terraces near the head of navigation for ocean-going ships, offered ready access to crude oil and natural gas, ample water for industrial processes, and in the river, a giant sink for wastes, in addition to the favorable winter climate.[20] The petrochemical industry exploited the vast reserves of oil, natural gas, and salt located within the region, begin-ning first in the upland areas of Louisiana and then, by the 1930s, moving into the coastal zone of south Louisiana. Once Standard Oil established itself in the region, several other petrochemical refiners and manufacturers followed suit by constructing industrial facilities south along the river between Baton Rouge and New Orleans. The industry received a boost dur-ing WWII as federal investment accelerated to maintain fuel supplies for the war effort. By 1947, there were 177 refineries and chemical plants in Louisiana; 126 were located along the lower Mississippi River by 1962.[71] The American plastics industry made the region home during this period, and the industrial zone stretching from Baton Rouge south along the river to New Orleans has been known as the "Petrochemical Corridor" or "Cancer Alley" ever since. Industrial development in the area was encour-aged by state tax policies that allowed incentives and exemptions for devel-oping and enhancing industrial plants in Louisiana. The laissez-faire environmental oversight by the state during the 1950s and 1960s added to the attractiveness of industrial development in the region, although at sig-nificant health and environmental costs.[22] In addition, the continual main-tenance and upkeep of the levee system along the River provided an

---

[20] Robert N. McMichael, *Plant Location Factors in the Petrochemical Industry in Louisiana.* (Ph.D. diss., Louisiana State University, 1961).

[21] Colten C.E. 2006. The Rusting of the Chemical Corridor. *Technology and Culture,* 47(1).

[22] Colten C. E. 2000. Too Much of a Good Thing: Industrial Pollution in the Lower Mississippi River. In *Transforming New Orleans and Its Environs.* C.E. Colton (Ed). University of Pittsburgh Press. pg. 148–49.

indirect subsidy from the U.S. Army Corps of Engineers to industry that located facilities in the Mississippi River floodplain. Baton Rouge was threatened by flooding in 1912 but the city, because of its high ground, escaped extensive damage to infrastructure then, and again during the historic Mississippi River flood of 1927.

In the post WWII era, Baton Rouge grew as petrochemicals became more important in the American consumer economy. The construction of Interstates 10 and 12 also buoyed the economy between Houston and Mobile. Baton Rouge is at a nexus of waterborne (the river and the Intracoastal waterway), rail, and highway transport. Much of the housing development around the city consisted of single-family homes in subdivisions expanding into former agricultural areas (i.e. suburban sprawl). Some of the most dramatic impacts on the city's population occurred in the wake of Hurricane Katrina in 2005, when Baton Rouge experienced an influx of an estimated 200,000 evacuees from New Orleans, many of whom stayed. It is estimated that half the homes in the City had evacuees from the storm. Population growth in the area has led to a housing construction boom frequently manifesting itself as sprawling suburban neighborhoods constructed in floodplains and former agricultural areas throughout the parish. The growth of State government, as well as the two universities in the city (Louisiana State University and Southern University) also contributed to the economy of the region.

Houston, Texas: Oil City, USA

Houston is situated towards the western end of the arc of coast in the north central Gulf of Mexico that includes major natural ecosystems with high freshwater input (Galveston Bay, the Mississippi delta, Mobile Bay), some of the nation's largest fisheries, several of the largest ports in the nation (Houston, Lake Charles, lower Mississippi River, Mobile), and major oil and gas production, refining, and petrochemical activity.

Population

Houston is the fifth largest city in the country. It began along Buffalo Bayou and grew tremendously throughout the last century as a result of energy, shipping, and aeronautics industries. Houston is the energy capital of the United States. In 1850, Houston's size was comparable to Baton

Rouge with a population of 2396, but by 1900, the population had grown
to 44,633. The city has been a hub for the international oil and gas industry ever since the most productive well in the world at the time,
Spindletop, came into production in 1901. Spindletop was discovered
near Beaumont, Texas just 80 miles east of Houston. A year later, the federal government appropriated funds to build the Houston ship channel.
The port of Houston handles more international cargo than any other port
in the country and ranks second in total tonnage. From 1910 to 1930, the
population increased by 250 %, and by 1950 there were 596,163 inhabitants. As the energy industry in the city continued to expand through the
twentieth century, so did its population. Houston is the largest city in the
state of Texas with approximately 2.1 million residents in 2010. The metropolitan area was over 4.6 million in 2000. The petroleum industry that
grew up around the port and ship channel buoys the economy and
employs more people than any other sector.

History

Buffalo Bayou originates in Katy, Texas and winds its way through the city
of Houston before discharging freshwater into the Houston ship channel
that flows into Galveston Bay. Prior to European settlement, Native
Americans hunted, fished, and farmed the area along the bayou. The landscape that provided sustenance to the native peoples consisted of tall grass
prairie, pine and hardwood forests, swamps, and rich estuaries. Prairie
deer and buffalo were abundant throughout the vast landscape, and the
natives farmed and traded maize. A rich woodland ecosystem lined the
banks of the Bayou.

European settlers did not permanently settle in the area until the 1820s,
although the French and Spanish maintained trade with the natives prior to
this time. Before Houston was established in the late 1830s, the town of
Harrisburg, located near the ship channel east of Houston, served as the
region's shipping hub along the Gulf. Many of the early settlers in the
region arrived via boat from Louisiana. In 1836, Harrisburg was burned to
the ground by Mexican General Santa Anna during the Texas Revolution.
A short time later, Sana Anna was defeated and Texas established itself as
a Republic. Two New York land speculators, the brothers John and
Augustus Allen, began searching for land to develop a colonial settlement.
They chose a site near the confluence of Buffalo Bayou and White Oak
Bayou to develop into a city based on shipping. The area was of strategic

importance for three reasons.[23] First, Buffalo Bayou flows east to west, it is enhanced by tidal flow throughout the four seasons, and it is augmented by rainfall. Thus, the bayou was largely unaffected by drought during dry seasons unlike many of the north to south flowing rivers in the eastern part of the state such as the Brazos, Colorado, Trinity, and San Jacinto rivers. In order to establish a "great commercial center", the brothers needed continuous water levels sufficient for shipping, which was what the tidal flow delivered. Continuous, relatively high water levels allowed for year round shipping of agricultural products from the prairie to both national and international markets. Second, the Allen brothers needed at least 10 feet of water to accommodate riverboats of the time. They found this at what is now known as Allen's Landing. Third, they needed a turning basin for 75-foot long paddle boats, and White Oak Bayou allowed that.

After purchasing 6642 acres for just over $9,428, they began marketing it to potential buyers as the future "great interior commercial emporium of Texas". The American publisher, surveyor, and inventor of condensed milk, Gail Borden Jr., drew up the first street maps for the city that would be named after Sam Houston, the hero of the battle of San Jacinto and the Texas Revolution.

The first ship to visit the city arrived in January 1837, less than 6 months after the land was purchased. The city was incorporated and made into the temporary state capital 6 months later. The city contained an overtly lawless frontier-style culture in its early years before a legitimate government was established to deal with this menace to potential business. In 1842, Congress designated funds to dredge Buffalo Bayou for the establishment of a port, but the effort proved inadequate for the establishment of a deepwater port to handle large-scale shipping traffic.

It would take some time for commercial ships to begin conducting business on a large scale since there were other natural obstacles, including two substantial sand bars in Galveston Bay that prevented boats with greater than a five-foot draft from accessing the Bayou to Houston. The shallowness of the Bay and Bayou prevented ocean-going vessels from entering the city, so shallow draft paddlewheel steamboats were the watercraft of choice for moving goods in and out of the city from the late 1830s to the 1850s. Boats carrying exports were forced to offload their cargo at Galveston, before transferring it to ocean going vessels. Imports required the same treatment, and businessmen quickly noticed their profit margins shrinking as a result of the transfer costs. The business community in Houston decided that a deepwater port was necessary to circumvent these losses and

[23] Aulbach, L. 2011. An Echo of Houston's Wilderness Beginnings. CreateSpace Independent Publishing Platform, 752 p.

increase commerce for the city. Several rail routes were established in connection with the shallow water port to move products from the port to markets beyond the city. Southern agricultural crops such as cotton and lumber were traded through the port in Houston.

During the Civil War, the economy of Houston, like much of the rest of the nation's, suffered. Federal troops blockaded Galveston in 1862, which shut down shipping. Social unrest resulting from the war and poor economy resulted in a military government being installed in Houston during Reconstruction. After the Civil War, shipping magnate Charles Morgan, owner of the Ship Channel Company, was able to get a nine foot deep canal dredged through the Galveston sand bars. The channel, however, was not dredged all the way to Houston and shipping was still limited. The Texas railroad network continued to expand from Houston to cities like Dallas, Fort Worth, San Antonio, and El Paso, boosting Houston's prominence as the major transport thoroughfare in the state. Shortly thereafter, the railroad expanded to New Orleans. In 1870, Houston became a port of entry, allowing foreign imports to enter. Businesses began to develop in the city throughout the decade as a result of increased commerce and jobs, and the city had two newspapers by 1880 when the population reached 16,513. Two hospitals were established in the city in the span of 6 years from 1887 to 1893.

In the mid 1890s, congress approved a 25-foot deep channel that became the precursor to the modern ship channel. In addition to the bottom dredging, the channel was straightened by the removal of three bends in the Bayou. The city then established the Port of Houston as barges and tugboats became the watercraft of choice. By the turn of the century, the city contained 44,633 inhabitants. In 1947, the modern ship channel was developed by the Army Corps of Engineers and dredged to a depth of 45 feet. To this day, Houston remains the largest inland port in the nation (Fig. 4.15).

Houston experienced rapid growth in the early part of the twentieth century as a result of two major events. The first was the Galveston hurricane of September 1900, when winds of 145 mph pushed a hurricane surge across Galveston leaving thousands of people dead and wiping out much of the city. A well-written history of the storm and the hubris of humans in the face of nature are detailed in *Isaac's Storm* by Erik Larson.[24] Many businesses invested in more protected inland real estate near Houston after the storm. The other major event that would affect the city and surrounding region was the aforementioned discovery of oil in 1901 near Beaumont to the east. Anthony Lucas, a Croatian born oil explorer, who was one of the first to understand the relationship between subsurface salt domes and

---

[24] Erik Larson and Isaac Monroe Cline. 1999. *Isaac's storm: A Man, a Time, and the Deadliest Hurricane in History*. Vintage.

Fig. 4.15 The Houston Ship Channel with downtown Houston in the distance. The channel is lined with oil refineries and petrochemical plants (http://commons.wiki-media.org/wiki/File:Houston_Ship_Channel.jpg By United States Coast Guard, PA2 James Dillard [Public domain], via Wikimedia Commons)

crude oil deposits, drilled the well that gushed oil hundreds of feet into the air and produced about 100,000 barrels per day. This kicked off the Texas oil boom and forever changed the southeast Texas economy and practically everything else. Spindletop was the most productive oil field in the world up until that time. Houston was transformed from a railroad hub town to a big city as the population in 1910 grew to 78,800, surpassing the population of Galveston for the first time.

The decade from 1910 to 1920 saw a large influx of Mexicans moving into Houston, partly as a result of the Mexican revolution south of the border and partly because of economic opportunity. With the increase in gasoline demand as the automobile became a widespread consumer product and the onset of World War I the Houston oil industry grew. Twelve oil companies were located there by 1913. A decade before, rice became a popular agricultural crop, thanks in part to Japanese immigrants who had come over to expand the agricultural base of the region. However, rice prices declined and much of the Japanese community had to take up other jobs or move out of the region altogether. By 1920, the population of Houston reached 138,276. In the 1920s, the first state highways were built, connecting the city with the hinterland. This was the beginning of automobile traffic in a city now renowned for its rush hour congestion. The institution that would become the University of Houston was established in the city during the decade. The population of Houston more than doubled by 1930, reaching 292,352. An influx of African Americans from southern states, most notably Louisiana, occurred during the Depression Era. Commercial air traffic also increased as the city transformed into a large urban area.

In December 1935, after 17 in. of rainfall, Houston experienced the worst flood in its history. Floodwaters inundated downtown, drowning several people, and causing $2.5 million in damage. Floods had been a part of Houston's history since a year after it was founded, when the naturalist John James Audubon and his party were "drenched to the skin" by rainfall during their visit in 1837. Part of Houston's vulnerability to floods lies in it's flat topography and its location within 75 miles of four different river basins. At least one newspaper reporter referred to the city as the "Venice of Texas". The Harris County Flood Control District was set up shortly after the 1935 flood to mitigate against future flooding. The result was the creation of the Barker and Addicks reservoirs, which allowed for water storage and managed release over time for flood prevention.

Houston grew rapidly in the 1940s as World War II gripped the country. The shipping and oil industry in the city was supplemented by aircraft and shipbuilding industries as federal coffers injected hundreds of millions of dollars into the Houston economy as part of the war effort. Steel, munitions, and petrochemical products including synthetic rubber and high-octane fuel were all in high demand, and Houston was one of the major production centers for these critical resources. The demand for labor in the city led to further population growth in the Mexican and African American populations throughout the decade. Policies to deal with this growth did not come easy. Voters rejected zoning ordinances when they went to the polls. The birth of suburban Houston sprawl occurred in the post WWII era. The now famous M.D. Anderson Foundation established the Texas Medical Center in 1945. Houston's population in 1950 was 596,163 (Fig. 4.16).

The 1950s were marked by the realization that the drawdown of city groundwater was causing subsidence, especially in the area around Texas City. Whole neighborhoods had to be abandoned as Galveston Bay waters flooded into homes. The city turned to surface water resources in Lake Houston, Lake Conroe, and Lake Livingston in an effort to deal with the problem.

Population and water use continued to grow throughout the decade until 938,219 people were living in Houston in 1960. The increase in international air traffic that led to the construction of a new International Airport in the 1960s and growth of the downtown health care complex were both signs of the city's emergence on the national and international scene in post

Fig. 4.16 View of downtown Houston (http://commons.wikimedia.org/wiki/ File:Houston_from_Westheimer.JPG By iodine127 (Own work) [Public domain], via Wikimedia Commons)

WWII America. In 1961, category 4 Hurricane Carla slammed into the Texas coast killing 43 people and causing at least 325 million dollars worth of damage. The low number of fatalities is attributed to the fact that over 500,000 people were evacuated from the hurricane's path in the largest peacetime evacuation the nation had ever seen to date. Carla ranks as the second most intense hurricane to strike the Texas coast.

More people and jobs came to Houston when NASA established its Manned Spaceflight Center in the southeastern part of the city. In 1962, Houston voters again rejected zoning ordinances and the city sprawl continued as population reached 1.2 million by 1970. The 1970s were marked by the construction of skyscrapers in downtown and a population increase at least partially attributable to the 1973 Arab oil embargo and the need for more labor in the east Texas and Gulf Coast oil fields. Rapid population growth in the city continued until 1986 when the price of oil declined from $27 per barrel to less than $10 per barrel, and recession set in on the Houston economy.

Hurricane Alicia struck Galveston and Houston in 1983 causing $2.6 billion in damage. The hurricane spawned 23 tornadoes that tore across the Texas landscape from Galveston to Tyler, Texas. Population continued to grow from more than 1.6 million in 1990 to over 1.9 million by the new millennium. During this time, Houston voters went to the polls to vote down zoning ordinances for a third time. Heavy investment into the aeronautics and health care industries after the recession helped to diversify the economy and decrease reliance on the oil industry. Tropical storm Allison struck Houston in June 2001, killing 17 people, flooding the central business district and disrupting services in the medical complex district. Five years later, Houston experienced an influx of people fleeing the flooded city of New Orleans after Hurricane Katrina. More flooding occurred in Houston during Hurricane Ike in 2008. By 2010, over 2 million people lived in the city of Houston and over 4.6 million in the metropolitan area.

Houston continues to be a world leader in the oil, natural gas and petrochemical industry sectors as well as one of the top two commercial ports in the nation. Like the Mississippi delta to the east, the continued degradation of the coastal environment threatens the city. Climate change will impact this area in terms of sea level rise and hurricanes, as well as potentially more drought. Oil and gas production will decline in coming decades, forcing Houston to continue to adapt again.

Cedar Rapids, Iowa: Small Metropolis on the Fertile Plain

Located over a thousand miles north of the Gulf Coast, Cedar Rapids is in the center of a region that supplies a enormous amount of grain that feeds the nation and flows down the Mississippi past Baton Rouge and New Orleans. It is the second largest city in Iowa after the state capital Des Moines. It is one of many small cities established in the rich agricultural region of the Midwest in the nineteenth century. The city is situated on both banks of the Cedar River, a tributary of the Mississippi River, which is located about 75 miles east of Cedar Rapids. The city's population was 1830 in 1860 and increased to 25,656 by 1900. It took another 30 years for the population to double, surpassing 100,000 by 1970. The population remained relatively stable during the 1980s and 1990s. In 2010, the population of the city was just over 125,000, while that of the three-county metropolitan area was 255,452. Grain processing forms the basis of the city's most important economic sector, reflecting Iowa's importance in the Midwest bread basket. Large agricultural companies that have offices in Cedar Rapids include Quaker Oats, Archer Daniels Midland, Cargill, and General Mills. The value of agricultural products in east central Iowa is $598 million from crops and $302 million for livestock. Other important employers are the defense and commercial avionics company Rockwell Collins and the insurance company Aegon. Partly because of Rockwell Collins, Cedar Rapids has more engineers per capita than any other city in the U.S. The city also has a higher percentage of exported products, per capita, than any other area of the country, mainly reflecting agricultural products.

History

Before colonization, the area was inhabited by tribes belonging to the Prairie-Plains Indian culture whose members lived both migratory and settled lifestyles. The Sauk and Meskwaki tribes dominated eastern Iowa. These tribes sold their lands to the federal government and were relocated to Kansas. Some returned but there are few Native Americans in Iowa today.

The area was settled just before the middle of the nineteenth century as part of the westward movement of settlers across the U.S. The city was established in 1838 at the rapids that existed at the site in the Cedar River. The city was first called Columbus but in 1841 it was renamed for the rapids and the large number of red cedar trees that grew along the banks of the river.

Like much of Iowa, Cedar Rapids was settled by European immigrants, and became incorporated in 1849. By 1870, farms and small towns covered much of the state, ending the frontier era. The population of Iowa was then just over 1,900,000. By 2005, the state's population had reached about 2,990,000. The state actually lost population in the 1980s but began growing again in the 1990s (Fig. 4.17).

From its founding, the economy of Cedar Rapids—like many towns and cities in Iowa, and much of the Midwest farm belt—has been tied to agriculture. And agriculture has changed over time. Prior to the Civil War, wheat, oats, barley, hay, and sorghum were important crops. Crops were first shipped to markets on the Mississippi River, but in the 1850s, rail transport began and, in 1867, the Chicago and North Western Railroad reached the western border of the state. Until well into the twentieth century, railroads dominated transportation in the state.

Following the Civil War, the main agricultural products shifted from wheat to corn and hogs. For the most part, corn was fed to hogs that were then sold on the market. The state ranked first or second in corn and hog production. By 1900, Iowa was also an important egg producer.

These activities were initially located along the Mississippi where there was regular steamboat traffic. But with the expansion of the railroads, business activity spread across the state. This is when Quaker Oats constructed

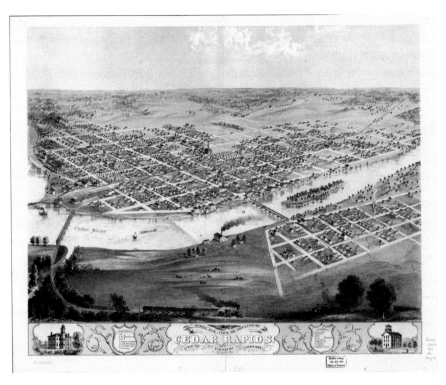

Fig. 4.17 Cedar Rapids in 1868 (http://commons.wikimedia.org/wiki/File:Cedar_Rapids_IA_1868.jpg. This map is available from the United States Library of Congress's Geography & Map Division. By Arthur Ruger [Public domain], via Wikimedia Commons)

an oat processing plant in Cedar Rapids. Meat processing became widespread in Cedar Rapids and the rest of Iowa. After 1900, non-agricultural industries began to establish themselves in the state. Coal mining became the second largest industry in the state in the late nineteenth and early twentieth centuries. This was tied to the development of railroads as coal provided fuel for the trains (Fig. 4.18).

Following World War II, agricultural production increased, becoming more mechanized and energy intensive as it utilized heavy equipment such as combines, corn pickers, and larger tractors. Farmers also began using herbicides and pesticides to control weeds and pests. Farms became larger and there was movement to corn and soybean production, as well as large numbers of hogs. The face of agriculture and the way farm families lived in Iowa and the rest of the nation was changing. In the first part of the twentieth century, farms produced much of the food they consumed typically from large flocks of chickens, large gardens, and small dairy herds and fruit trees.

Fig. 4.18 "Ridge-till and strips of corn and soybeans in northwest Iowa field" (http://commons.wikimedia.org/wiki/File:NRCSIA99311_-_Iowa_(3315)(NRCS_Photo_Gallery).jpg By Lynn Betts/Photo courtesy of USDA Natural Resources Conservation Service, via Wikimedia Commons)

After the Second World War, most farm families bought their food rather than produced it. Rural electrification meant that farm homes could include modern amenities and be as modern as those in urban areas.

Also, Iowa and the rest of America were becoming urbanized. In 1880, nearly 85 % of Iowans lived in rural areas. But by 1956, more than half of Iowans lived in urban areas and by 2000, less than one in ten Iowans lived on a farm. The industrialization of agriculture led to the transformation of rural areas. Rural institutions began to disappear including churches, local schools, and small town businesses. Over half of farm families had members who worked off the farm. The number of farms in Iowa shrank but individual farms became larger. Large-scale production of hogs and chickens in what are called CAFOs (Confined Animal Feeding Operations) became common.

Cedar Rapids has followed the path of many small cities in the Midwest. Urban areas have increased as farming has become more mechanized. But the city economy is still dominated by the agricultural economy. Rich soils and a favorable climate will continue to sustain the area and perhaps make it more sustainable than some of the other places that we have visited.

Amarillo, Texas: Cattle and Oil on the Semi-arid Plain

Amarillo, the principal city in the Texas Panhandle, has a metropolitan area population of about 236,000. The population of Amarillo did not surpass 1000 until 1900 but grew steadily throughout much of the twentieth century, reaching almost 75,000 people by 1950. After the only decade of population decline in the 1960s, Amarillo started to grow again but at less than 20 % per decade until the present.

The city got its start based on ranching and cattle production, and marketing these continues to be a major economic activity today.[25] Amarillo is one of the largest meat packing areas in the U.S. The U.S. military is also an important economic driver in the area. The V-22 Osprey aircraft assembly plant is located in Amarillo, as is Pantex, the only nuclear weapons assembly and disassembly plant in the country. An iconic American attraction, the Cadillac Ranch located west of Amarillo on Interstate 40, is a monument of painted Cadillacs buried in the ground hood first. The famous Route 66, popularized in the American musical canon by Nat King Cole and Chuck Berry, passes through the city (Fig. 4.19).

Fig. 4.19 The Cadillac Ranch located along old Route 66 just outside of Amarillo (http://commons.wikimedia.org/wiki/File:Cadillac_Ranch.jpg By Richie Diesterheft from Chicago, IL, USA Tipping Painted Cars, Uploaded by PDTillman [CC BY 2.0] http://creativecommons.org/licenses/by/2.0, via Wikimedia Commons)

---

[25] P.H. Carlson. 2006. *Amarillo: The Story of a western Town*. Texas Tech University Press, pg. 283.

History

Amarillo was founded late in the nineteenth century. Nomadic Native American tribes lived in the area for centuries and this was one of the very last Native American strongholds in the country. The Kwahadi Comanche chief Quanah surrendered to the US Army in 1875 in the Red River War, opening the area to settlement by Anglos for the first time. Fort Elliott was established on Sweetwater Creek in 1875 to keep the Indians on Oklahoma reservations. Hunters then killed most of the buffalo that had grazed the grasslands for centuries. During the beginning of the last quarter of the nineteenth century, the Texas Panhandle was a vast, sparsely populated area.

With the removal of the Indians, settlers began moving into the Panhandle. Within 10 years, the entire region was divided into very large ranches, some of them covering areas larger than some individual New England states. These large ranches prospered with the introduction of barbed wire and water pumping windmills that provided a reliable source of water in this semi-arid region. Exceptionally cold winters in 1885 and 1886 killed vast numbers of cattle when herds were trapped against barbed wire fences and froze to death. This caused the breakup of the large ranches into more manageable smaller ranches and farms, and led to a dramatic increase in the population of the Panhandle.

Amarillo began in 1887 as "Ragtown", a tent city of about 500 railroad construction workers. The arrival of the railroads and the selection of the town as the county seat of Potter County made the town a rapidly growing center for cattle marketing. Heavy rains in the spring of 1889 flooded the city, and it was moved to higher ground a mile east. Buildings were placed on logs and teams of horses rolled them to the higher site. Amarillo (Spanish for "yellow") was named for the yellow flowers that grew in great profusion along Amarillo Lake and Creek. The original Spanish pronunciation, Ah-mah-REE-yoh, evolved into Ah-mah-RI-loh with the arrival of the English speaking railroad workers.

The arrival of railroads in 1887 and 1888 also spurred population increase and the early growth of the city. With the beginning of rail service, Amarillo became a center for cattle shipping; by the 1890s, more than 100,000 head of cattle were being shipped each year. The cattle business led to local production of wheat and other grains, the construction of grain elevators, and the city became a milling and feed manufacturing center. This led to rapid population growth. From 482 people in 1890, the city grew by almost 200 % to 1442 in 1900 and by an additional 500 % to nearly 10,000 in 1910. Reflecting the rapid growth of the city and an evolving urban culture, the Grand Opera House was built in the first decade of the twentieth century.

Natural gas was discovered in 1918, and oil 3 years later. In 1924, it was discovered that the Cliffside gas field had an unusually high percentage of helium associated with it. It was purchased by the Federal government in 1927 and the Amarillo Helium plant, which opened 2 years later, became the sole producer of commercial helium in the world for several years. The U.S. National Helium Reserve is still located in the Cliffside field. Amarillo is near the Panhandle Field, a productive gas and oil area covering 200,000 acres spread over eight counties.

The depression of the 1930s was especially hard on the region because of the Dust Bowl, caused by a multi-year drought. This was a time of dust storms in the American prairie caused by a combination of severe drought and unwise farming practices. Farming in the plains at the time did not include such conservation measures as crop rotation, wind breaking trees, fallow fields, and cover crops. Deep plowing of the original soils in the decade preceding the dust bowl years led to loss of deep-rooted natural grasses that maintained soil moisture and vegetation cover. And because this is an area of high winds, the denuded soils were easily eroded when vegetation was lost. Most of the plains areas in the affected region receive less than 20 in. of rain and some less than 10 in. per year.

During the drought that lasted through most of the 1930s, much of the region had 15–25 % less precipitation than average for and some years had less than 50 % of normal levels. During this decade, large areas of the plains were effectively deserts, receiving less than 10 in. of rain in some years, and for several years running in some places.

During the 1930s drought, dust sometimes reached all the way to the east coast. Many regions lost 75 % of their topsoil. During the dust storms, visibility was sometimes reduced to less than a few feet. Over 100,000,000 acres of the plains were affected, most strongly in the panhandles of Texas and Oklahoma, and also in adjacent areas of New Mexico, Colorado, and Kansas. Hundreds of thousands of people, including many residents of Oklahoma, often referred to as Okies, were displaced. The Great Depression made the plight of these people even worse (Fig. 4.20).

The human suffering and environmental degradation associated with the Dust Bowl is seared into the American psyche. John Steinbeck's novels *Of Mice and Men* and the *Grapes of Wrath*, and Woody Guthrie's folk songs both gave vivid accounts of the plight of the Okies, and "Migrant Mother" by Dorothea Lange and Arthur Rothstein's depiction of a farmer and his two sons seeking shelter are among the most famous photos of the era. More recently, the Ken Burns documentary entitled "The Dust Bowl", which recounted the human cost of this difficult period in tragic detail, shows the enduring impact of the Dust Bowl on the national consciousness.

Fig. 4.20 Farmer and sons walking in the face of a dust storm. Cimarron County, Oklahoma, USA (April 1936) (http://en.wikipedia.org/wiki/File:Farmer_walking_in_dust_storm_Cimarron_County_Oklahoma2.jpg By Rothstein, Arthur, 1915–1985, photographer [Public domain], via Wikimedia Commons)

The Texas drought of 2011 has evoked grim memories of the Dust Bowl. Total precipitation in the area was 7 in. in 2011, less than half of normal. Since 2000, the region of the Texas Panhandle has been drier than normal. People are beginning to wonder if a new Dust Bowl is beginning; we will return to this topic.

The end of the drought and the close of World War II led to economic recovery. The establishment of the Amarillo Army Airfield and the Pantex Army Ordnance Plant provided employment for many. Both of these facilities closed after the war, but the Pantex plant reopened in 1950 and produced nuclear weapons throughout the cold war. It presently serves as America's only nuclear weapons assembly and disassembly facility. The airbase was reopened as Amarillo Air Force Base, and became home to a Strategic Air Command B-52 wing. These bases led to increased economic activity, and the population increased from 74,443 in 1950 to 137,969 in 1960. The closure of the airbase in 1969 led to a decrease in population to 127,000 by 1970. In the 1980s a number of manufacturing plants were established in the city, including several food processors. In 2010, the population of the metropolitan area was 236,113.

Amarillo is the regional economic center for the Texas Panhandle, eastern New Mexico and the Oklahoma Panhandle. Agriculture continues to be the most important economic activity of the region with approximately 14 million acres (57,000 km²) of agricultural land surrounding the city. The meat packing industry is a major employer, with about one-fourth of U.S. beef processed in the area. Amarillo is the headquarters for the Texas Cattle Feeders Association. This is cattle country. Reflecting this, the Big Texan Steak Ranch is famous for offering visitors a free 72 ounce steak if it can be eaten in less than an hour. The city's largest employer in 2005 was Tyson Foods, with 3700 employees. Corn, wheat, and cotton are the major crops. Sorghum, silage, hay, and soybeans are also grown. Beginning in 2000, the Panhandle has become a fast growing milk producing area with a number of large, modern dairies with center point irrigation providing fodder crops for the dairy herds. The rise of the dairy industry in the region is in part a result of the water shortages in California, where much of the industry was previously based, as well as less stringent environmental regulations in Texas.[26]

Climate has always played a critical role in the economic health and soul of the region. The regional climate is semi-arid with annual precipitation averaging about 20 inches. Summers are hot and dry while in winter there are cold fronts from the north and northwest with occasional blizzards. The record low was −16 °F in 1899 and the record high was 111 °F in 2011. The Texas Panhandle is in Tornado Alley and Amarillo is the third windiest city in the U.S. Because the area is semi-arid with frequent droughts, irrigation is necessary for agriculture.

Amarillo gets it drinking water supply from the Ogallala Aquifer and from Lake Meredith, formed when the Canadian River northeast of Amarillo was dammed. The 2011 drought caused lake levels to drop so much that it could no longer be used as a municipal water source. The Ogallala Aquifer also supplies water for agriculture in the Panhandle.

---

[26] William Ashworth. 2007. *Ogallala Blue: Water and Life on the High Plains*. The Countryman Press Woodstock, VT, pg. 59.

Without water from the aquifer, agriculture would not be possible. We will revisit the Ogallala Aquifer later in the book. Like Cedar Rapids, the foundation for the economy is agriculture. However, climate change may threaten the base of the economy in coming decades.

Las Vegas, Nevada: Betting on the Uncertain Future of Water

Las Vegas is the epitome of twentieth century tourist cities in the U.S. The city began as a small oasis town in the early 1900s and grew to become a desert city with just under a half million inhabitants by the year 2000. The storybook history of Las Vegas is a remarkable tale of a western town strategically located along an early trade route that began as a small settlement operating as a collection of independent businesses, eventually developed by outside interests before finally emerging as a modern corporate-dominated international tourist destination with a reputation for illicit conduct.

As was the case with many towns in the Southwest, population growth did not occur at a large scale in Las Vegas until the second half of the twentieth century. Population increased from 25 residents in 1900 to almost 25,000 in 1950. Over the next 60 years, the population of Las Vegas increased to over 580,000 people, 20 times as many people as lived there in 1950, and almost 25,000 times more than 1900!

Las Vegas is a world-renowned gambling and entertainment center built largely on imported water and the disposable income of the tourists who notoriously flock to "Sin City" for vacations, gambling, weddings, and conferences. Las Vegas competes with Orlando for the largest number of tourists annually. However, Las Vegas is more of a theme park for adults than for children. The dry Las Vegas climate requires the city to draw on distant lakes and rivers to meet the water demand of the indigenous and transient population. Presently, a large percentage of Las Vegas residents work in the service industry in hotels, casinos, and restaurants. A half-century ago, many post WWII residents stayed in the city after serving at nearby military-based industrial facilities and military bases north of the city. The present population has seen a considerable downturn in the economy as a result of the housing crisis, increases in energy prices, and the loss of discretionary income that makes travel less affordable for tourists. The future of Las Vegas is uncertain since many of the basic requirements of the city are under severe stress as energy and basic resources become scarce and more expensive, and as mass tourism faces a decline.

History

In 1829, the Spanish Mexican explorer Antonio Armijo and his commercial caravan party were attempting to establish a trade route from New Mexico to Los Angeles. The route they took is known as the Old Spanish Trail. It had been used by Spanish explorers since the 1500s. Camped approximately 100 miles northeast of Las Vegas, the party sent a scouting team out to search for water. One of the scouts, Rafael Rivera, wandered into the Las Vegas Valley, thereby becoming the first documented non-Native American to set foot in the valley. The valley contained artesian wells that were fed by the runoff from snowmelt atop Mt. Charleston, situated to the west of the valley. The abnormally high amount of water in the valley made it a natural oasis that included extensive green meadows not found elsewhere in the dry, arid landscape; Las Vegas means the "meadow" or "fertile plain" in Spanish. It is difficult to imagine the lush green meadows of pre-European settlement when viewing the city through the lens of the twenty-first century (Fig. 4.21).

Las Vegas was an oasis amidst the vast deserts of central Nevada. The Tudinu, or "Desert People", who were ancestors of the Paiute Indians, inhabited the region. The Paiutes were displaced from the region following

Fig. 4.21 The desert outside of Las Vegas (http://commons.wikimedia.org/wiki/File:The_desert_outside_Las_Vegas_(8294515450).jpg By Bert Kaufmann from Roermond, Netherlands (The desert outside Las Vegas Uploaded by russavia) [CC BY-SA 2.0] http://creativecommons.org/licenses/by-sa/2.0, via Wikimedia Commons)

the policy of "Manifest Destiny" that led to widespread western expansion of the U.S. population during the middle part of the nineteenth century. The tribe currently owns a popular tribal smokeshop in downtown and a golf resort southwest of the city.

Approximately 15 years after the Armijo scouting party stumbled upon the Las Vegas valley, John C. Fremont and his associates became the first American explorers to visit the valley. They were on an expedition to map and document areas of the west for what was then known as the Corps of Topographical Engineers, now the U.S. Army Corps of Engineers. Fremont documented his travels across the land and produced the most comprehensive information source for others interested in regional settlement.

In 1855, Mormon missionaries came to the Las Vegas valley in order to convert the local Paiute populace. Most of the missionaries left a couple years later when the Utah War began, leaving behind a fort that had been constructed in the present day downtown area. Las Vegas was a strategic outpost for the missionaries because of the water resources and its location in the "Mormon Corridor" between Salt Lake City, Utah and San Bernardino, CA. For 50 years, Las Vegas remained a small, sparsely inhabited settlement. Around 1900, a Montana mining baron and U.S. senator named William Andrews Clark was advancing his plans for the San Pedro, and the Salt Lake Railroad that would once again utilize Las Vegas as a strategic transport stopover and as a source of water for the steam powered locomotives. The train line from Los Angeles to Salt Lake City served as a shortcut to the Chicago and New York markets. Clark had purchased 2000 acres and the water rights to the land from a local ranch owner and set about developing the area into a small town. On May 15, 1905 the first Las Vegans began purchasing commercial and residential land at auction in the Clark Las Vegas Townsite. The snowmelt runoff from Mt. Charleston that fed the local aquifers before draining into the Colorado River supplied the necessary water for the railroad town. The population of Las Vegas in 1910 was 800.

From the beginning, the Las Vegas economy was conducive to traveler services and the nightlife that would eventually make it a world-renowned tourist attraction. Early on, travelers could exit the train during a break in service and head to the 'red light district' or Block 16 for bowling, billiards, brothels, and bars selling booze without licensing restrictions. Even during the prohibition era from 1920 to 1933, liquor flowed freely through the speakeasies of Block 16. By 1920, the population of Las Vegas had grown to 2304 with most of these residents employed at the railroad yard or in the local traveler services industry.

The Las Vegas economy was dealt a major blow in 1921 when the railroad labor board decided to cut wages for workers and the approximately

1.5 million nationwide railroad union workers went on strike. This included workers in Las Vegas who, by 1921, were working for the Union Pacific railroad company. Soon thereafter, Union Pacific decided to move operations from Las Vegas to Caliente, NV, leaving the Las Vegas economy in the lurch. Much of the 1920s progressed as a time of economic hardship for Las Vegas, although the federal government eased the pain when Congress announced construction plans for the Hoover Dam in 1928. A massive 24 h per day, 7 days per week labor pool began moving into the area just 25 miles southeast of Las Vegas in Boulder City, NV. The damming of the Colorado River for the intended purpose of generating hydroelectric power while providing flood control and irrigation water to downstream and nearby municipalities, allowed the city of Las Vegas to grow beyond the natural limitations of the local ecosystem resources, most importantly water. The dam provided electricity to the city that had become a major stopover point for the quarter million annual tourists who visited the dam by the mid 1930s when it was completed. Government leaders legalized casinos and loosened the marriage and divorce laws earlier in the decade, giving even more reason for tourists to visit.

By 1940, the population of Las Vegas was 8422. World War II was instrumental in the growth of Las Vegas, and for the second consecutive decade, federal government contracting led to a rise in the city's population. Most of the new arrivals were associated with the magnesium (the "miracle metal") production facility located in nearby Henderson. The Basic Magnesium Plant supplied magnesium for incendiary munitions casing and airplane components that were being manufactured in southern California. In addition, veterans stationed at the nearby gunnery school just north of Las Vegas added to the population, which doubled by 1945. The growing population required access to more water. The solution came in the form of a pipeline from the Basic Magnesium Plant that was already plumbed into the source at Lake Mead.

Beginning in 1947, Las Vegas was once again a modern transport stopover town, as recently paved Highway 91 brought in travelers from southern California. The historic U.S. 91 ran straight through the city and became the Las Vegas strip. Plenty of traffic flowed into Las Vegas in the post war era, and local and not-so-local developers saw the potential for resort hotels where travelers could relax and enjoy the nightlife of Las Vegas. Some of these tourists were lucky enough to take in the hundreds of nuclear tests on display in the desert 65 miles northwest of Las Vegas. Spectators were entertained not only by above- and below-ground nuclear tests, but also by spectacles such as the Miss Atomic Bomb pageant and refreshments such as the atomic cocktail. The huge number of people working on the nuclear tests also contributed to the burgeoning Las Vegas population.

Modern transport infrastructure improvements continued through the 1950s with the development of Interstate 15 to California and the construction of the Las Vegas airport. This allowed for fast, easier access for more tourists from outside the region. Local leaders made a point of targeting the growing business convention market as evidenced by the construction of the Las Vegas convention center in 1959. This market was also aided by the recently passed income tax deduction for conference attendees, known locally as the "three-martini lunch". During this decade, much of the funding for casino development and management was financed by well connected, not always above board, outside management entities. The exploits of these groups and their lasting legacy in Las Vegas have been made it into a host of Hollywood movies, including "Bugsy" in 1991, and the 1995 feature film "Casino".

By 1960, the population of Las Vegas reached 64,405 and the popularization of the desert city in the American mind was being cultivated through movies such as "Ocean's 11", starring the Rat Pack, who would go on to entertain a multitude of tourists in the city for the next thirty plus years. It was the 1960s, when Howard Hughes moved into the top floor of the Desert Inn in Las Vegas and slowly began buying up five more hotels over the next 4 years, thereby ushering in the era of corporate hotels in Las Vegas. As the population doubled again during the decade of the 1960s, more water became a necessity. This time, the south Nevada water project was drawn up to provide water directly from Lake Mead. The fabric of Las Vegas was torn beginning around 1969 when a 2-day riot broke out, fueled at least in part by racial strife, and crime began to rise into the 1970s. In the 1970s, Las Vegas was characterized by critiques on the American Dream by Hunter S. Thompson, shows by Elvis and Liberace, the mob feeling the heat by the corporate interests, and competition from the opening of Atlantic City, the New Jersey gambling paradise on the East coast that further exacerbated the economic slowdown beginning around 1976. The gas crisis and the cancellation of the Trans World Airlines non-stop flight from New York around the same time did not help the situation.

By 1980, 164,674 people called Las Vegas home. The 1980s in Las Vegas included more disasters as a series of hotel fires, an air show accident, and several large explosions at the nearby Pacific engine plant made the headlines. However, the economy improved in the late 1980s with the construction of such mega hotels as the 3000 room Mirage and 5000 room MGM Grand, which opened to a public hungry for theatrical attractions and large scale productions that went beyond the simple bells, whistles, and winnings of the slot machine. This was also the era of the master planned communities, which, simply put, was "suburban living at its best."[27] These communities led to a large amount of real estate development surrounding the downtown area and pushing out into the peripheral desert environment. Towards the end of the 1990s the investment in theatrical tourism and mega-resorts began to pay off as Las Vegas became the number one tourism destination in the world, surpassing Orlando in 1999 and the holy city of Mecca in 2001. The city and tourism infrastructure in Las Vegas consists of more than 130,000 hotel rooms, 200,000 slot machines, 15,000 miles of lighted neon tubing, all consuming approximately 22 million Mwh of electricity per year. Over the 20-year period from 1990 to 2010, the population of Las Vegas increased from 258,295 to 583,756, ranking as the 30th most populated city in the U.S. (Fig. 4.22).

However, the Las Vegas economy, especially the real estate market, was hit hard when the housing mortgage crisis struck in 2007. Many have

Fig. 4.22 The famed Las Vegas Strip (http://commons.wikimedia.org/wiki/File:Las_Vegas_strip.jpg By Jon Sullivan. [Public domain], http://www.pdphoto.org/About.php)

---

[27] According to: http://www.lasvegasmove.com/las-vegas-planned-communities.asp; "The master plan is the be-all end-all to community living in the 21st Century. You will appreciate a complete community to support your every need from schools and churches to shopping and the arts. Housing ranges from the low priced up to the very expensive, and in any desired size or floor plan. Examples of these communities include: Summerlin, Green Valley, Anthem, Desert Shores, The Lakes, Seven Hills and many more!".

suggested that Las Vegas and its master planned communities—essentially tract housing plopped down in a desert—represented the epicenter of the crisis as home prices in southern Nevada dropped an average of 65 % between 2008 and 2012. A rough estimate by a University of Nevada Las Vegas real estate expert suggests total real estate losses at $113 billion for the local market. Accompanying this loss in real estate dollars was a loss of tens of thousands of jobs in construction, real estate, finance, and other real estate-related professions. At the nadir of the housing market, those buyers who bought existing homes had a 50/50 chance of foreclosure, and those who bought a new home had a 1 in 4 chance of seeing a foreclosure notice in the next 5 years. In addition, many planned projects in the downtown area including hotels, casinos, and other tourist attractions have been put on hold. As we have mentioned, the economic slowdown in Las Vegas is partially a result of a decline in the discretionary income that tourists have to draw upon for vacations and entertainment travel, as well as a highly inflated real estate market built with cheap credit and energy during the housing market bubble. As the megatrends of climate change, energy scarcity, and decreasing discretionary income continue to impact the tourist economy, Las Vegas will continue to function as the "canary in the coal mine", a twentieth century cheap energy desert urban oasis attempting to maintain relevance. Ladies and gentlemen, place your bets.

Los Angeles, California: Life in the Desert Fast Lane

Los Angeles is the epitome of explosive growth in the twentieth century in the arid Southwest. The population of Los Angeles has increased every decade since it was established in the first half of the nineteenth century. The first Census Bureau record of Los Angeles in 1850 reported a total 1610 people living in the city. From 1890 to 1910, the city's population increased by 650 % to over 300,000 people. By 1930, there were over a million people living in L.A. and this number reached nearly two million by 1950. Population growth continued in the second half of the twentieth century both for the city and the region. The population of L.A. was 1.9 million in 1950 and 3.8 million by 2010. L.A. County grew to 4.1 million in 1950 and 9.8 million in 2010. Southern California had 18.9 million people in 2010 in the counties of Los Angeles, Orange, Ventura, Riverside, and San Diego. John C. Fremont, one of the first non "Anglo" explorers of Southern California, said that it would never be settled by Europeans. He said, basically, a nice place, but it's too dry for European–American settlement.

History

The history of Los Angeles, the second largest metropolitan area in the country, is characteristic of many large urban areas in the arid southwest. The most important environmental resource issue is water. Los Angeles depleted local water resources early on and then began reaching further and further afield in an increasingly desperate attempt to obtain sufficient water.

About 5000 Native Americans lived in the Los Angeles basin when Europeans first arrived in the area in the eighteenth century. What was later named the Los Angeles River flowed year round, and there was rich agricultural production by Native Americans throughout the basin. The early survival and success of Los Angeles depended greatly on the presence of the nearby and prosperous Native American Gabrielino village called Yaanga. Its residents provided the colonists with seafood, fish, bowls, pelts, and baskets. For pay, they would dig ditches, haul water, and provide domestic help. They often intermarried with the region's Mexican colonists.

Los Angeles had its beginnings between 1765 and 1771 as part of the settlement of missions in what the Spanish called Alta California. The military also established forts, and one role of the missions was to supply the troops with food and other goods. The mission of Los Angeles was located in the valley of the Los Angeles River in an area of black and loamy soils. The official date for the founding of Los Angeles was in September 1781 when the town was called "El Pueblo de Nuestra Señora de los Angeles de Porciuncula", or "the town of Our Lady of the Angels of Porciuncula." Porciuncula was the name of the river before it was renamed the Los Angeles River. The name Porciuncula comes from the small church near the town of Assisi, Italy where the Franciscan religious order, which ran the missions, got its start in the thirteenth century. By 1821, Los Angeles was a self-sustaining farming community, the largest in southern California. The Los Angeles River, which still flowed all year, provided water for the city. The area had vineyards of the Mission variety that were introduced by the Franciscan Brothers. Later vines were introduced from the Bordeaux region to improve the quality of wine. The area was also an important producer of cattle.

When Mexico gained independence in 1821, the people of Alta California became citizens of Mexico. This brought economic growth and by 1841, the population of Los Angeles was nearly 1700. Immigration from the U.S. and Europe increased, and these immigrants would play an important role in the takeover of the area by the U.S. The Mexican American War broke out in 1846. In August of that year, Commodore Robert Stockton landed in the coastal town of San Pedro and took control of Los Angeles after a series of battles in January 1847, ending Mexican control of the area.

With the discovery of gold in the Sierras in 1849, Los Angeles developed a thriving cattle industry to supply beef and other foods to miners in San Francisco, the Sierra foothills, and other areas in the central part of the state. However, settler population growth was slow, as the population increased from about 1600 in 1850 to approximately 5700 in 1870. Growth was much more rapid for the rest of the nineteenth century and the population reached slightly over 100,000 by 1900.

The construction of railroads helped spur growth. The transcontinental railroad, which reached Los Angeles in 1876, changed Southern California forever. The southern hub of the Central Pacific railroad was Los Angeles, and this led to further rapid growth of the city. The CP Railroad stretched from Sacramento, CA to Ogden, UT and was part of the first transcontinental railroad in North America. In 1871, dredging led to the establishment of a harbor at San Pedro that would eventually become part of the Port of Los Angeles. The Southern Pacific railroad reached the port of Santa Monica in 1876. These developments led to Los Angeles becoming an important port, a role it retains to this day. Oil was discovered in 1892, and Los Angeles grew into an important center for oil production. By 1923, the area was producing one fourth of total world oil production. After this, the economy of the area diversified. L.A. became a center for the movie industry when East coast producers relocated to the area to get away from the Edison Manufacturing Company's New York City headquarters and the ongoing film patent wars that prevented independent filmmakers from using unlicensed equipment. The proximity to Mexico and the temperate climate was also attractive to early moviemakers as it allowed for filming to be conducted year round. The city also developed a garment industry in the years just before World War II. At first devoted to regional merchandise such as sportswear, the industry eventually grew to be the second largest center of garment production in the United States.

The history of Los Angeles is intimately tied to water. Unlike the wet east where rainfall and rivers were abundant, southern California was, and still is, arid. In the early twentieth century, Los Angeles' relationship to water changed. For much of the nineteenth century, the Los Angeles River flowed year round. The river originated from the aquifer under the San Fernando Valley, supplied from the runoff of the surrounding mountains. The underlying geology caused water from the aquifer to surface at the Glendale Narrows. From there it flowed about 20 miles to the sea. There were also a number of other smaller water sources in the area. However, rainfall in the area is erratic, with droughts interspersed with strong storms. This led to great variability in the flow and course of the river.

Although the Los Angeles region is arid with an average annual rainfall of about 15 in., the area also has very intense rainstorms. Some of the most intense storms occur in the winter months in the San Gabriel Mountains north of Los Angeles. For example, more rain fell in January 1969 in the mountains in 9 days than normally occurs in New York City in a year. And in 1978, nearly a foot of rain fell in 24 h. These storms are the cause of massive debris flows described by John McPhee in his book *The Control of Nature*. McPhee also discusses L.A.'s vulnerability to wildfires. After extremely heavy rains in 1938, the federal government initiated a large program that lined the Los Angeles River and its tributaries with cement. A number of dams were built and catchment basins were constructed to capture debris flows caused by heavy rains.

Until late in the nineteenth century, the Los Angeles River supplied enough water for the city. But when groundwater was withdrawn from the San Fernando Valley, the river did not flow year round any more, and it dried up in the 1920s. Today's Los Angeles River is a cement lined ditch that serves mainly for flood control.

By the turn of the twentieth century, the city began to look elsewhere for water. The business of importing water to L.A. transpired amid many shady deals fueled by the rampant greed and corruption of those seeking control of the resource. Plans were drawn up for aqueduct projects, and insider knowledge provided large returns for those in the know. For example, L.A. insiders undermined water rights from residents of Owens Valley, and the water wars that played out during this time were famously documented in the film "Chinatown" (1974) and in Marc Reiner's book *Cadillac Desert*.[28] Over time, the city looked farther afield and eventually tapped three major water sources, the Owens Valley via the Los Angeles Aqueduct, the Colorado River via the Colorado River Aqueduct, and northern California via the California Aqueduct. This last aqueduct was part of a gigantic water supply system called the Central Valley Project that supplied the major cities of northern California (like the San Francisco region, Sacramento, and Bakersfield), the vast agricultural region of the Central Valley, and last but not least, Los Angeles (Fig. 4.23).

The first major non-local source of water was the Owens River, located approximately 250 miles northeast of the city on the Nevada border. J.B. Lippencourt, who worked for the U.S. Reclamation Service and also secretly received a salary from the City of Los Angeles, convinced farmers and water companies in the Owens Valley to give up water rights to over 200,000 acres of land. Lippencourt's studies of water availability in Owens Valley were used in the design of an aqueduct to bring water to Los

---

[28] Marc Reisner. 1993, *Cadillac desert: The American West and its disappearing water*. Penguin.

Fig. 4.23 The three major aqueducts that supply water to Los Angeles. These include the Los Angeles Aqueduct from the Owens Valley, the Colorado River Aqueduct from the Colorado River, and the California Aqueduct from Northern California (Map courtesy of Dr. Lynn Ingram, from Lynn Ingram and Frances Malamud-Roam. 2013. The West Without Water. University of California Press, Berkeley. 256 p. Used by Permission University of California Press)

Angeles. In 1905, L.A. city officials drained water into sewers to create an artificial drought to convince voters to approve bonds for over $22 million to build the aqueduct. The opening of the aqueduct provided the city with four times more water than it needed at the time. But further supply would be needed for continued population growth throughout the twentieth century.

The Los Angeles aqueduct consists of two separate pipelines. The first Los Angeles aqueduct is a gravity-fed, 223-mile long, 12-foot diameter steel water conveyance pipeline that was built over a 5-year period from 1908 to 1913 at a cost of $24.5 million. The maximum rate of flow of water through the pipeline is slightly over 13 m³/s (about 300 million gallons per day). Nowadays, the gravity fed water is harnessed for hydropower with the added benefit of electricity production as it travels south. Water that was originally used to irrigate the Owens Valley for agriculture was redirected south to the city to quench the urban population's growing thirst for water. Several locals in the Owens Valley took offense to the diversion of local water resources and sabotaged the pipeline during the California Water Wars of the 1920s. The Owens Valley became much drier and less productive as a result of the aqueduct while the population of Los Angeles was able to grow as a result of the diverted water. Prior to the post WWII suburbanization phenomena in L.A., the aqueduct had the effect of changing agricultural practices in the L.A. region from wheat production to more water intensive irrigated crops such as corn, beans, squash, and cotton because for a while water was plentiful.

The second Los Angeles aqueduct pipeline begins just south of Owens Lake at the Haiwee Reservoir and runs parallel to the first aqueduct for approximately 140 miles before merging with the first pipeline. This project took 5 years, from 1965 to 1970, at a cost of $89 million. Unlike the first pipeline, the second pipeline is not gravity fed and therefore requires pumping to move the water south to the city. The pipeline has the capacity to move water at a rate slightly more than 8 m³/s (nearly 200 million gallons per day).

The Colorado River Aqueduct was constructed in the 1930s as the largest depression era public works project in southern California. It employed about 30,000 people during its 8-year construction period. The purpose of the aqueduct project was to deliver water from the Colorado River at Parker Dam, located approximately 242 miles east of L.A., to the city's growing population. William Mulholland, the self taught engineer and overseer of the Los Angeles aqueduct, concocted a plan to dam the river in the 1920s and harvest the water from Lake Havasu on the California–Arizona border to provide a new source of water for L.A. In 1931, a $220 million bond was approved for construction to begin and, by 1939 the waters of Lake Havasu began flowing into L.A. The aqueduct is composed of 63 miles of canals, 92 miles of tunnels, and 84 miles of buried conduit, along with several siphons, five pumping stations, and a couple of reservoirs. The 9000 HP pumps are something to behold; each can fill an Olympic-sized swimming pool in about 20 s The aqueduct has a much larger conveyance capacity (45 m³/s or about a billion gallons of water per day) than the Los Angeles Aqueduct.

The California Aqueduct consists of three branches known as the east, west, and coastal branch. The west branch provides water to western Los Angeles at its terminus in Castaic Lake. The east branch supplies an artificial reservoir known as Lake Palmdale before terminating at Lake Perris in Riverside County. The coastal branch terminates at an artificial lake known as Lake Cachuma in Santa Barbara County. The California Aqueduct involves a series of canals, tunnels, and pipelines that begins at the San Joaquin-Sacramento River Delta where water is pumped southward at a maximum capacity of 370 m^3/s. Approximately 40 % of the freshwater inflow into the San Joaquin-Sacramento River Delta is diverted. This has had a profound effect on the ecosystem health of the delta. Several pumping stations are required to convey water along the length of the 710-mile aqueduct.

The water subsidies conveyed by the three major aqueducts allowed for continued rapid expansion of the city. The city covered 29 $mile^2$ in 1890, 90 $mile^2$ in 1910, and 450 $mile^2$ by 1932. World War II led to dramatic growth as the city was a center for production of airplanes, ammunition, and other war materials. Automobile culture began to claim the city in the 1940s post war era after the construction of the Pasadena Freeway, now the Arroyo Seco Parkway, which was the first freeway in California and the western U.S. The freeway represented the transitional phase between older style parkways and modern freeways. Building on the industrial base created by the war, post-war L.A. became a center for the production of automobiles, tires, furniture, and clothing. After World War II, the city became a center for the aerospace and defense industries. It also became the national hub for movies, radio, and television. The television industry and recording industry reinforced the city's reputation as the entertainment capital of the country. The city spread rapidly, with development in the San Fernando Valley and the building of more freeways. The car culture that swept the country at the time led to severe air pollution problems, because the city is located in a bowl surrounded by mountains. Smog plagued the city throughout the 1940s and early 1950s before measures were taken to curtail polluting emissions generated from so much happy motoring. The city continues to experience poor air quality conditions on many days and the war against smog is an ongoing battle. Air pollution regulations instituted during the 1970s and improved public transportation has led to some advances, but these measures have largely been offset by the increasing population (Fig. 4.24).

In 1961, wildfires destroyed nearly 500 structures and scorched almost 16,000 acres of land in the BelAir, Brentwood, and Topanga Canyon neighborhoods. The city remains vulnerable to wildfires due to the arid landscape that surrounds the city and the movement of people into fire-prone areas. Further drying as a result of climate change may heighten this vulnerability, especially in those areas where suburban housing developments sprawl amongst the hillsides on the outskirts of the city.

Fig. 4.24 The famed sign for Hollywood. Note the semi-arid scrub in the foreground (http://commons.wikimedia.org/wiki/File:Hollywood_sign.jpg By Jon Sullivan. [public domain], http://www.pdphoto.org/About.php)

In addition, earthquakes pose a threat to the city. One of the more recent quakes, the Northridge earthquake of 1994, measured 6.8 on the Richter scale and caused considerable damage to buildings across the city. Larger ones appear inevitable, for California was formed by massive plates smashing into each other, and the plates are still moving.

In the second half of the twentieth century, Los Angeles experienced major economic changes. Most major manufacturing has moved elsewhere to other states or developing countries. By the end of the century, there were no longer any major auto plants, steel mills, furniture manufacturers, or agricultural operations. Even the aerospace and entertainment industries had declined. The aerospace industry declined due to the end of the cold war, and many entertainment industry companies had moved to states such as Louisiana, offering cheaper production facilities and tax incentives. During the last few decades the L.A. economy experienced growth mostly in the real estate and financial sectors.

One of the economic sectors that has grown dramatically since the mid twentieth century is international trade. The combined ports of Los Angeles and Long Beach handle approximately 47 % of all containers imported into the U.S. Five of the top ten ports are on the west coast. This dramatic growth in trade reflects the internationalization of trade in the twentieth century based on abundant supplies of cheap energy and natural resources that allowed the development of a far flung industrial system connected by an extensive trade network.

Top ten container ports in the U.S. (Fig. 4.25)

Port	Number of inbound containers in the first quarter of 2012[a]
Los Angeles	946,063
Newark/New York	673,723
Long Beach	673,172
Savannah	271,526
Seattle	195,326
Norfolk	188,802
Oakland	180,445
Houston	156,326
Charleston	155,951
Tacoma	118,188

[a]http://www.logisticsmgmt.com/images/site/LM1205_TopPorts.pdf. Containers are generally given in TEUs. A TEU is the 20-foot equivalent unit based on the volume of a typical 20-foot-long container

The changes in the LA economy have led to socio-economic changes, since many new jobs tended to be low-wage. The Hispanic population of the region has grown, and the proportion of poor families has increased. The city will soon be majority Latino. The growing gap between rich and

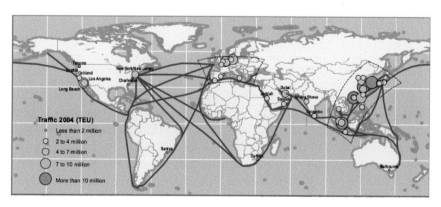

Fig. 4.25 Map of the top container ports in the world in 2004. The largest ports in the world are located in the manufacturing capitals of Asia; these ports ship massive volumes of consumer goods to wealthy economies of North America and Europe (Reprinted from Rodrigue, Jean-Paul, and Michael Browne. 2008. International Maritime Freight Transport and Logistics. In Transport Geographies: An Introduction. R. Knowles, J. Shaw, I Docherty (eds) pgs. 156–178. Whiley-Blackwell Publishing, Used by Permission)

poor makes Los Angeles one of the most economically divided cities in the nation. This mirrors a wealth gap in much of the nation. Predicted future water and energy scarcity pose very serious challenges for the city and, indeed, for the entire State of California and the Southwest.

Portland, Oregon: Green Paradise, Or Is It?

Portland is the second largest city in the Pacific Northwest after Seattle. The climate in Portland is classified as temperate oceanic with warm, dry summers and mild, damp winters, where the sun does not shine for weeks at a time. On average, only 13 % of the annual total average rainfall of 36 in. falls in the months of June, July, August, and September.

In 1850, the population of the city was just 821. However, with the expansion of western settlement over the next 50 years, the population grew to 90,426 by 1900. Population growth accelerated up until the Great Depression when it slowed, but growth began again in the 1940s. There were 373,628 people living in Portland by 1950. Population declines occurred in the 1960s and 1980s before the present trend in growth began in the 1990s. By 2010, the population of the city had increased to more than half a million people, and the metropolitan area included about 2,225,000 individuals. Portland has emerged as the archetypical "green" city, and is often considered the greenest city in the U.S., especially by those who live there. We will consider what this means later.

History

Portland was established at the confluence of the Willamette and Columbia Rivers. The area was settled by British, Canadian, and American traders and trappers, and was incorporated on February 8, 1851. This area was rich in natural resources, and the Columbia River was navigable by ocean-going vessels up to Portland. The rich natural resources of the area supported bands of the Upper Chinook Indians as well as the early economic activity of Portland. The Clatsop Indians also inhabited the coastal region near the mouth of the Columbia. A malaria epidemic in 1825–1826 largely decimated the Chinooks and Clatsops. The Lewis and Clark Expedition (1804–1806) was the first major scouting party of white settlers from the east to survey and map the Oregon Territory. The Expedition was largely a result of Thomas Jefferson's hope of finding a waterway to the Pacific, but the Expedition discovered that there was no all-water route across the continent. The Oregon Territory mapped during the Expedition included pristine mountains, valleys, rivers, lakes, and forests that provided a large resource base for the area's Native American population, as well as settlers from the east. The Oregon Territory was the expanse of land stretching from California to Canada along the Pacific coast and eastward, encompassing parts of what are now the states of Washington, Idaho, and Montana. When the Lewis and Clark Expedition reached the Oregon coast, just downriver from Portland, they established Fort Clatsop using the abundant grand fir trees that grew in the cool climate and wet evergreen landscape. Fort Clatsop was initially a salt making camp for the Expedition, whereby seawater was boiled in large kettles until it evaporated, the residual salt was scraped from the sides of the kettle and then used in the preservation of food. The difficulty of food preservation in such a moist environment is also evidenced by the fact that the first structure erected at Fort Clatsop was a smokehouse. The expedition survived predominantly on elk they hunted and dried fish and roots purchased from the Clatsop Indian tribe. They were occasionally able to purchase dogs to supplement their diet and Meriwether Lewis preferred them to deer or elk meat, though Clark found it less amenable to his own palate.[29] When Lewis and Clark returned back east in 1806, their reports provided the most detailed description and maps of the opportunity awaiting those adventurous souls that wished to pursue a new life west of the Rocky Mountains. However, it would be another 40 years until Portland was established as a sizeable frontier settlement.

Portland's location at the confluence of the Willamette and Columbia rivers provided a natural advantage over other potential settlement sites. The depth of the channel at Portland allowed access for ocean-going vessels, making it a convenient thoroughfare for travelers, traders, trappers, and settlers. These vessels were too large to reach Oregon City, Oregon's largest settlement in the 1840s. Many of the travelers passing through Portland at the time were heading for Fort Vancouver, the nearby fur-trading outpost along the river that served as the Columbia District headquarters of the Hudson Bay Company. The settlement got its name in 1845 when Francis Pettygrove and Asa Lovejoy agreed that the winner of a coin toss would name the new town after their own place of residence. Pettygrove, a Mainer, won the toss and the town was named Portland. Had the toss gone the other way, we might be referring to Portland as Boston. At the time, Portland was a male dominated frontier village of 821 inhabitants often referred to as "Stumptown" because of the numerous stumps from fir trees that had been logged during land clearing in the early settlement years. By 1850, Portland

---

[29] Stephen Ambrose. 1997. *Undaunted Courage: Meriwether Lewis, Thomas Jefferson, and the Opening of the American West.* Simon & Schuster. 521 p.

was the largest settlement in the Pacific Northwest, and the city served as the major regional port for most of the nineteenth century. The population of Portland grew dramatically with trade and commerce centered on port traffic. This changed in the 1890s when railroads from the east reached Seattle. This allowed railway access to the deepwater Puget Sound harbor in Seattle and eliminated the need for sailors to navigate the dangerous Columbia River mouth bar, a major impediment to ship traffic in a stretch of coastal ocean known commonly as the "Graveyard of the Pacific". Seattle then took over as the largest port in the northwest (Fig. 4.26).

The Lewis and Clark Centennial Exposition world's fair was held in Portland in 1905. This brought more recognition to the city and contributed to more than a doubling of the population from 90,426 in 1900 to 207,214 in 1910. Between 1880 and 1910, growth soared between 95 and 163 % per decade. Growth then slowed as it took another 30 years to reach a population of 305,000. After the Great Depression years of the 1930s, the city experienced an economic boom and grew by 23 % reaching nearly 374,000 by 1950. This growth was spurred by two factors. The first was nearly $2 billion in spending by the Bonneville Power Administration that included a number of large dams on the Columbia. This made electricity plentiful and cheap in the Northwest. In addition, the onset of World War II led to a major expansion of shipbuilding in the region. Three shipyards in the area built aircraft carrier escorts and liberty ships. These shipyards employed about 150,000 workers.

During the 1950s, 1960s, and 1970s, population growth was less than 3 % per decade, and in the 1960s and 1980s, the city's population actually declined. From the mid-to-late 1990s, the city prospered with the booming dot-com industry. There was an influx of people in their 20s and 30s, attracted to the graphic design and internet industries as well as trendy sportswear companies like Nike and Adidas. When the dot-com bubble burst, there was a large pool of creative young people in the city, and their numbers grew with an influx of like-minded people from Seattle and San Francisco. The 2000 census listed over 10,000 artists in Portland. During this time, the city gained a reputation as a trendy place to live and work, as portrayed by the satirical series "Portlandia", with abundant natural areas that also served as boundaries to urban growth.

The urban growth boundary surrounding Portland provides a limit to urban sprawl that has allowed the city to culture a green appeal, largely as a result of the progressive politics and planning of Tom McCall (R), Governor of Oregon from 1967 to 1975. Governor McCall's first major victory after being elected was the passage of the "Beach Bill" which granted the state the power to zone the beaches, thereby preventing overdevelopment by private developers along the Oregon coast. Next, he proposed that

Fig. 4.26 Portland, Oregon Harbor in the early 1900s (http://en.wikipedia.org/wiki/File:Portland_Oregon_harbor_early_1900s.jpg, [Public domain], via Wikimedia Commons)

all local governments complete comprehensive land use plans in 1969 and then enacted the country's first mandatory bottle bill in 1971. A nickel each for a bottle or can requires the consumer to pay up front for waste while creating a monetary incentive to recycle. During Governor McCall's terms, Portland enacted the nation's first urban growth boundary that is still maintained and is now also required for other cities in Oregon. It has since been adopted by cities in Washington and Tennessee. Governor McCall's urban growth boundary vision was backed by enough of the state's citizens, including environmentalists and farmers, to become a statewide mandate, requiring all cities and counties to have a long-range plan for urban growth and the protection of the state's natural resources and landscapes. Based on these initiatives, Portland touted itself as the greenest city in the nation. There is no doubt that these initiatives have made Portland a desirable place to live.

But what does being green mean? Green is often used as a feel good moniker appealing to the emotions of college educated, earth conscious adults and their children. The appeal of being green is to people of conscience about the relationship of consumption, pollution, and the environment. The use of the term green hints at our notion of the color we perceive when we see the summer leaves on trees and fertile green pasture grasses. "Greenwashing" is a term that is applied to tactics used by companies and marketing agencies to pull at the heartstrings of consumers, directing their choices and preferences towards a product that supposedly contains some morally or ethically charged notion of goodwill. These campaigns have stepped up their efforts considerably in the last two decades as environmental awareness has entered into mainstream consumer culture. The notion of caring about the environment through the consumption of green products has become a cliché status symbol for a certain demographic of consumer society. Whether items or people labeled "green" do, in fact, have less impact on the environment than others that are similar but not called green is something that rarely has been studied objectively. We address this below (Fig. 4.27).

The west coast states of California, Oregon, and Washington are often considered to be at the forefront of the green environmental movement. The history of this movement dates back at least to the 1960s where progressive politics surrounding free speech, civil rights, anti-Vietnam war, and flower power movements, signaled that it was time for change in the U.S. California was the hot bed of this movement, and the coincidental location of the electronics heartland in the Bay Area at least partially influenced the progressive thinkers who would become leaders in the development of the computer age. Apple and Microsoft were developed in California and Washington respectively, and the I-5 corridor that connects these two areas

Fig. 4.27 The Portland skyline with Mount Hood in the background. The verdant landscape helps further the city's "Green" image (http://commons.wikimedia.org/wiki/File:Portland%26MtHood.jpg, Photo by David E. Wieprecht [Public domain] Courtesy of United States Geological Survey, via Wikimedia Commons)

includes Portland. The interconnectedness of the region and the rise of the software industry led to ever increasing population and cultural dispersion among these coastal cities over the last three decades. However, the notion that one city in an industrialized nation is more "green" than another is often simply an aesthetic notion rather than any sort of reality that is going to lead to a sustainable society living in harmony with the planet. The energy and material flows required by the populations of these so-called "green" cities are hardly different from other cities that are often deemed less progressive.

For example, Dr. James Brown and colleagues at the University of New Mexico calculated energy and material use by Portland, Oregon, because it is considered one of the most sustainable, "green" cities in the U.S.[30] They reported that "each year the Portland metropolitan area consumes at least 1.25 billion liters of gasoline, 28.8 billion megajoules of natural gas, 31.1

---

[30] Burger, J. R., et al., 2012. The macroecology of sustainability. *PLoS Biol*, 10(6), e1001345.

billion megajoules of electricity, 136 billion liters of water, and 0.5 million tons of food, and the city releases 8.5 million tons of carbon as CO_2, 99 billion liters of liquid sewage, and 1 million tons of solid waste into the environment. Total domestic and international trade amounts to 24 million tons of materials annually". These figures for consumption and generation of wastes are about what would be expected for a city the size of Portland. This is not to say that Portland is not a nice place to live, only that it is dependent on economic activities and ecosystem services in other areas just like other cities. If we look at Portland in more detail, however, a more nuanced picture emerges. The work of Jones and Kammen (see Chap. 3) shows significant differences in CO_2 emissions in different parts of Portland. CO_2 emissions of residents living in the city center of Portland are actually low compared to many other places (~22–24 metric tons CO_2 per capita direct plus indirect emissions). However, outside of the city center, emissions more than double to near 60 metric tons per capita. According to Jones and Kammen, housing size, transport, and income affect these differences. So we must consider both what is happening in a particular city as well as the broader societal context of energy and materials production and consumption.

There are several aspects of Portland that give it the appearance of a green city and an appeal different from other similar-sized cities in the U.S. Portland's green image is based partially on the city's location at the intersection of two large rivers, the numerous small green spaces located within the metro area, a metro area delimited by protected greenspace of an urban-region ring, an urban-growth boundary that sharply reduces the rate of outward urbanization, and the especially wide and long green wedge of Forest Park that serves as a recreational space and wildlife habitat for the city. In addition, there are market gardening areas located in close proximity to the city, citywide recycling and composting programs, many miles of bike lanes, and a commuter rail system that provides various public transportation options to visitors and residents. These features provide aesthetic as well as ecological benefits that make Portland a desirable place to live. However, the city is still a modern industrial urban environment that requires enormous material and energy throughput on a level that is on par with many similar-sized American cities. The April 2014 issue of National Geographic magazine provided a diagram of the average hours per year spent in traffic for a number of cities in the U.S. and other industrialized nations.[31] At just over 40 h, the average Portlander spent less time in traffic than residents of Los Angeles, New York City, or Houston but more time

---

[31] *National Geographic Magazine*. April 2014. Commuter Science. p. 20.

than residents of Baton Rouge, Orlando, or New Orleans.[32] As noted above, this time spent in traffic is likely largely a function of Portlanders living in suburban areas of the city. Portland receives a suboptimal amount of sunlight year round; therefore, the potential to expand solar electricity generation is less than sunnier areas. The potential for wind power generation in Oregon is considerable but is located along the coast. Portland's location about 80 miles inland from the coast makes it difficult for cost effective wind power transmission and distribution to the city. Portland benefits from the enormous hydro electricity generated by dams on the Columbia River. Portland relies heavily on fossil fuels for materials and energy produced elsewhere and the energy required to transport these.

Portland garners a large amount of attention when it comes to being green and sustainable, but at the end of the day the city is probably not a whole lot more or less sustainable than any other industrialized city in the U.S. Any green future that would be combined with existing living standards is un-achievable in isolation from the surrounding hinterlands and larger global economy that provide the energy and materials required for a thriving twenty-first century American city. Later in the book we will discuss how the cities we have been discussing rank in terms of "real" sustainability.

Summary

In this chapter, we have chronicled the development of 12 cities and 10 regions in the U.S. from just before European colonization until the early twenty-first century. These cities and regions represent a cross section of American communities from the largest urban areas (New York City and Los Angeles) to smaller regional cities (Asheville, Cedar Rapids, and Amarillo). The initial settlement and growth of all of these cities were based on local resources such as timber, wild game, fur-bearers, fisheries, rich soils for agriculture, and coastal and river trade routes. In all cases, Native Americans living in these areas were largely pushed aside by disease, killing, and relocation. Las Vegas is an exception because it sprang largely *de novo* from the fossil fuel age. Growth in the eastern US took off first in the late seventeenth and eighteenth centuries, based on the rich natural resources of eastern North America. Growth in the western US generally began in the nineteenth century and did not take off until the twentieth

---

[32] These were the only cities of the ones we discuss in this book that were included in the National Geographic article.

century with the advent of cheap energy to overcome the lack of water and lower level of ecosystem services. This was key to growth in the West, as large public works projects made water available to large areas where little existed before. For example, by the mid nineteenth century New York had over a half million residents, and New Orleans about 116,000. By contrast, Los Angeles, now the second largest city in the U.S., had a population of 1610 in 1850 and Las Vegas did not even exist. In the next five chapters, we discuss some of the mega trends that will influence the sustainability of all these cities and regions in the twenty-first century.

Chapter 5

The Wealth of Nature Is the Wealth of Nations: Ecosystem Services and Their Value to Society

Everyone is aware of nature. It is all around us. Nature is the weather, sunshine, winds, rain, flowing rivers, and the tides on the coast. It is also the living world, including forests, grasslands, wetlands, and farm fields. There is also much that we don't see, from microscopic organisms such as protozoans, bacteria, and fungi to the many chemical reactions that take place in nature. If we live in the country, we may experience much of nature very directly. Even in the middle of cities, there are parks, trees along streets, weeds in the cracks in sidewalks, and for many the green spaces around homes. But for most people, especially in a developed country like the U.S., nature is out there. It is something we enjoy, or not, as our mood dictates. Unless we work in one of the occupations where people interact intimately with nature, such as farming, forestry, field biology (the authors of this book), or fishing (Alice Monro's drowned fisherman), our contact with nature is casual and somewhat intermittent. For most people living in cities, jobs have little directly to do with nature. This is important since the majority of the world's people now live in cities, reaching more than 80 % for the U.S. Weather is a factor in our lives, but it is generally externalized as a potential source of pleasure or inconvenience, not the matter of economic survival that it is for a farmer. Nature is something we can choose to experience or not, often with—at best—a vague realization that others working in primary economic production such as farming, fishing, logging, and mining are dependent on nature for their direct economic well-being, which in turn assures access for office workers, doctors, teachers and carpenters to the essential products of these activities.

© Springer Science+Business Media New York 2016
J.W. Day, C. Hall, *America's Most Sustainable Cities and Regions*,
DOI 10.1007/978-1-4939-3243-6_5

In reality, nature has everything to do with our lives. Simply put, we cannot *not* live with nature because we are part of it. We can live without computers, smart phones, cars, microwaves, air conditioning and the many other artifacts of modern civilization. After all, our ancestors lived without these things for millennia. What they could not live without was the natural world. Nature gives us the climate, clean water, good soil, plants and animals that we eat, wood and stone for our houses, bees to pollinate our crops, and metals for our implements. The writer John Michael Greer calls this the wealth of nature in his book of the same name.[1] Ecologists call what nature gives us *ecosystem goods and services*. Since these goods and services are essential to any notion of sustainable cities and regions, let's take a closer look.

What Are Ecosystem Goods and Services and Why Are They Important?

Ecosystem goods and services are all the things that nature provides and that humans depend on for a functioning economy and, indeed, for our very survival. These services are provided by nature at no cost. To begin to think about the services of ecosystems and their value, consider this example. When a modern shopper goes to the supermarket and buys some seafood for 20 dollars (it could be a salmon from Puget Sound, a red drum from the Mississippi delta, or blue crab and oysters from Chesapeake Bay), what is it that these 20 dollars are paying for? They are paying for the labor of the fishers, the upkeep of the fishing vessel (including maintenance, payments to the bank, taxes, insurance, and other expenses), the fuel used to power the boat, the cost of getting the seafood to market, and the cost of the store, including the energy needed for heat and refrigeration, to sell it to us. What the 20 dollars are not paying for is all the complicated and intricate work that the sun and ecosystems do to produce the fish and shellfish. These include the hydrologic cycle that delivers freshwater to these three coastal ecosystems, the tide that raises and lowers water levels, the growth of plants that directly and indirectly feed the fish, oysters, and crabs and provide them habitat, as well as the natural processes that cleanse the water to keep the ecosystem healthy. There are many more such services that we will discuss later, but suffice it to say that without the natural world, there would be no fish, and indeed there would be no human economy. Life itself

---

[1] John Greer. 2011. *The Wealth of Nature*. New Society Publishers, British Columbia, Canada.

would not exist. The search for planets out in the universe that might contain life like ours emphasizes the rather narrow and exacting range of conditions that must exist for life, as we know it, to exist. Because our earth has an atmosphere and water and an appropriate temperature range, life evolved. We are not just talking about the millions of individual species that exist on earth, but all the interactions among these species and their interactions with the soil, water, and atmosphere. These species and all these interactions make up the ecosystems of the natural world. *Homo sapiens* are just as dependent on these natural systems as all other species are. We are unique only in the ability of certain individuals to distance themselves from intimate contact with these natural systems, at least much of the time.

We know that thousands of years ago, ancestors of modern humans caught and consumed salmon, drum, crabs, and oysters (they also consumed giant turtles, birds, and mammoths that are now extinct, and may have been made extinct by humans). Archeologists find the bones and shells of these organisms at sites where our ancestors lived and ate. And images of these organisms are found on artifacts and in ancient stories that come down from early humans. The beautiful, and strikingly modern paintings of deer, bear, bison, and other organisms on the caves at Lascaux in France and other ancient sites attest to the importance that these organisms played in the lives of ancient peoples, just as they do in our lives today. A drum or a salmon eaten thousands of years ago provided the same high quality nutrition that these fish do today. And studies have shown that consumption of high quality seafood was an important factor leading to the development of early civilization.[2]

An important point is that our ancestors who caught fish thousands of years ago received the same nutritional benefits but paid no money for them. The natural environment did the same work to produce the seafood. Thus the basic workings of the human economy evolved long before modern concepts of economics (we will talk more about the "science" of economics in a later chapter). Humans can, and did, live without formalized concepts of economics, but our modern economy simply cannot exist without the complex web of ecosystem services provided by the wealth of nature. All of the materials and energy we consume comes from the natural world. Human ingenuity (or technology) is absolutely critical in obtaining and processing the goods and services of nature so that we can use them, but technology can't create something from nothing. We also have to

---

[2] Day, J., J. Gunn, W. Folan, A. Yanez, and B. Horton. 2012. The influence of enhanced post-glacial coastal margin productivity on the emergence of complex societies. *The Journal of Island and Coastal Archaeology*, 7, 23–52.

consider how these ecosystem services vary over the landscape, and how humans have degraded them.

Robert Costanza of the Australian National University is one of the leading scientists studying ecosystem goods and services. He defines ecosystem services as the ecological characteristics, functions, and processes that directly or indirectly contribute to human well-being—the benefits people derive from functioning ecosystems.[3] By characteristics, functions, and processes, he means all that goes on in the natural world that makes them function well and provide benefits to humans. Ecosystem processes and functions may contribute to ecosystem services, but they are not synonymous. Rather these processes and functions describe biophysical relationships, and exist regardless of whether or not humans benefit. Ecosystem services, on the other hand, exist only if they contribute to human well-being and cannot be defined independently of humans. This is true from a human perspective, and indeed, humans need to think much more about what we are doing to natural systems and the way they support us. But if salmon, drum, crabs, or oysters could think and write about these things, they would also describe how these ecosystem services contribute to salmon, drum, crab, and oyster well-being. No doubt they would be incensed about how humans are affecting *their* natural world (notwithstanding the television commercials showing ants having a beer party or of animals agog at a sleek, new SUV). When natural ecosystems are degraded, ecosystem services are diminished, and the consequences have real impacts on human well-being. A major point we make in this book is that the value of ecosystem goods and services is not constant over the landscape. Some natural systems, or environments, provide much higher levels of these goods and services than others.

If nature were able to sell her services, the price might be very high. Imagine a kiosk at the foot of the Catskill Mountains where Mother Nature is hanging out. The director of the water works of New York City approaches, hat in hand, and asks for some water. "That might be very expensive", says Mother nature. "But I need it for my eight million people in New York," says the DWW. "I am willing to pay 500 million dollars." "What use do I have for money?" asks Mother Nature. "But maybe we can cut a deal. I would like you to restore the 300 square miles of oyster beds that used to be in the Hudson River. I would like for you to remove all the

---

[3] Costanza et al. 1997. The value of the World's Ecosystem Services and Natural Capital. *Nature* Vol. 387. 15 May 1997. See also Millenium Ecosystem Assessment, 2005. *Ecosystems and Human Well-being: Synthesis*. Island Press, Washington D.C. The subject of ecosystem goods and services and how we should value them is beyond the scope of this book but we provide a number of references in the literature section.

dams that are keeping my shad and salmon from using the rivers. I would like for you to reforest Eastern New York State, and I would like you to reduce your population and greenhouse emissions by 90 percent."

Do you think we would pay the price?

What Are the Types of Ecosystem Services?

Ecosystem services can be separated into four broad categories: supporting, provisioning, regulating, and cultural (Fig. 5.1). Supporting services refer to all the interactions that go on in an ecosystem that are necessary for all the other ecosystem services. Soil formation is an important supporting service. Any farmer or gardener can attest to the value of good soil for growing food. All of the chemical reactions that go on in an ecosystem are critical for chemical availability, such as the nitrogen and phosphorous cycles that maintain soil fertility. So too is the growth (or primary production) of plants of all kinds, because they provide the organic material that supports all other life. Provisioning services refers to services provided by nature that sustain our basic needs such as food, water, and shelter. The fishes, oysters, and crabs discussed earlier are examples of provisioning services. The wood cut from forests and plants growing in a field are provisioning services, as are the genetic resources that provide the incredible diversity of living organisms on which human life depends. Oil pumped from the ground is ancient plant material that has been processed over millions of years. Regulating services refer to the benefits realized from the regulation of ecosystems. Examples include maintaining the earth's climatic system, disease regulation, and water supply and purification. An example of the latter is the New York City water system, which receives water from forest preserves in the Catskill Mountains. This forest watershed provides abundant and clean water for the city. Without this, New York would have to spend billions of dollars obtaining and processing water from the Hudson and other polluted sources. Storm protection provided by coastal wetlands is another type of ecosystem service. The previous loss of these wetlands in the Mississippi delta exacted a heavy toll when Hurricane Katrina hit the coast.[4] Cultural ecosystem services include nature's ability to fulfill the spiritual and recreational interests of humans. The gods of early societies

---

[4] Shaffer, G., J. Day, S. Mack, P. Kemp, I. van Heerden, M. Poirrier, K. Westphal, D. FitzGerald, A. Milanes, C. Morris, R. Bea, and S. Penland. 2009. The MRGO navigation project: A massive human-induced environmental, economic, and storm disaster. *Journal of Coastal Research*, SI 54, 206–224.

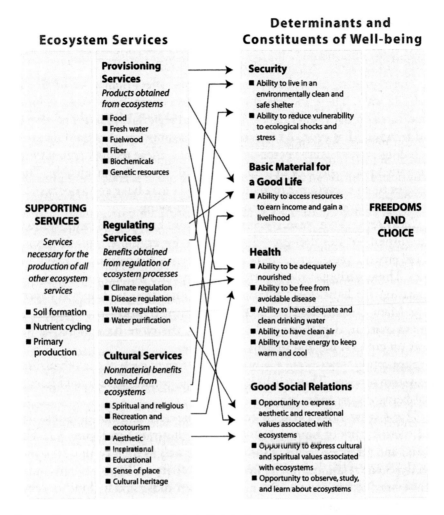

Fig. 5.1 Ecosystem services and their link to human well-being. From Ecosystems and Human Well-being: A Framework for Assessment by the Millennium Ecosystem Assessment. Copyright © 2003 World Resource Institute. Reproduced by permission of Island Press, Washington, DC

were often specifically related to values of nature. Most people have experienced the contentment and calming effect of just sitting in a forest, or on a lake or at the seashore. All the authors of this book started their careers as field biologists. One of the important things that drew us to this work was the opportunity it afforded us to be a part of nature, and to enjoy it regularly and intimately.

All of these services combine and interact to support humans in a variety of ways. They provide security, basic materials that we depend on for a good life, health, and good social relations. Thus, it is clear that nature is not just some adjunct to society. Nature is absolutely fundamental to humans, for we cannot get along without the services of nature; it enriches our lives immeasurably. Thus, the term ecosystem service implies nature serving mankind. In this respect, healthy ecosystems may be viewed as a resource that can be used by humans to maintain our way of life. This means that humans are a *part* of nature rather than *apart* from nature. Regardless of the philosophical or moral underpinnings of our relationship with nature, the term services in this case should be recognized as nature providing essential accommodations to the human species. Using energy and technology allows us to exploit these services more intensively, often to the point of degrading or depleting them, and to transform materials into new forms that did not appear in nature.

Ecosystem services within the four major categories are dependent upon one another. For example, a change in climate due to increase in carbon dioxide in the atmosphere (a regulatory service) can have negative impacts on food production (a provisional service). Alteration of ecosystems can result in changes, both positive and negative, in ecosystem services. The activities of modern society are, unfortunately, much more prone to degradation than enhancement as humans continue to scale up a resource-intensive lifestyle and culture that is unsustainable given known resource constraints and our growing knowledge of the adverse effects of large-scale, industrial economies on the natural environment.

It is important to note that ecosystems provide essential components of virtually all goods and services to humans and absorb the wastes generated by humans. For example, nature supplies both food (current solar energy transformed into plants), fiber (wood and paper) and fossil fuels (ancient solar energy transformed over millions of years into coal, oil, and natural gas) that humans use. As a result, both solid and liquid wastes (in the case of food) and carbon dioxide (in the case of burning fossil fuels) are produced. Ecologists refer to these as the source and sink functions of the biosphere. For most of human history, at least at a global level, these sources and sinks were essentially infinite. But now both the human population and its consumption of resources have grown so large that this is no longer the case. The earth can no longer supply all the materials and energy that the human population needs, nor absorb all of the wastes it generates. Growing carbon dioxide in the atmosphere and resulting climate change is perhaps the best-known example of the inability of the earth to absorb these wastes.

How Are Ecosystem Services Valued, and How Much Are They Worth?

The value of nature was recognized by ancient cultures throughout history and was often incorporated into their spiritual view of the world. However, it was only in the last few decades that the term ecosystem services developed as a major area of study and use within ecology. Economists and ecologists began putting economic values on ecosystem goods and services as a way of demonstrating the importance and value of ecosystems to the economy. Prior to this time, ecosystem services were rarely mentioned, overlooked externalities of the human economy. Financial and economic markets in the human economy did not—and still largely do not—include the value of these services in the provision of essential goods to the human economy.

Ecosystem valuation is the concept of putting a value (often monetary) on the different services that ecosystems provide to humans. One of the important papers that brought the concept of ecosystem services to prominence was a report by Robert Costanza and colleagues, published in 1997 in the scientific journal *Nature*.[5] They identified a number of important ecosystem services such as water supply, climate regulation, erosion control, soil formation, water treatment, and pollination (Table 5.1). More importantly, they placed a monetary value on these different services. The total value of the world's ecosystem services was about 33 trillion dollars per year. This was roughly equivalent to the world gross domestic product in 1997. Thus the work nature does for humans is roughly equal to the value of the money

Table 5.1 Average global values of annual ecosystem services. (Data From Costanza 1997)

Natural habitat type	Value ($ per acre per year)
Coastal	
Estuaries	9244
Seagrasses and Algal Beds	7694
Coral Reefs	2460
Forests	
Tropical Forests	813
Temperate Forests	122
Grass and Range Lands	
Wetlands	
Tidal Marsh and Mangroves	4045
Swamps and Floodplains	7927
Lakes and Rivers	3440

---

[5] Costanza et al. 2007, ibid.

economy. Moreover they reported that the services of different ecosystems were vastly different. For example, estuaries, aquatic vegetation, and wetlands had values in excess of $4,000 per acre per year, while the open ocean, grasslands and some forests had values less than $200 per acre per year. High mountains and deserts have very low values based on biological processes. But mountains collect and store water as snow and release it slowly. Thus, Montana provides ecosystem services to the Mississippi delta. Mountains and desserts also provide services in the sense of individual, social, and cultural well being. Overall, this clearly means that some parts of the landscape are much more valuable in terms of ecosystem services than others. We will come back to this later when we discuss how sustainability changes for different parts of the landscape.

We give here a few examples of how different ecosystems have been valued. Breaux Bridge, Louisiana, a small town in southern Louisiana, locally known as the Crawfish Capital of the World for its annual crawfish festival, saved nearly $3 million dollars on municipal sewage treatment by using wetlands as part of the treatment process compared to conventional treatment alone.[6] One of the authors of this book, David Pimentel, and coworkers reported that bee pollination in the U.S. is worth $16 billion annually.[7] Other studies have estimated the valued of goods and services of Galveston Bay, Texas, at over 5 billion dollars per year[8] and the average value of global wetlands at over $5,846 per hectare per year.[9] Ecosystem service evaluation has also contributed to the development of the concept of natural capital.[10] Just as in economics, capital can be defined as a stock that yields a flow of valuable goods or services into the future. Natural capital is the stock of natural ecosystems that yield ecosystem goods and services into the future. The value of only five ecosystem services (flood control, water purification, biodiversity, nursery habitat, and water supply) in Alberta, Canada, one of the largest inland deltas in the world, is reported as $846

---

[6] Ko Jae-Young, et al. 2012. Policy adoption of ecosystem services for a sustainable community: A case study of wetland assimilation using natural wetlands in Breaux Bridge, Louisiana. *Ecological Engineering*, 38, 114–118.

[7] Pimentel D. et al. 1997. Water Resources: agriculture, the environment, and society. *BioScience* 47, 97–106.

[8] Ko Jae-Young. 2007. *The Economic Value of Ecosystem Services Provided by the Galveston Bay/ Estuary System.* Final Report. Texas Commission on Environmental Quality Galveston Bay Estuary. Available online: files.harc.edu/Projects/Nature/GalvestonBayEconomic Value.pdf.

[9] Costanza et al. 2007, ibid.

[10] Costanza, R. and Daly, H. E. 1992. Natural Capital and Sustainable Development. *Conservation Biology*, 6, 37–46.

million Canadian dollars.[11] In a study of the Mississippi Delta, David Batker and coworkers estimated the value of ecosystem goods and services such as hurricane and flood protection, water supply, water quality, recreation and fisheries.[12] They reported that Delta ecosystems provide at least $12-47 billion in benefits annually. If this natural capital were treated as an economic asset yielding an annual flow of value, the delta's minimum capital asset value would be worth $330 billion to $1.3 trillion. They examined three restoration scenarios. If the deterioration of the delta continues, there will be an additional $41 billion in lost goods and services over the next several decades. If, however, the delta is restored, there will be an increased benefit of $62 billion. These values are conservative since they include only partial values of 11 ecosystem services and do not include the value of increased protection for levees, avoided catastrophic impacts such as levee breaching, the benefit of reduced displacement of residents, reduced FEMA relief and recovery costs, lower insurance rates, lower national oil and gas prices, less litigation, or the benefits of an expanding coastal economy, greater employment, and stability gained for existing communities and residents.

How are these values obtained, and why is there such a large range of estimates as reported, for example, for the Mississippi delta? Economists and ecologists estimate these values in a number of ways. For example, they can ask people what they are willing to pay to catch a fish, or go canoeing or hunting, or take a vacation in a national park. They can estimate how much stronger, and thus more expensive, a levee would have to be if there were no wetlands to reduce hurricane storm surge. They can calculate what it would take to do nature's work if humans had to build bigger treatment plants to treat sewage or drinking water, as the cities of Breaux Bridge and New York did. These different techniques give a range of values, and the market price at a given time is a very inexact estimator of how much energy and materials it takes to produce a product. However, any of the estimated values placed on the work that nature performs for humans is enormous. Additionally, these examples show that as humans degrade natural ecosystems, the value of the services they provide decreases.

So in summary, it is important to realize that ecosystem services provide goods and services to humans at little to no financial cost to society, and include such things as maintaining a benign climate, protection of natural habitats, waste absorption and breakdown of pollutants (for example, cost

[11] Timoney, K. 2013. *The Peace-Athabasca Delta: Portrait of a Dynamic Ecosystem*. University of Alberta Press, 608 p.

[12] Batker, D., et al. 2014. The importance of Mississippi Delta Restoration on the Local and National Economies. In *Perspectives on the Restoration of the Mississippi Delta: The Once and Future Delta*. Eds. Day J.W., Kemp G.P., Freeman A.M., Muth D.P. pp. 155–173. Springer, Dordrecht, The Netherlands.

of treatment using wetlands versus a conventional plant), prevention of soil erosion by forests and grasslands, and the service of bees pollinating crops. Money does not exchange hands when these processes occur. But if these processes are lost due to human degradation of ecosystems, there is a large cost to replace them. The processes are the product of healthy, functioning ecosystems that are threatened by a burgeoning human population that demand resources in greater abundance as population and consumption of resources increases. In areas where such services are less abundant due to low rainfall, poor soils, or other factors, ecosystem services must be replaced, if possible, at high ongoing energy costs.

Human activity has greatly altered natural systems and the goods and services they produce. This is demonstrated in many ways. Humans dominate, in one way or another, approximately two-thirds of the land area of Earth's land surface[13] and divert—directly or indirectly—from 40 to 50 % of the plant growth of the earth to their own ends.[14] By this we mean that agriculture and forestry have altered naturally occurring plant communities. Plant communities are disappearing due to human activities such as forest fires and desertification. Large areas of vegetation have been paved over by roads, buildings, airports, and industry. Many fish stocks are overfished and are near collapse.[15] Plants and microorganisms naturally convert nitrogen gas in the atmosphere into ammonia or organic forms of nitrogen in a process called nitrogen fixation. Humans increased the rate of this "fixation" by a factor of 10 times in a little over a century.[16] Most of this fixed nitrogen is incorporated in fertilizers. Some of this excessive nitrogen is eventually transported to rivers and coastal waters, leading to poor water quality and low oxygen conditions.[17]

It is estimated that about 50,000 species of plants, animals, and microbes have been introduced into the United States since Columbus discovered America. Several of these species, especially our crops and livestock, are valuable introductions. However, many invasive species are serious pests,

---

[13] Vitousek PM, Mooney HA, Lubchenco J, Melillo JM. 1997. Human domination of earth's ecosystems. *Science*, 277, 494–499.

[14] Vitousek PM, Ehrlich PR, Ehrlich AH, Matson PA. 1986. Human appropriation of the products of photosynthesis. *BioScience*, 36, 368–373.

[15] For example, Pauly, J., Christensen, V., et al. 1998. Fishing down marine food webs. *Science*, 279, 860–863.

[16] Vitousek et al. 1997 ibid.; Galloway, J.N., Aber, J.D., Erisman, J.W., Seitzinger, S.P., Howarth, R.W., Cowling, E.B., Cosby, B.J. 2003. The nitrogen cascade. *BioScience*, 54, 341–356.

[17] Nixon S.W., et al. 1996. The fate of nitrogen and phosphorous at the land sea margin of the North Atlantic Ocean. *Biogeochemistry*, 35, 141–180.

causing an estimated \$120 billion in damage and control costs each year.[18] These invasive species also cause an estimated 40 % of all species extinctions in the United States.[19] Scientists report that the earth is now experiencing the sixth great extinction of the last half billion years.[20] Unlike the first five great extinctions, this one is caused by human activity.

One measure of the impact of humans on nature is called the ecological footprint.[21] Ecological footprint is a measure of how much of the earth's resources is required to support humans. More specifically, William Rees defined ecological footprint as follows.[22]

> The area of land and water ecosystems required, on a continuous basis, to produce the resources that the population consumes and to assimilate (some of) the wastes that the population produces, wherever on Earth the relevant land/water is located

The size of a population's eco-footprint (EF) depends on four factors: population size; average material standard of living (i.e., the EF reflects consumption); productivity of the land/water base; and the efficiency of resource harvesting, processing and use at the time of the analysis. In other words, Ecological Footprint Analysis provides an area-based 'snap-shot' of a population's demand on the ecosphere.[23]

Table 5.2 gives the Ecological Footprint of the world and for a number of countries.[24] Wealthy, mainly urban, consumers clearly impose a larger average per capita demand on the ecosphere than people living in poorer countries. The citizens of high-income countries, such as Americans, Canadians, and Europeans, have average EFs ranging from four to over ten hectares (defined as average global hectares—gha), or up to 18 times larger

---

[18] Pimentel et al. 2005. Update on the environmental and economic costs associated with alien-invasive species in the United States. *Ecological Economics*, 52(3), 273–288.

[19] Pimentel et al. 2005. Ibid.

[20] See Elizabeth Kolbert. 2014. *The Sixth Extinction*. Henry Holt, New York. 319 p.

[21] Wackernagel, M., Schulz, N., Deumling, D., Callejas Linares, A., Jenkins, M., Kapos, V., Monfreda, C., Loh, J., Myers, N., Norgaard, R., & Randers, J., 2002. Tracking the ecological overshoot of the human economy. *Proceedings of the National Academy of Science* 99(14), 9266–9271.

[22] Rees, W.E. 2006. "Ecological Footprints and Bio-Capacity: Essential Elements in Sustainability Assessment." In Jo Dewulf and Herman Van Langenhove (eds) *Renewables-Based Technology: Sustainability Assessment*, pp. 143–158. Chichester, UK: John Wiley and Sons.

[23] For fuller details of the method, including inclusions, exceptions and limitations, see Rees, 2006, ibid. World Wide Fund for Nature, Gland, Switzerland WWF (2010) Living Planet Report 2010. and www.footprintnetwork.org/atlas.

[24] William Rees. 2012. Cities as dissipative structures: Global change and vulnerability of urban civilization. In M. Weinstein and R. Turner (eds), *Sustainability Science*. Springer, New York.

Table 5.2 The eco-footprints, bio-capacities and overshoot factors of selected nations (estimated from 2007 data in WWF 2010)

Country	GDP per capita (in 2010 international dollars)	Per capita eco-footprint (gha)	Per capita domestic bio-capacity (gha)	Overshoot factor
World	10,700	2.7	1.8	1.5
United States	47,284	8.0	3.8	2.1
Canada	39,057	7.0	15.0	0.5
Netherlands	40,765	6.1	1.1	5.5
France	34,077	5.0	2.7	1.9
United Kingdom	34,920	4.8	1.5	3.2
Malaysia	14,670	4.8	2.6	1.8
Japan	33,805	4.7	0.7	6.7
Hungary	18,738	3.0	2.3	1.3
Mexico	14,430	3.0	3.5	0.9
Brazil	11,239	2.9	8.9	0.3
Thailand	9,187	2.3	0.6	3.8
China	7,519	2.2	1.0	2.2
Indonesia	4,394	1.2	1.4	0.9
India	3,339	0.9	0.5	1.8
Malawi	827	0.8	0.8	1.0
Bangladesh	1,572	0.6	0.4	1.5

than the EFs of the citizens of the world's poorest countries such as Afghanistan and Bangladesh.

The final column of Table 5.2 shows each country's 'overshoot factor.' The overshoot factor is the ratio of the national average per capita eco-footprint to *per capita* domestic bio-capacity. Countries with overshoot factors larger than one impose a greater burden on the ecosphere than could be supported by their domestic ecosystems. That is, these countries are at least partially dependent on trade and on exploitation of the global ecological commons to maintain their current lifestyles. The Netherlands and Japan are often used as examples of how we can put many more people onto the land, as their population density is very high. But in fact these two countries use the ecological services of 5.5 and 6.7 times as much productive land/water elsewhere in the world than is contained within their respective borders. All countries in overshoot are running 'ecological deficits' with the rest of the world.

Ominously, the world as a whole is in a state of overshoot. At the rate that humans are consuming resources, it would take 1.5 worlds to sustain the current rate of consumption. Thus, human demand exceeds Earth's regenerative capacity by about 50 %.[25] We are able to do this by living, in

---

[25] WWF 2010, ibid.

part, by depleting and dissipating as waste, the enormous stocks of potentially renewable natural capital (fish, forests, soils, etc.) that have accumulated in ecosystems over millions of years. Thus, we are not living on the "income" of the earth but also depleting its natural "capital".

Variation in Ecosystem Goods and Services across the Landscape

Ecosystem services vary dramatically across the North American continent. But why? It's a function of a number of things. Elevation is important; little can grow in very high mountains with extreme cold and barren soil. Soil type is very important. Some soils, like those of the Midwest, are almost perfect for growing many crops without the need for soil amendments (as we shall see in the chapter on food).

But the most important factor affecting ecosystem productivity and ecosystem services is water. "Water, water every where, nor any drop to drink." Samuel Taylor Coleridge wrote this famous line in 1798 in his poem *The Rime of the Ancient Mariner*. Over the last two hundred plus years it has become one of the most often quoted references to the preciousness of fresh water to human life. Areas with sufficient rainfall, generally at least 20–30 in., engender high growth rates of plants, both natural and cultivated. The eastern U.S. generally has abundant rainfall and much of the area was forest when Europeans arrived in the seventeenth century. The Midwestern prairie supported rich grasslands watered by frequent rains and which over centuries developed some of the richest soils of the world. Moving west, though, rainfall decreases and natural vegetation becomes sparse. This is the Arid Region of the U.S. as depicted on John Wesley Powell's map that is reproduced in Chap. 6. Without irrigation, agriculture in much of the west would not be possible. Only in a narrow coastal strip in the Northwest, from San Francisco north to Canada, is there generally sufficient rainfall for agriculture. In general, ecosystem services are high in the narrow coastal strip in the Northwest and in the eastern U.S. and low in the Arid Region.

But it is not just how much precipitation falls on a particular spot of land but also whether the precipitation falls as snow or rain, and how water moves across the landscape. Winter precipitation generally falls as snow, especially high up in mountains. In the west, this is extremely important because slowly melting snow provides summer water in much of the west. If there were no snow pack, most of the human settlement patterns and lifestyle in the west as they exist now, even in the wet Northwest, would not be possible. Across the world, more than a billion people depend on slowly melting snowpack for their water. So snowpack is an important ecosystem service that nature provides. But as we discuss in Chap. 6 on climate, the snow pack in much of the west is projected to decrease due both to less total precipitation and more winter precipitation falling as rain, which runs off quickly or soaks into the ground. This has severe and worrisome implications for the region.

Ground water is another ecosystem service that benefits humans. Groundwater is used for direct human consumption but much more is used for irrigation. If the groundwater withdrawals are matched by recharge replenished by annual rains, then there is no net loss of ground water. But in many areas, water withdrawals exceed recharge rates and there is depletion of the aquifer.

A classic case of groundwater drawdown is the Ogallala aquifer.[26] This enormous aquifer that underlies about 174,000 mile^2 in parts of eight states of the Great Plains (Texas, Oklahoma, New Mexico, Colorado, Kansas, Nebraska, South Dakota, and Wyoming) is fossil water (see Fig. 6.9). This means that it was filled as the last glacial period ended about 10,000 years ago. During this time, the west was much wetter and runoff filled the enormous aquifer completely. The saturated thickness of the aquifer ranges from less than 50 to more than 1000 feet. Beginning in the mid twentieth century, farmers on the plains began pumping water from the Ogallala at unsustainable rates. Ogalla water feeds many of the large circular green fields visible on flights over the plains, using water pumped by what is called center pivot irrigation systems (Fig. 5.2). Twenty-seven percent of the irrigated land in the U.S. is fed by water pumped from the Ogallala. The aquifer also provides drinking water to 82 % of the people who live within its borders. Natural recharge rates of the aquifer are currently much, much less than the pumping rates and ground water levels in the Ogallala have dropped by hundreds of feet in some areas. If fully drained, it would take about 100,000 years to refill given today's rainfall patterns. Thus pumping is clearly unsustainable. Some natural systems such as streams and wetlands that were fed by seeps where the aguifer was close to the surface have dried up. This reduces the level of ecosystem services for the region. Experts project further drying due to climate change for the Great Plains that will lead to further reliance on the Ogallala. The pumping of groundwater from the Ogallala and many other aquifers is analogous to pumping oil. Both oil and water in the Ogallala region are being pumped out at rates that greatly exceed renewal or recharge and must be considered non-renewal resources at current rates of use.

River basins are examples of how water flowing down stream enriches the lower parts of the basin. In the upper basin, runoff flows into small

---

[26] William Ashworth, 2006. *Ogallala Blue: Water and Life on the High Plains*. Published by the Countryman Press, Woodstock, VT.

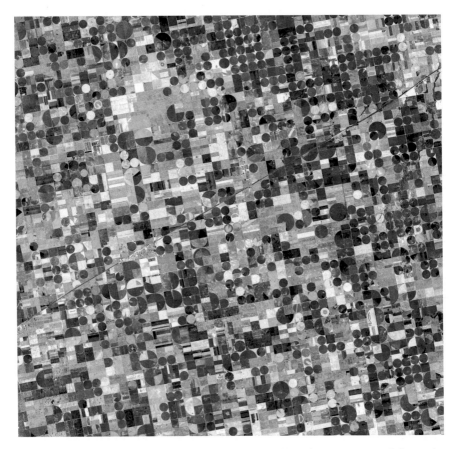

Fig. 5.2 Center Pivot Irrigation Systems in Kansas. Water is pumped from the Ogallala Aquifer to irrigate these crops. (By NASA [Public domain], via Wikimedia Commons)

tributaries that have small to non-existent floodplains. As the tributaries come together and the streams grow in size, the flood plain becomes larger and larger. In large rivers like the Mississippi, the flood plain can become truly enormous. The lower Mississippi River alluvial floodplain stretches for over a thousand miles from Cairo, Illinois to the Gulf of Mexico and can be more than 100 miles wide. For example, the eastern edge of the flood-plain at Baton Rouge is demarcated by what is called the Pleistocene Terrace, a rise that is 20–30 feet higher than the adjacent floodplain. The campus of Louisiana State University sits mostly atop the terrace and would not flood even if there were no levees along the river. The western edge of the floodplain is 60 miles to the west at Lafayette; it also sits high and dry on the terrace. It is clear that a lack of flooding threat strongly influenced the establishment of both of these cities. The huge alluvial floodplain encompasses nearly ten million acres and was once home to one of the largest bottomland hardwood ecosystems in the world. These were huge old growth forests teaming with wildlife. They were largely cut by the first quarter of the twentieth century.

The huge and productive bottomland forests of the Mississippi alluvial floodplain were, and those still existing are, still very productive. The reason is that for years, the river fertilized, watered, and enriched the floodplain. Annually the river would rise up and spread out over the flood plain, delivering water, sediments, and enriching nutrients. Many organisms followed the water as it spread out including fish, birds, and mammals. The floodplain ecosystem is an incredibly important habitat for these organisms. Scientists call this repeated flooding the *flood pulse* concept.[27]

When the waters of the Mississippi reached the Gulf of Mexico, they built and sustained the huge Mississippi delta, one of the most important coastal ecosystems of the world. And just as in the upstream alluvial floodplain ecosystems, the river enriched and sustained the delta and supported its incredible richness. Most coastal ecosystems are rich because they receive drainage from the land as well as input of organisms and salt water from the sea. The famous ecologist, Eugene Odum, called coastal areas tidally subsidized fluctuating water level ecosystems. Coastal areas are renowned for their high level of fisheries. All these ecosystems are subsidized by their entire drainage basin. At least, they once were, until the massive human impacts of the twentieth century.

So water flowing together from all the tributaries, the Snake, Missouri, Arkansas, Upper Mississippi, Ohio, Tennessee and thousands of others in the basin enriches downstream ecosystems, just as wealth flowing from all over the country and world enriches the financial sector in New York City. The enormously productive Mississippi delta could not exist without its enormous watershed. Likewise, the rich financial sector in New York and similar sectors in other cities could not exist without their enormous "wealthshed" which is ultimately enriched, in turn, by natural wealth flowing from healthy ecosystems all over the world. This speaks volumes about both the future of the delta and NYC. Properly managed, the Mississippi Basin can regain its once world-class status as a continental sized river basin ecosystem, providing a wealth of wide ranging ecosystem services. But mega trends of the twenty-first century will inevitably reduce the wealth flowing into NYC and other similar areas. We will come back to these ideas in Chap. 10.

---

[27] Junk, W. J. et al. 1989. The flood pulse concept in river-floodplain systems. Pages 110–127 in D. P. Dodge, (ed). Proceedings of the International Large River Symposium. *Canadian Special Publications Fisheries Aquatic Sciences*, 106.

Ecosystem Services Across the Landscape

Thus far, we have shown that ecosystem services are an important input to the human economy, and that they vary across the landscape. They are generally higher where there is sufficient water availability, either through precipitation or runoff, and where it is not too cold—in other words, places that are moist and with moderate temperatures. Another way of saying this is that if you look out of your window in the spring or summer where you can see part of the natural world (and not the side of a building) and it is green and there are trees, the place you are living probably has higher than average ecosystem services. Such a place has high amounts of what scientists call primary productivity, or the rate at which plants grow. A map of primary productivity is a good indication of how ecosystem services vary across the landscape. This is shown in Fig. 5.3. Green colors indicate higher rates of plant growth while tan indicates low rates. The scale of productivity

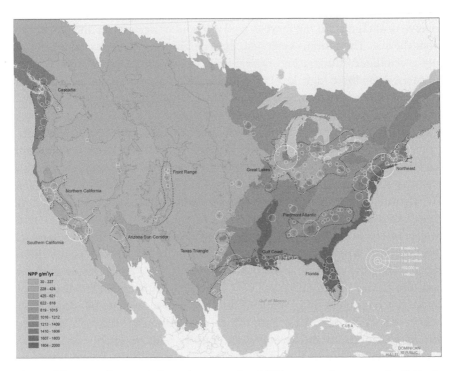

Fig. 5.3 The growth rate of plants (expressed as NPP or net primary productivity in grams per meter square per year) across the central part of North America. The tan areas have very low productivity, and dark green areas are highly productive. Ecosystem goods and services are higher in areas with higher plant growth rates such as the eastern United States and the Northwest. Also shown are large urban areas and the 11 megaregions discussed in Chap. 3

ranges from less than 30 to more than 2000 g of dry plant material produced per square meter per year. This pretty much encompasses the range of productivity that occurs in nature. Some areas of the landscape have near zero primary productivity in areas such as northern ice fields, the tops of high mountains, and extremely dry deserts. Life has difficulty existing in such environments. At the other end of the range are the most productive ecosystems on earth. These include healthy bottomland hardwood forests such as once grew in the Mississippi alluvial floodplain, or in rich tidal marshes and other coastal ecosystems. The middle of this range—at about 500 g per square meter per year—is typical of productive agricultural land, or about 150 to 200 bushels of corn per acre per year. Productivity values less than 100 g per square meter per year are typical of very dry areas. Much of the western U.S. is naturally arid or semi arid and thus has very low natural levels of productivity. These are the scrublands and deserts. The importance of water in controlling the productivity of land is clearly shown where there is irrigation. Those circular green fields we spoke of earlier are there because of irrigation. The agricultural bounty of California's Central Valley would not happen without irrigation water pumped from the ground or flowing from the mountains and captured behind dams. The importance of water is especially seen in the drought years of the past few years as we write in 2015.

Areas that have high rates of primary productivity generally have high levels of ecosystem services.[28] Wetlands, estuaries, flood plains, forests, rich grasslands, lakes and rivers have high ecosystem services. It is a measure of the importance of these services to humans that most people lived in and adjacent to these ecosystems for most of history. Ancient (and modern) Egypt is a good example of this. The area of Egypt is almost 387,000 mile^2, larger than the Northeast of the U.S. But almost everyone in Egypt lives in a narrow strip of land along the Nile River and in the Nile delta that constitutes less than 5 % of the area of the country. A view from space clearly shows this as a narrow strip of green that contrasts sharply with the barren desert on either side. The people of Egypt fed themselves from food produced along the river and in the delta and from fish caught in the Mediterranean offshore from the delta where the sea is fertilized by river water. This whole productive region was sustained by the *flood pulse* that we spoke of earlier. Each year the river would inundate its flood plain. The Egyptians learned how to control this pulse by canals and structures such as small dams that delivered the water to the rich fields. It is a

---

[28] Costanza, R. et al. 1998. Special Section: Forum on Valuation of Ecosystem Services. The value of ecosystem services: putting the issues in perspective. *Ecological Economics* 25, 67–72.

measure of how rich this land was, and still is, that the ancient Egyptians were not only able to feed themselves but also to export large quantities of grain to the Romans.

So, let's sum up. How, specifically, do ecosystem services vary across the central part of the North American continent? The map of productivity sums it up clearly. Coastal regions and river flood plains have the highest levels of ecosystem services. The Lower Mississippi Valley stands out in this image because of its large size. But flood plains support rich ecosystems and productive agriculture throughout the country, especially in the eastern half of the country. Productive plant communities like salt marshes and estuaries thrive in coastal areas. The Mississippi delta, the sounds of North Carolina, Chesapeake and Delaware Bays, Long Island Sound, and Puget Sound are clearly visible. These coastal systems support rich fisheries not only inside of estuaries but also in coastal waters.

Much of the eastern U.S. is green. Forests covered most of this area before European colonization. A considerable area of forests has been cleared for agriculture because rainfall was sufficient. Much of the farmland, especially in the Northeast has either reverted to forest as the region became dependent on other areas for much of its food, or was engulfed by growing urban areas. The rich grasslands of the Midwest and eastern prairies form one of the most important farming regions in the world. It is no coincidence that most people lived in the eastern U.S. prior to the twentieth century. This is where the "wealth of nature" of the nation was. Most of the western U.S. contrasts with the east in terms of ecosystem services. The west is mostly dry and lacks extensive forests and rich prairies. Only in higher mountains or in the Northwest is there sufficient water to support forests. Without irrigation, farming would be limited and many fewer people could live in the West.

It bears repeating that the explosive growth in the west during the twentieth century was fueled by cheap fossil fuels. This allowed the development of extensive water distribution systems that included many dams, canals, and aqueducts as well as the extensive use of ground water. Water from the Colorado is key to the entire Southwest, as its water is parceled out to more than 30 million people and tens of thousands of acres of farmlands. Dams on the Columbia and its tributaries generate electricity and impound water that supports agriculture in drier regions of the northwest. The Ogallala and California Central Valley aquifers help support huge areas of farms. California has the most extensive and sophisticated water system in the world. The Colorado River, Los Angeles, and California aqueducts parcel out water to millions of Californians and Mexicans, along with hundreds of thousands of acres of farmlands. These farmlands feed many people in the country. What these systems have done is convert huge areas—which under natural conditions had low levels of ecosystem services—to areas with high levels of ecosystem services. It is a clear demonstration of the importance of fresh water. As it currently exists, California could not exist without this water distribution system. And it was cheap energy that allowed this to happen.

As society moves through the twenty-first century, the great megatrends discussed in this chapter and in Chaps. 6–9 will have serious implications for ecosystem services. During the twentieth century, ecosystem services were both degraded and enhanced. In the west, arid and semi-arid land was made much more productive by the addition of water, high levels of agricultural technology, and cheap fuels. The result was an agricultural bonanza, especially in California, that provides about half of the nation's vegetables, fruits, and nuts (the edible variety). As we discuss in the chapters on climate and energy, much of the West is projected to dry, and energy will become much more expensive. This suggests that the artificially high levels of ecosystem services in the West will likely decline. It will be difficult to maintain the economy of the west if this happens.

Climate change will threaten the integrity of coastal ecosystems and the high levels of ecosystem services they provide, especially along the Atlantic and Gulf coasts. Sea level is projected to rise by a meter or more by the end of the century as land based ice masses melt and warming sea water expands. Strong hurricanes will become more common, threatening both natural systems and human infrastructure in the coastal zone. Hurricanes Katrina and Sandy and the damages they wrought may be harbingers for the twenty-first century. Increasing sea level and greater number of more intense storms will lead to the loss of large areas of coastal wetlands. One of the important ecosystem services these wetlands provide is storm buffer, reducing storm surge. So impacts on coastal development will likely increase as the twenty-first century progresses.

The Mississippi delta is a case in point for the impacts of megatrends on ecosystem services. As we discussed earlier, there is an ambitious 50-year plan to restore the degraded delta. The plan is expensive and energy intensive. In addition, the climate change impacts of accelerated sea-level rise and more intense hurricanes will reduce the effectiveness of restoration efforts. One other climate impact that will impact the delta is a projected increase in the Mississippi River discharge. It was the River that built and sustained the coast over the past several thousand years. If used properly and judiciously, more years with larger flood-increased discharge represents a growing resource to rebuild the coast. However, growing energy scarcity will limit all of our options.

Chapter 6

Global Climate Change: A Warmer and More Unpredictable Future

Introduction

On May 9, 2013, the earth's climate system reached a notable milestone. The concentration of carbon dioxide, or CO_2, at the Mauna Loa Observatory in Hawaii passed 400 parts per million (ppm) when averaged over a whole day. This is the first time in more than 3 million years that the concentration has been this high. The CO_2 concentration has increased by 40 % since 1890 when levels were about 280 ppm, an unprecedented rate of increase. Over the past million years, CO_2 ranged between about 170 and 300 ppm as ice ages waned and waxed. According to Professor Clive Hamilton of Charles Sturt University in Canberra, Australia, "if you are not frightened by this fact," and it is a fact, "then you are ignoring or denying science."[1] The rapidly increasing CO_2 is a harbinger of dramatic climate changes to come in this century. But nowhere are these changes going to be felt more dramatically than in the American Southwest and along the coasts of the country.

In his book, "A Great Aridness," William deBuys chronicles the dramatic impacts that climate change is having and is projected to have on the American Southwest.[2] This, the driest and hottest region of the United States, is projected to become drier and hotter with more extremes of both rainfall and droughts. Hotter temperatures mean more evaporation and drier means

---

[1] *International Herald Tribune,* May 27, 2013.

[2] William deBuys. *A Great Aridness.* Oxford University Press, Oxford. 2011. 369 p.

© Springer Science+Business Media New York 2016
J.W. Day, C. Hall, *America's Most Sustainable Cities and Regions,*
DOI 10.1007/978-1-4939-3243-6_6

less water both to evaporate and to use. More winter precipitation will fall as rain rather than snow and thus will run off earlier. This is important because many water distribution systems in the west are designed for a slowly melting snowpack. Hotter and drier conditions are leading to more tree death, super forest fires, loss of species, and more dust. And these are not just projections for the future; the changes are already taking place. News reports of giant wildfires consuming large areas of vegetation and many homes have become common. The wildfire that destroyed over 400 homes near Colorado Springs, and the extremely high temperatures, some of which topped 120° in June 2013, are examples of what is becoming a regular part of the summer news in the West. Large wildfires continued throughout the 2013 and 2015 fire season, including one that burned into Yosemite National Park.

Al Gore referred to global climate change or "global warming" as "an inconvenient truth." We present a number of inconvenient truths is this book, but this chapter focuses on climate science and how it portends varying consequences for different parts of the landscape. We will focus on North America, primarily the United States, but will compare what is predicted to happen here with other parts of the world.

While there is considerable evidence that these changes are caused by human changes to the atmosphere, most notably an increase in CO_2, the climate has always changed and always will. So we must always keep a keen eye on humanity's role in climate. Humans are always in a precarious situation relative to climate. We must also recognize that burning fossil fuels has not only led to climate change but has also allowed us to be less affected by climate change by buffering our extreme personal temperatures, through the impact of irrigation and fertilization on agriculture, and through increasing transportation. So it is important to understand that energy availability is another aspect of how energy affects our ability to live in different landscapes.

The Southwest

A good place to begin our consideration of the American Southwest is the beautiful colored map of the "Arid Region of the United States", produced by John Wesley Powell in a report to the Geological Survey in 1890 (Fig. 6.1). The one-armed Powell is famous for one of the first boat trips down the Colorado through the Grand Canyon. The arid Southwest includes almost all land roughly west of the 100th meridian running through west-central Texas, south through the Rio Grande Valley, all the way to the Gulf of Mexico and north to Canada through Oklahoma, Kansas, Nebraska, South

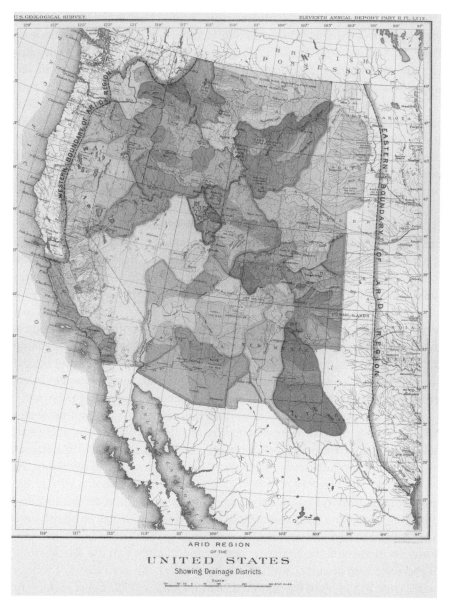

Fig. 6.1 Arid Region of the United States. This map was produced by John Wesley Powell in 1890 based on his travels in the region. *Different colors* represent different drainage basins (John Wesley Powell. Map of the Arid Region of the United States. U.S. Geological Survey, Eleventh Annual Report Part II PL LXIX)

Dakota, and North Dakota. Only a narrow band of land in the west between the Pacific and the Cascades from San Francisco to Canada lies outside of Powell's Arid Region. The Arid Region includes all of the Colorado, Rio Grande, Yellowstone, North and South Platte, and Snake River basins, the upper Missouri and Arkansas basins, and a number of smaller streams.

The Colorado is the most important river in the Southwest. Its watershed covers almost a quarter million square miles, or about 8 % of the lower 48 states. The discharge of the Colorado could decrease by 5–20 % by 2050, and much more by the end of the century according to some estimates.[3] This bodes ill for the 30 million people in this region of the U.S. and Mexico, including the populations of Los Angeles, Phoenix, San Diego, and Las Vegas, who depend on the river for 50–100 % of their water. This population is projected to grow to 38 million by the third decade of this century. Indeed, Lake Meade behind Hoover Dam is at less than 50 % capacity now, and a recent study by scientists from the Scripps Institution of Oceanography in La Jolla predicted that the lake had a 50 % chance of drying up in two decades or less due to the combined effects of reduced river discharge and increasing rates of water evaporation and consumption.[4] In fact, it is likely that Lake Mead will rarely, if ever, refill to capacity again. Yet another critical issue is that water flowing through the turbines of Hoover Dam and other large dams in the west produces large amounts of electricity. One of the largest single uses of this electricity is to pump water around. So as lake water levels drop, there will be both a water shortage and an energy shortage.

Part of the impending water shortfall has nothing to do with projected future declines of river flow due to climate change and future increased demand.[5] The Colorado River Compact, which was signed in 1922, had as its goal the portioning out of the water of the river. In order to divide up the water of the river, members of the Colorado River Commission who wrote the Compact had to know how much water was in the river. In deciding this, they used flow data from the first two decades of the twentieth century. As it turns out, these decades were among the wettest in the last millennium. In the end, the compact commission thought that the average flow of the river was about 17 million acre feet (maf). An acre foot is 1 foot of water covering an acre (or about 325,850 gallons or 1233 cubic meters) or about what four typical homes use in a year. To be conservative, the commission decided to use 15 maf. However, reconstructions of ancient river

---

[3] Brad Udall, director of the University of Colorado's Western Water Assessment Program. Cited in deBuys, 2011, ibid.

[4] Barnett, Tim P., Pierce, David. 2008. When will Lake Mead go dry? *Water Resources Research*. Vol. 44 Issue 3. March 2008. Available online: http://onlinelibrary.wiley.com/doi/10.1029/2007WR006704/abstract

[5] Much of this discussion is based on DeBuys, 2011, ibid.

flow based on tree ring studies concluded that the average flow at Lee's Ferry, just downstream of the Glen Canyon dam, was somewhere between 13 and 14.7 maf. In addition, these studies indicated that going back over a thousand years, droughts more severe than those during the historical record occurred regularly. For example, in the eleventh to thirteenth centuries, multi-decadal droughts likely contributed to the collapse of a number of Native American civilizations, including the Anasazi and Fremont cultures in the eleventh and twelfth centuries, and the Lovelock culture in the thirteenth century.[6] Therefore, from the beginning, more water was allocated from the Colorado than the river had to give. When we add it all up, total allocations of Colorado River water (including evaporation from reservoirs) are almost 2.5 maf greater than what the river has to give.

Dr. Virginia Burkett of the U.S. Geological Survey, a member of the U.S. delegation to the Intergovernmental Panel on Climate Change, said that the U.S. National Climate Assessment projects declines in rainfall in the southwest that, when combined with continued increase in temperature, portend more widespread and intense droughts that make the historical conflicts look like child's play.[7]

Several solutions have been proposed to deal with growing water shortages in the Southwest. The first is to shift water from agriculture, which consumes the most water in the west, to urban and industrial uses. In California, almost 80 % of water use is for agriculture, and the picture is similar for the other western states. One "simple" solution to the water shortage is to stop farming on some or most of the current farmland in the region. But how much agricultural land is critical? California alone produces about 50 % of the nation's fruits, vegetables, and nuts. And most potatoes consumed in the U.S. grow on irrigated farmland in Idaho. Any significant reduction in farming would have serious implications for the nation's food supply. We will have more to say about food and agriculture later in the book, but it is obvious that the competition between agriculture and other uses for water is a very complicated issue. Given projections for progressively worsening water shortages, shifting away from agriculture can work for only so long.

A second solution is to bring water to the Southwest from other, more water rich areas. There has been talk—and even plans—to bring water from

---

[6] Benson, L. V., Kashgarian, M., Rye, R. O., Lund, S. P., Paillet, F. L., Smoot, J., Kester, C., Mensing, S., Meko, D., and Lindstrom, S.: 2002, 'Holocene multidecadal and multicentennial droughts affecting Northern California and Nevada', *Quaternary Science Reviews* 21, 659–682.; Benson L. et al. (2007) Anasazi (Pre-Columbian Native-American) Migrations During The Middle 12th and Late 13th Centuries – Were they Drought Induced? *Climatic Change*, 83(1), 187-213.

[7] Dr. Virginia Burkett as told to Dr. John Day.

as far north as the Yukon and from as far east as the upper Mississippi and Missouri basins, the Arkansas River, and even the Great Lakes.

Such schemes would cost tens if not hundreds of billions of dollars and consume prodigious amounts of energy to pump water over vast distances and up significant elevation gradients, in some cases over the continental divide. Assuming the energy needed is available (and that's a big if), climate projections are for rainfall in the Ohio, Upper Mississippi, and parts of the Missouri to be wetter. However, surplus is in the eye of the beholder. For example, people in Louisiana are dealing with disastrous loss of wetlands and other habitats in the Mississippi delta, a problem that will be greatly aggravated by climate change in the form of accelerated sea-level rise. River floods, such as occurred in 2011, are a central resource in dealing with this problem, because "excess" freshwater and sediments help to rebuild the delta and compensate for low-flow years. Any plan for mitigating water supply problems in the Southwest by importing water from other areas faces major and potentially unacceptable trade-offs, both politically and in terms of economic and energy costs.

Other parts of Southwest ecosystems will also be altered, with predictable consequences. The high temperatures and low water availability are stressing forests so much in the Southwest and Rocky Mountain region that trees are becoming more susceptible to borer beetles.[8] The beetles bore into the bark of the tree, where they mate and lay their eggs in the living inner bark. When the larvae hatch, they tunnel through the inner bark, feeding on living tissue. If trees are not stressed they can control the beetles by producing resin. But if trees are stressed by high temperatures and lack of water, they cannot fight off the beetles. This is leading to widespread tree death throughout the mountain region from Mexico to Canada. As the trees die, forest fires are burning larger and larger areas. Some have predicted that by the end of the century, forests in the arid region of the U.S. will almost completely die out because of drought, beetles, and fire. In 2012, the largest fire in New Mexico history occurred in the Gila National Forest in the southwestern part of the state, consuming over 350 mile^2. Another massive blaze near Colorado Springs, CO in June 2012 led to the evacuation of tens of thousands of people from their homes and, as mentioned earlier, fires in June 2013 destroyed more than 400 homes.[9] The trend towards higher

[8] deBuys, 2011. ibid.

[9] United States Geological Survey. Responding to Whitewater-Baldy Fire, USGS Installs Early Flood Warning Network. Available online: http://www.usgs.gov/newsroom/article. asp?ID=3265#.VRBLF03wvcs. Tillery, A.C., Matherne, A.M., and Verdin K.L., 2012, Estimated probability of postwildfire debris flows in the 2012 Whitewater–Baldy Fire burn area, southwestern New Mexico: U.S. Geological Survey Open-File Report 2012–1188, 11 p., 3 pls.; Cole, C.J., Friesen, B.A., and Wilson, E.M., 2014, Use of satellite imagery to identify vegetation cover changes following the Waldo Canyon Fire event, Colorado, 2012–2013: U.S. Geological Survey Open-File Report 2014–1078, 1 sheet.

temperatures, less rainfall and earlier snowmelt across the southwest will amplify conditions that lead to larger wildfires and longer fire seasons.

Evidence for Climate Change

Before moving on to other regions, we should ask ourselves if climate change is real. Some politicians refer to "sound science", but others are calling the evidence into question. Thousands of scientists have carried out research and published on the subject. Several studies show that 97–98 % of scientists who publish about climate agree that the warming of the atmosphere has accelerated since the 1880s and that humans are influencing the climate.[10] The main scientific group addressing climate change in a coordinated manner is the Intergovernmental Panel on Climate Change (IPCC). The World Meteorological Organization and the United Nations Environment Programme established the IPCC in 1988. The goal of the IPCC is to synthesize a growing body of scientific findings on climate change into a format that is accessible to policy makers and scientists across disciplines, as well as to the general public. The panel consists of three "working groups" that assess, respectively, (1) the scientific aspects of climate change, (2) the effects of climate change on socio-economic and natural systems and, (3) options for reducing greenhouse gasses and mitigating the effects of climate change. The IPCC produces a variety of special reports and technical papers, but the most anticipated and widely read publications have been the comprehensive Assessment Reports published in 1990, 1996, 2001, 2007, and 2013–2014. These peer-reviewed reports, the product of thousands of experts from all over the world and from all three working groups, assess and synthesize the current scientific and socio-economic literature concerning climate change. After years of consideration and analysis the IPCC, in their 2013 assessment report, issued their strongest statements to date[11]:

---

[10] Doran, P., and Zimmerman M.K. 2009. Examining the scientific consensus on climate change. EOS, 90, 22–23.; Cook, J., Nuccitelli, D., Green, S. A., Richardson, M., Winkler, B., Painting, R., … & Skuce, A. 2013. Quantifying the consensus on anthropogenic global warming in the scientific literature. Environmental Research Letters, 8(2), 024024.; Anderegg, W. R., Prall, J. W., Harold, J., & Schneider, S. H. 2010. Expert credibility in climate change. Proceedings of the National Academy of Sciences, 107(27), 12107–12109.

[11] IPCC. 2013. *Climate Change 2013: The Physical Science Basis*. Contribution of Working Group I to the Fifth Assessment Report of the Intergovernmental Panel on Climate Change [Stocker, T.F., D. Qin, G.-K. Plattner, M. Tignor, S.K. Allen, J. Boschung, A. Nauels, Y. Xia, V. Bex and P.M. Midgley (eds.)]. Cambridge University Press, Cambridge, United Kingdom and New York, NY, USA, 1535 pp.

"Warming of the climate system is unequivocal, and since the 1950s, many of the observed changes are unprecedented over decades to millennia. The atmosphere and ocean have warmed, the amounts of snow and ice have diminished, sea level has risen, and the concentrations of greenhouse gases have increased." "Human influence on the climate system is clear. This is evident from the increasing greenhouse gas concentrations in the atmosphere, positive radiative forcing, observed warming, and understanding of the climate system...Human influence has been detected in warming of the atmosphere and the ocean, in changes in the global water cycle, in reductions in snow and ice, in global mean sea level rise, and in changes in some climate extremes. This evidence for human influence has grown since the 2007 report. It is extremely likely that human influence has been the dominant cause of the observed warming since the mid-20th century." What led the scientists of the IPCC to these somber conclusions?

Over the past million years or so, there have been repeated ice ages separated by warmer interglacial periods. Each cycle lasted about 100,000 years. On average, each ice age lasted for about 90,000 years and the warmer interglacial lasted for about 10,000 years. This is shown in the graph below. The temperature and carbon dioxide curves follow very similar patterns with CO_2 concentrations highest during the warm interglacial periods. At no time over the past million years did CO_2 levels exceed 300 ppm before the twentieth century, but in a little more than one century, they surged to over 400 ppm in 2013. Ancient concentrations of CO_2 are measured in gas bubbles extracted from ice cores, and temperatures are based on such things as the ratio of different forms of oxygen in shells of ancient shellfish.

A primary cause of these variations in temperature and CO_2 is a phenomenon called the Milankovitch cycle, named after the Serbian mathematician Milutin Milankovitch. Milankovitch first developed the idea that describes the effects of earth movements on climate.

Changes in orbital shape (more or less elliptical), wobble in the earth's axis of rotation, and the tilt of the earth's axis lead to slight changes in the total amount of solar radiation reaching the earth's surface, and to variations across the globe. You can see that increases in CO_2 may lag a bit behind temperature in Fig. 6.2 because changes in orbital dynamics are the major driver of glacial cycles. Climate skeptics have used these facts to argue that CO_2 increases in the twentieth century are not a primary driver of temperature increase. But CO_2 increases within each cycle are closely coupled to warming temperatures. These changes give rise to the regular changes in CO_2 and temperature shown in the figure. However, over the

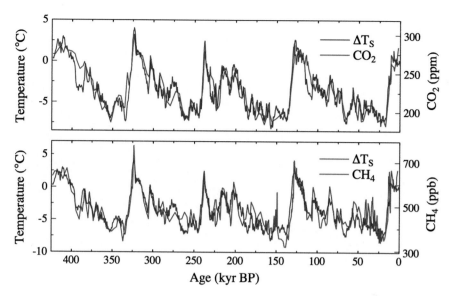

Fig. 6.2 Global temperature, CO_2 (carbon dioxide), and CH_4 (methane) levels based on ice cores taken at Vostok, Antarctica. The *red line* in the *upper panel* is carbon dioxide, and in the *lower panel* is methane. The *blue line* in *both panels* is temperature. It is important to note that the current concentration of CO_2 line is over 400 ppm even though the highest value shown on the graph is 300 ppm (Petit, J.R. et al. 1999. Climate and atmospheric history of the past 420,000 years from the Vostok Ice Core, Antarctica. *Nature* **399**:429–436)

past two centuries, human activities have had an increasing impact on CO_2 and temperature.

The most widely known information showing human impact on the atmosphere and on climate change is the famous carbon dioxide record from the Mauna Loa Observatory located at 11,135 ft on Mauna Loa volcano on the Big island of Hawaii. Carbon dioxide concentrations have been measured continuously there since the 1950s. The remote location and minimal local impact of vegetation or human activities make it an ideal location for monitoring global-scale changes in CO_2 in the atmosphere.

Carbon dioxide levels in the atmosphere increased from preindustrial levels in the late nineteenth century of about 280 ppm to 400 ppm in May 2013. Prior to direct measurements of CO_2 in the atmosphere, atmospheric scientists measured the concentration of CO_2 in bubbles trapped in ice in Greenland and Antarctica. The World Meteorological Organization also reported that the rate of CO_2 increase is accelerating. Between 2004 and 2014, CO_2 increased by 2.07 ppm, which is more than double the increase in

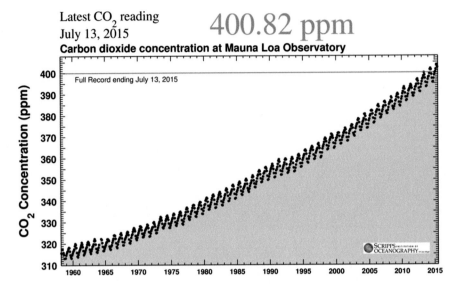

Fig. 6.3 Carbon dioxide measured at the Mauna Loa Observatory in Hawaii. (Courtesy of Scripps Institution of Oceanography, UC San Diego. https://scripps. ucsd.edu/programs/keelingcurve/)

the 1960s.[12] The annual variation in CO_2 shown in the graph is due to seasonal uptake and release of CO_2. During summer in the northern hemisphere, CO_2 decreases due to plant uptake, while in the winter, CO_2 increases due to respiration by biological organisms. The overall increase is due mainly to the burning of fossil fuels (Fig. 6.3)[13].

---

[12] http://co2now.org. The CO_2 now website contains much interesting information on carbon dioxide levels in the atmosphere. World Meteorological Organization. 2014. Global Atmosphere Watch. WMO Greenhouse Gas Bulletin. The State of Greenhouse Gases in the Atmosphere. No. 10 6 November 2014. A recent scientific publication showed that CO_2 levels in the atmosphere are increasing at rates consistent with the highest IPCC scenarios. P. Friedlingstein, et al. Persistent growth of CO_2 emissions and implications for reaching climate targets. *Nature Geoscience*, 7, 709–715.

[13] Hall, C.A.S., C. Ekdahl and D. Wartenberg. 1975. A fifteen-year record of biotic metabolism in the Northern hemisphere. Nature 255: 136–138.

Temperature

"U.S. average temperature has increased by 1.3 °F to 1.9 °F since 1895, and most of this increase has occurred since 1970. The most recent decade was the nation's and the world's hottest on record, and 2012 was the hottest year on record in the continental United States."[14] (Fig. 6.4). And as stated earlier, the IPCC concluded that this warming is unequivocal and is likely due to the increases in greenhouse gases.

The temperature record in Fig. 6.4 shows a clear increase in temperature since the mid nineteenth century. But there are large year-to-year variations and natural decadal scale oscillations in the climate system. Therefore, a sufficiently long record is necessary before any trend can be established.

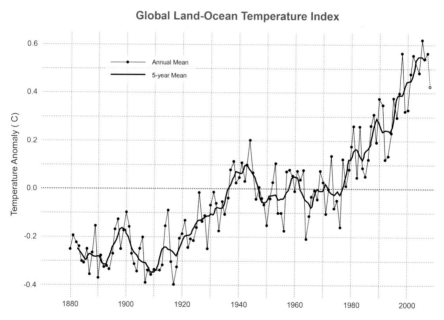

Fig. 6.4 Global mean surface temperature change, 1880–2010 (U.S. Global Change Research Program, 2009. United States Global Change Research Program. Global Climate Change Impacts in the United States. Available online: http://library.global-change.gov/products/assessments/2009-national-estimate-assessment)

Some of the recent controversy over lack of increasing temperature in recent years has to do with the rate of warming. The IPCC stated that "The rate of warming over the past 15 years (1998–2012; 0.05 [–0.05 to +0.15] °C per decade) is smaller than the trend since 1951 (1951–2012; 0.12[0.08 to 0.14] °C per decade)."[15] But the last decade was the warmest on record.

The U.S. National Climate Assessment of 2014 reported that "U.S. average temperature has increased by 1.3 °F to 1.9 °F since record keeping began in 1895; most of this increase has occurred since about 1970. The most recent decade was the nation's warmest on record."[16] The hottest month on record for any state was for Texas in August 2011. The IPCC WGII concluded in 2013 "It is extremely likely that human activities caused more than half of the observed increase in global average surface temperature from 1951 to 2010."[17]

The increase in temperature in the twentieth century has had a number of demonstrable effects. These include decreases in Arctic sea ice and Greenland ice cap, worldwide retreat of glaciers, melting of permafrost in the Arctic, and sea-level rise. As seas warm and permafrost melts, large amounts of methane (natural gas) are released. This is very troubling since each molecule of methane causes about 25 times as much warming as a molecule of carbon dioxide. We will come back to some of these as we move across the country.

The IPCC predicts that global temperatures will likely increase by as much as 1 to 5C during the twenty-first century compared to the increase of approximately 1C during the twentieth century.[18] A 1° increase in temperature doesn't sound like much when daily temperatures often change by 20° or more. However, the rapidity of this increase is brought into focus when compared to past increases in average global temperature. For example, temperature increased about 15 C from 15,000 to 5000 years ago (or about 1.5° per millennium) as the last ice age ended, glaciers melted, and sea level rose. Thus modern temperature may increase by about a third of the amount in a single century (the twenty-first) of what took 10,000 years previously. Temperature directly affects precipitation patterns, sea-level rise, the intensity and frequency of tropical storms, as well as biological processes.

[15] IPCC, 2013, ibid.
[16] Melillo et al. (eds) 2014, ibid.
[17] IPCC, 2013, ibid.
[18] IPCC, 2013, ibid.

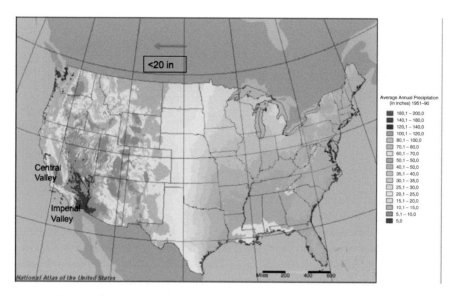

Fig. 6.5 Historical precipitation patterns in the United States. *Dark oranges* and *reds* are generally less than 10 in. per year while *blues* and *purples* are more than 60 in. per year. Areas west of the 100th Meridian receive less than 20 in. of precipitation and are unsuitable for agriculture (From the National Atlas of the United States. National Atlas of the United States, 2013. Precipitation (accessed on 22.07.2013). This map series 1997–2004 has was retired in September, 2014. Small scale maps are now available on nationalmap.gov/small_scale)

Precipitation

Let's take a look at "normal" precipitation patterns in North America.

Figure 6.5 shows average annual precipitation in the United States. In general, the 100th meridian running through west-central Texas north to Canada divides the U.S. into a moist eastern half and a drier western half as was illustrated in Powell's map discussed earlier. Exceptions to dryness in the west are in high mountains and in the northwest from northern California through Washington. Less than 20 in. (50 cm) of rain falls on most of the western U.S. each year. As a general rule of thumb, most agriculture in the West is not possible without irrigation. This is significant, as the southern Great Plains region (especially Nebraska and south) produces most of the beef in the U.S. while about 50 % of vegetables, fruits, and nuts are produced in the rich soils of the Central and Imperial Valleys of California. The Central Valley averages about 10 in. per year, while the Imperial Valley averages less than 5 in.

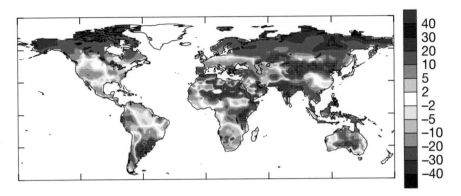

Fig. 6.6 Expected climate change impacts on water runoff by the end of the twenty-first century. *Blue* denotes relative increase and *red* denotes relative decrease, in percentages (Reprinted from From Macmillan Publishers Ltd. on behalf of Nature, Milly, P. C., Dunne, K. A., & Vecchia, A. V. 2005. Global pattern of trends in stream-flow and water availability in a changing climate. *Nature*, **438**(7066), 347–350, Copyright 2005)

What, then, do these patterns show us of expected future changes associated with global climate change? The southern Great Plains and the southwest are projected to have decreased freshwater availability of 10–50 % due to decreased precipitation and greater evaporation. The upper Mississippi and Ohio valleys and the northeast are projected to have increases in precipitation of 5–20 %. Precipitation in the southeast is not expected to change significantly on average. Globally, the area around the Mediterranean, southern Africa, the southwestern coast of South American, and much of Australia are expected to become drier. Higher latitudes in the northern hemisphere will be wetter (Fig. 6.6).

The Temperature–Precipitation Feedback Loop

As mentioned earlier, increasing temperatures in many areas mean that more winter precipitation falls as rain rather than snow, and more evaporates before it runs off. More rapid runoff leads to higher river levels in the winter and spring, and less stream flow in summer months. This affects both natural ecosystems and human uses of water for direct consumption and for irrigation.

The switch from snow to rain in mountains particularly impacts human populations that depend on melting snow for summer fresh water consumption. This is the case for the west coasts of both North and South America. La Paz, Bolivia is an extreme example of this situation. La Paz is situated in a desert with very little rain. Glaciers above the city provide its main sources of fresh water. The 18,000 year old Chacaltaya glacier disappeared in 2009, 6 years earlier than predicted. As the glaciers melt, La Paz will lose its source of water.[19]

Along the west coast of the U.S., summer water use is dependent almost entirely on snow melt stored in reservoirs. These reservoirs are sized for slow melting of the snow pack over the spring and summer. As total winter precipitation amounts fall and evaporation increases, and as more winter precipitation occurs as rain, total runoff becomes less; a greater proportion of this reduced runoff is occurring early in the year. This is a problem for all of the west coast, but it becomes increasingly more acute to the south because of increasing aridity and large water demand. In California, the demand is both from agriculture and from the large population. As stated earlier, the Central Valley of California is one of the most important agricultural areas in the U.S. But the Central Valley is very arid and is absolutely dependent on reliable supplies of water for irrigation. Reservoirs in much of the Southwest have been running at much below capacity for many years. Increasing temperatures, lower rainfall, higher evaporation, and high demand point to chronic water shortages for decades to come. Many think the severe drought of the last several years, which became acute in 2014 and 2015, will become more common in the coming decades. Dr. Benjamin Cook of Columbia University and colleagues report that droughts in the Southwest Great Plains will likely be worse than any in the last thousand years (Figs. 6.7 and 6.8).[20]

The Great Plains

The great plains between the Southwest and the Mississippi Valley straddle Powell's dividing line between the arid region of the west and the wetter east—but summers are expected to get much drier in this region over the coming decades. The Great Plains stretch from Texas to

---

[19] Francou B. et al. 2000. Glacier Evolution in the Tropical Andes during the Last Decades of the 20th Century: Chacaltaya, Bolivia, and Antizana, Ecuador. Ambio. Vol. 29, No. 7, Research for Mountain Area Development: The Americas. pp. 416–422.; Francou, B. et al. 2003. Tropical climate change recorded by a glacier in the central Andes during the last decades of the twentieth century: Chacaltaya, Bolivia, 16° S. *Journal of Geophysical Research* 108.; International Bank for Reconstruction and Development/The World Bank and International Cryosphere Climate Initiative. 2013. *On Thin Ice: How Cutting Warming Can Slow Pollution and Save Lives*. A Joint Report of the World Bank and The International Cryosphere Climate Initiative.

[20] Cook, B. I., Ault, T. R., & Smerdon, J. E. 2015. Unprecedented 21st century drought risk in the American Southwest and Central Plains. *Science Advances*, 1(1), e1400082.

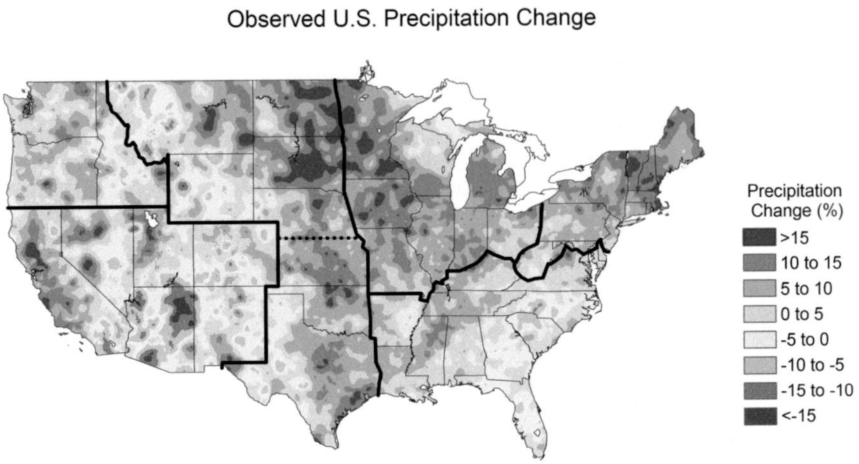

Observed U.S. Precipitation Change

Fig. 6.7 Observed precipitation change, from the 2014 National Climate Assessment. The *colors* on the map show annual total precipitation changes for 1991–2012 compared to the 1901–1960 average, and show wetter conditions in most areas. California was very dry in 2013 and 2014 (From Climate Change Impacts in the United States: The Third National Climate Assessment. 2014. Data from National Oceanic and Atmospheric Administration National Climatic Data Center/Cooperative Institute for Climate and Satellites North Carolina)

Canada and support wheat farming, cattle ranching, and assorted other agricultural activity. Much of the area is underlain by the famous Ogallala aquifer that extends over parts of eight states (Fig. 6.9).[21] The aquifer contains "fossil" water left over from a wetter epoch as the last ice age ended. Very little water is added to the aquifer now, but vast amounts of water are pumped out, and without this water, much of the agricultural richness of the Great Plains would be lost. The Ogallala contains enough water to fill Lake Erie nine times. Annually, five trillion gallons are pumped from the aquifer to support $20 billion in agricultural production. The extent of this irrigation is visible on almost any flight from the east to the west in the form of numerous large green irrigated circles of center pivot irrigation systems (see Fig 5.2). A paper in the prestigious Proceedings of the National Academy of Sciences predicted that nearly 70 % of the aquifer could dry up within 50 years.[22]

---

[21] An excellent book on the aquifer is William Ashworth. 2007. *Ogallala Blue: Water and Life on the High Plains*. The Countryman Press Woodstock, VT.

[22] Steward, D. R., Bruss, P. J., Yang, X., Staggenborg, S. A., Welch, S. M., & Apley, M. D. 2013. Tapping unsustainable groundwater stores for agricultural production in the High Plains Aquifer of Kansas, projections to 2110. *Proceeding of the National Academy of Sciences*. doi:10.1073/pnas.1220351110.

Projected Precipitation Change by Season

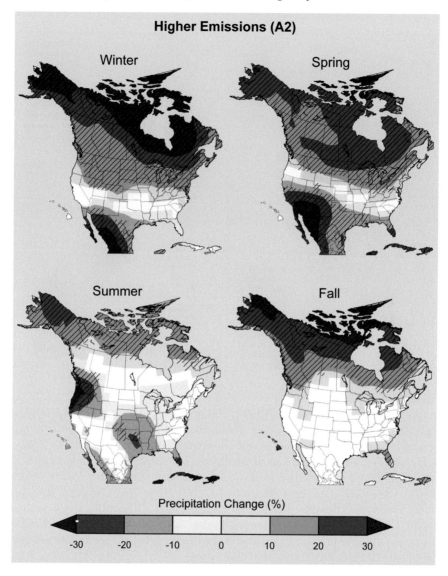

Fig. 6.8 Projected precipitation change by season, from the 2014 National Climate Assessment report. Maps show projected percent change in precipitation in each season for 2071–2099 (compared to the period 1970–1999) under an emissions scenario that assumes continued increases in emissions (A2). *Teal* indicates precipitation increases, and *brown*, decreases. Hatched areas indicate that the projected changes are significant and consistent among models. *White areas* indicate that the changes are not projected to be larger than could be expected from natural variability. Wet regions are generally projected to become wetter while dry regions become drier (This figure appears in chapter 2 of the Climate Change Impacts in the United States: The Third National Climate Assessment report. 2014 Cooperative Institute for Climate and Satellites – NC Kenneth Kunkel. http://nca2014.globalchange.gov/report/our-changing-climate/precipitation-change/graphics/newer-simulations-projected-precipitation)

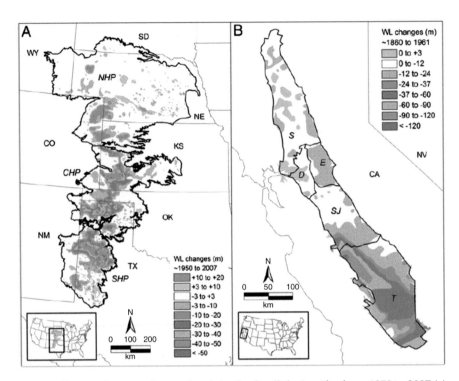

Fig. 6.9 Changes in groundwater levels in the Ogallala Aquifer from 1950 to 2007 (**a**) and Central Valley Aquifers from 1860 to 1961 (**b**). Dropping groundwater levels are especially pronounced in the Texas Panhandle and in the southern half of the Central Valley (Figure From: Scanlon, B. R., Faunt, C. C., Longuevergne, L., Reedy, R. C., Alley, W. M., McGuire, V. L., & McMahon, P. B. 2012. Groundwater depletion and sustainability of irrigation in the US High Plains and Central Valley. *PNAS* **109**(24), 9320–9325. Data From: McGuire V.L. 2009. *Water Level Changes in the High Plains Aquifer, Predevelopment to 2007, 2005–06, and 2006–2007*, Scientific Investigations Report 2009–5019 US Geological Survey, Reston, VA. and; Williamson AK, Prudic DE, Swain LA (1989) *Ground-Water Flow in the Central Valley, California*, Professional Paper 1401-D, US Geological Survey, Reston, VA)

The difference in precipitation between the Great Plains and Southwest on one hand and the Mississippi Valley on the other is striking. Almost everywhere west of a line from west-central Texas north to Canada has annual precipitation of less than 20 in. North and east of Texas, annual precipitation is greater than 30 in. almost everywhere. For precipitation greater than 30 in, agriculture without extensive irrigation is possible. With less than 20 in, it is not. Precipitation is expected to decrease over much of the Great Plains due to climate change. This will make the area even more dependent on irrigation water pumped from the Ogallala aquifer. We will come back to the implications of this later.

The Mississippi Valley

Now let's look to the east of the Arid Region to another and decidedly different drainage basin, the Mississippi. Ole Man River is the granddaddy of rivers in North America, and one of the largest in the world. In contrast to the Colorado basin and the Southwest, which are dry, the Mississippi basin, in general, is wet and likely to get wetter. It is a river of superlatives. In almost every way that rivers are measured, the Mississippi is in the top ten globally. At 2466 miles, the Mississippi-Missouri is the world's third-longest river. The river discharges an average of about 18,000 cubic meters of water per second (cms) into the Gulf of Mexico (the eighth-largest discharge in the world). By comparison, the mean discharge of the Colorado is about 550 cms. The Mississippi drains about 40 % of the continental U.S., including all or part of 31 states and 2 Canadian provinces. With an area of 3.2 million km^2, it is the world's fifth-largest drainage basin. The Mississippi basin is about five times as large as the Colorado basin, but discharges 32 times as much water, a testament to the higher rainfall over the Mississippi drainage.[23]

Like the Colorado, the Mississippi is a river greatly changed by human activity. Since the beginning of European colonization, humans have tried to tame the mighty Mississippi to meet their needs. The river is the busiest inland waterway in the world, and the Mississippi basin is one of the richest agricultural areas on Earth. For its last 150 miles, the river flows through the Mississippi delta. The delta is the largest contiguous coastal ecosystem in the United States—and one of the largest in the world. It covers about 25,000 km^2, or nearly 10,000 miles^2. This is as large as the lower third of the Florida peninsula, one-and-a-half times as large as the Olympic peninsula in Washington, nearly ten times as large as Grand Canyon National Park, comparable to a swath of land 25 miles wide stretching from Washington, D.C., to Boston.

The delta formed over the past five millennia as the river changed course repeatedly, depositing sediment that built new land and nourished wetlands. The river sustained the delta for thousands of years but during the twentieth century over a quarter of the wetlands were lost primarily as a result of human activities, especially levees that prevented the river from spreading out over the delta.[24] Because of this, the delta is weakened and highly susceptible to climate change.

---

[23] Coastal Protection and Restoration Authority of Louisiana. 2012. *Louisiana's Comprehensive Master Plan for a Sustainable Coast.* Coastal Protection and Restoration Authority of Louisiana. Baton Rouge, LA.

[24] Day, J., D. Boesch, E. Clairain, P. Kemp, S. Laska, W. Mitsch, K. Orth, H. Mashriqui, D. Reed, L. Shabman, C. Simenstad, B. Streever, R. Twilley, C. Watson, J. Wells, D. Whigham. 2007. Restoration of the Mississippi Delta: Lessons from Hurricanes Katrina and Rita. Science. 315, 1679–1684.; Day, J., P. Kemp, A. Freeman, and D. Muth. (eds). 2014. *Perspectives on the Restoration of the Mississippi Delta: The Once and Future Delta. Springer,* New York. 194 p.

Coastal Areas

Climate-change is already affecting the coast, and the effects will grow more severe during the twenty-first century. One of the most important effects is acceleration of the rate of sea level rise, which will impact coastal ecosystems and the people living along the world's coasts who depend on them. Because a large proportion of the human population lives near the coast, sea-level rise will disproportionately impact even those urban dwellers whose lives are not so clearly dependent on the productivity of coastal ecosystems. Sea level rises when the volume of the ocean increases. This is due to two primary factors: first, additional water is added to the ocean as land-based ice masses melt, and second, water in the ocean expands as it warms. As glaciers and other ice masses such as the Greenland ice cap melted at the end of the last ice age, sea level rose over 100 m from about 15,000–5000 years ago.

It is interesting to note that during the last interglacial period, when global mean temperatures were no more than 2 °C above pre-industrial values, maximum global mean sea level was, for several thousand years, 5–10 m higher than present with substantial contributions from the Greenland and Antarctic Ice Sheets. This is an important point, since there is much discussion among climate scientists about the potential for sea level rise due to melt water from Greenland and Antarctica.

Worldwide, the rate of sea level rise since the mid-nineteenth century has been larger than the mean rate during the previous two millennia (Fig. 6.10). Global mean sea level rose at an average rate of 1.7 mm per year from 1901 to 2010 and at a faster rate, 3.2 mm per year (based on highly accurate satellite measurement) from 1993 to 2010. There is a substantial anthropogenic contribution to the global mean sea level rise since the 1970s.

In its 2013 report, the IPCC predicted a rise of up to 1 m by 2100.[25] Some estimates based on so-called semi-empirical projections based on past relationships between temperature and sea level rise project a total rise of more than a meter by 2100.[26] Melting of ice on sea and land has an insidious effect called a reduction in 'albedo" or surface reflectivity. Because ice and snow are white, they reflect much solar radiation back into space. As melting

---

[25] IPCC, 2013, ibid.

[26] See for example, Vermeer, M., Rahmstorf, S., 2009. Global sea level linked to global temperature. *Proc. Natl. Acad. Sci. U.S.A.* 106, 21527–21532.

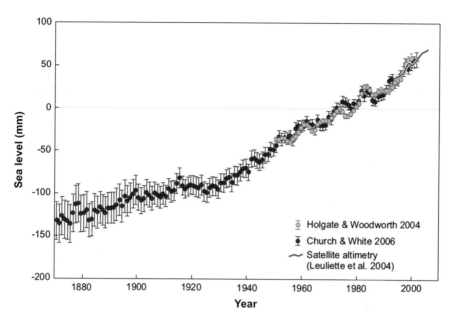

Fig. 6.10 Annual averages of global mean sea level from IPCC (2007). The *red data* are updated from Church &White (2006); the *blue data* are from Holgate &Woodworth (2004), and the *gray curve* is based on satellite altimetry. Zero represents the 1961–1990 averages for *red* and *blue* data (Figure from: IPCC. 2007. Climate change 2007-the physical science basis: Working group I contribution to the fourth assessment report of the IPCC (Vol. 4). Solomon, S. (Ed.). Cambridge University Press; Church, J. A., & White, N. J. 2006. A 20th century acceleration in global sea-level rise. *Geophysical Research Letters*, **33**(1); Holgate, S. J., & Woodworth, P. L. (2004). Evidence for enhanced coastal sea level rise during the 1990s. *Geophysical Research Letters*, **31**(7); Leuliette, E. W., Nerem, R. S., & Mitchum, G. T. (2004). Calibration of TOPEX/Poseidon and Jason altimeter data to construct a continuous record of mean sea level change. *Marine Geodesy*, **27**(1–2), 79–94)

occurs, the ground and water underneath is dark, absorbing sunlight and leading to even more heating.

If global warming exceeds a certain threshold resulting in near-complete loss of the Greenland Ice Sheet over a millennium or more, global mean sea level would rise about 7 m. A recent study suggests that the melting of the Antarctic ice sheet is irreversible and will lead to several meters of sea level rise over the next several centuries.[27]

---

[27] Rignot, E., Mouginot, J., Morlighem, M., Seroussi, H., & Scheuchl, B. 2014. Widespread, rapid grounding line retreat of Pine Island, Thwaites, Smith, and Kohler glaciers, West Antarctica, from 1992 to 2011. *Geophysical Research Letters*, 41(10), 3502–3509.

Because of the evidence of accelerated sea-level rise, many in the federal government and most coastal states are planning for an increase of about a meter. They are developing management plans to adjust and adapt to this rise. But North Carolina has taken a novel approach. It has, in effect, outlawed accelerated sea-level rise. The North Carolina State Legislature passed a bill that would prohibit state agencies from preparing for a 1 m rise. Democratic Gov. Bev Perdue let it become law by doing nothing. State legislators ignored a report by a panel of experts from N.C. that recommended that the state prepare for a 1-m rise, and ordered it stricken from the public record. The bill stated "The Division of Coastal Management shall be the only State agency authorized to develop rates of sea-level rise and shall do so only at the request of the Commission. These rates shall only be determined using historical data, and these data shall be limited to the time period following the year 1900. Rates of sea-level rise may be extrapolated linearly to estimate future rates of rise but shall not include scenarios of accelerated rates of sea-level rise". Word is not yet in on whether the warming sea will abide by the law.

Even before the bill became law, scientists from the U.S. Geological Survey reported that sea level rise along the portion of the East Coast between North Carolina and Massachusetts is accelerating at three to four times the global rate. The report predicted that sea level along the coast of that region, which it called a "hotspot," would rise up to 11.4 in. higher than the global average rise by the end of the twenty-first century.

Perfect Storms and Other Weather Events

According to the 2014 US National Climate Assessment, "the intensity, frequency, and duration of North Atlantic hurricanes, as well as the frequency of the strongest (Category 4 and 5) hurricanes, have all increased since the early 1980s. The relative contributions of human and natural causes to these increases are still uncertain. Hurricane-associated storm intensity and rainfall rates are projected to increase as the climate continues to warm."[28]

Recent evidence also indicates a trend toward stronger hurricanes that make landfall in the US Gulf or South Atlantic regions. For example, eight of the nine costliest Atlantic hurricanes in the U.S. occurred since 2000. Although no single weather event or flood can be attributed solely to

---

[28] Melillo et al. (eds) 2014, ibid.

climate change, recent extremes are strikingly consistent with the predictions of climate change models.

Three recent hurricanes exemplify what may be the future of the Gulf and Atlantic coasts of the U.S.: Katrina in 2005, Irene in 2011, and Sandy in 2012. Katrina is the most notorious of the three, and the story of its disastrous impact on the Gulf coast and New Orleans is well known. Katrina first swept east to west across south Florida, killing 11 and causing hundreds of millions of dollars in damage. Then, as it passed over the hot waters of the Gulf of Mexico, it grew into a Category 5 hurricane, with winds greater than 155 mph. It never deviated from a path that took it almost directly over New Orleans. On Sunday morning, August 28, a mandatory evacuation was ordered, and roads out of New Orleans were clogged with traffic.

Early morning on Monday, August 29, the storm, weakened to a Category 3, crossed over the mouth of the Mississippi and passed just east of New Orleans. Within 24 h, a major city was nearly destroyed by hurricane winds, driving rain and a failed levee system that, through various iterations, had protected it, more or less, for well over a century. The path of Katrina was very similar to that of Hurricane Betsy in 1965, a storm that also flooded much of New Orleans.

The statistics are well known. The passage of Katrina resulted in winds hitting the coast in excess of 150 mph, generating a tidal surge southeast of New Orleans of nearly 20 feet, and nearly 30 feet on the Mississippi state coast just east of the border with Louisiana. Levees in several parts of the city, especially in the eastern areas, were over-topped or failed. Eighty percent of the metropolitan area was flooded, and an estimated 90 % of homes were destroyed or damaged, leading to over 1500 deaths. Nearly 100 % of the population either left in advance of the deluge or was dramatically evacuated in its aftermath. It is interesting to note that had the levees not failed around New Orleans, Katrina would be remembered only as one of a number of strong storms that have hit the central Gulf coast. But it certainly would not have the notoriety it has now. Many believe that a disaster like Katrina was inevitable, and that another like it will happen again. We will come back to this question again.

Hurricane Irene formed in the tropical Atlantic in mid August 2011. Its path took it north of Cuba, over the Bahamas, and up along the east coast. It reached Category 3 strength but weakened to a Category 1 as it made landfall on August 27 on the Outer Banks of North Carolina. With winds of 85 mph, Irene hovered over land for 10 h, dumping large amounts of rain on North Carolina. It re-emerged into the Atlantic in southern Virginia near the mouth of Chesapeake Bay. It moved inshore again on August 28 along the New Jersey shore as a tropical storm but soon moved over water again and made a third landfall near Coney Island in New York City. The storm then tracked northeast over New England, Quebec, and finally passed into the Labrador Sea. The New York metropolitan area suffered flooding, and 350,000 homes and businesses were without power in Nassau and Suffolk counties.

But the most dramatic flooding occurred in New England. In Massachusetts, heavy rain caused widespread flooding of Connecticut River tributaries. In a few hours, the Westfield River rose almost 20 feet and the Deerfield rose over 15 feet, reaching flood stages not seen since the record flood of 1955. Widespread flooding drenched Vermont. Flood levels on the Deerfield River were higher than during the 1938 hurricane. Irene was only the second tropical cyclone to make a direct hit on Vermont. Throughout the state, many covered bridges, some over 100 years old were damaged or destroyed. There was extensive damage to roads and many small communities were isolated. Similar stories were common throughout New England and eastern upstate New York.

Hurricane Sandy, or Superstorm Sandy, was a big mess for the Big Apple and a big wake-up call. It was the deadliest (286 killed), most destructive, and most expensive hurricane of the 2012 Atlantic hurricane season, with damages estimated at over $68 billion.[29] In terms of cost, it was second only to Katrina in U.S. history. Hurricane force winds occurred over a diameter of 1100 miles, the largest Atlantic hurricane ever. Sandy developed from a tropical wave in the western Caribbean in late October. It then passed over the Caribbean, tracked up the east coast, and made landfall on October 29 just northeast of Atlantic City as a post-tropical cyclone with hurricane-force winds.

Normally hurricanes on Sandy's track would turn easterly and move out into the Atlantic. Instead, Sandy took a strong left turn and plowed into the New York City region. The storm was blocked by a high pressure ridge over Greenland that caused the jet stream to double back on itself just off the East Coast and flow toward the coast. Sandy was caught up in this westerly flow. The high pressure over Greenland also stalled an Arctic front that combined with Sandy to create a perfect storm for New York.

The storm affected an enormous area including all of the east coast and parts of the Midwest. It caused flooding of streets, along with vehicle and subway tunnels, and caused extensive power outages in and around New York City. Large areas of lower Manhattan flooded when the East River overflowed its banks. The water level at Battery Park reached 13.88 feet, over

---

[29] http://en.wikipedia.org/wiki/Hurricane_Sandy. Bennington J. B. and Farmer C. 2014. *Learning from the Impacts of Superstorm Sandy*, 1st Edition. Academic Press 140 pp.

2 feet higher than all other high water marks at the park since 1900 (Fig. 6.11). The Ground Zero construction site was flooded. The storm caused the release of over 10 billion gallons of raw and partially treated sewage. The storm severely damaged or destroyed about 100,000 homes on Long Island.

In the aftermath of Sandy, there has been a great deal of discussion about how to plan for similar events in the future. These discussions include rebuilding wetlands and dunes and making buildings more storm resistant. But given the densely populated nature of the area the potential for natural protection is limited. The central element of planning seems to be a storm barrier system similar to that of the Netherlands or New Orleans.

This is a complicated and difficult matter because of the geography of the Greater New York area. First, large areas of Staten Island and Long Island are low-lying and situated directly on the sea front. Thus there are not extensive wetlands and other natural systems to protect the area as there are, or were, for New Orleans. Second, storm surge can enter the New York area via New York Bay, the Arthur Kill River, and Long Island Sound. Any kind of effective barrier system has to deal with all three of these. A structure has been proposed for the entrance to New York Bay at the Verrazano Narrows. The preliminary cost is estimated at $6.5 billion.[30] Additional structures would be needed for the Arthur Kill ('kill' is derived from the Dutch word for creek) and Long Island Sound to protect the central core of New York City. The total costs of such a system would run into the tens of billions of dollars. Construction of such a system would almost certainly worsen flooding on the seaward side of the barriers in areas of Long Island Sound and Connecticut.

There has been a great deal of discussion about whether Sandy was related to global warming and climate change. Some have suggested that the shape of the jet stream was related to melting of Arctic ice. Others have maintained that the combination of the two storms was due to the natural variability of the weather. But sea level in the New York and New Jersey area has increased by about a foot since 1900 and this contributed to the 13 foot height of the surge (see Fig. 6.11). Sea level is projected to increase more in this region than for the oceans in general. Some have suggested that this will be the "new norm on the Eastern seaboard". Sea level rise is projected at a faster rate than the ocean in general for the rest of this century.[31]

---

[30] Aerts, J.C.J.H., Botzen, W.J. and De Moel, H. 2013. Cost Estimates for Flood Resilience and Protection Strategies in New York City. *Ann. N.Y. Acad. Sci.*

[31] Sallenger Jr. et al. 2012. Hotspot of accelerated sea-level rise on the Atlantic coast of North America. *Nat. Clim. Change*, 2, 884–888

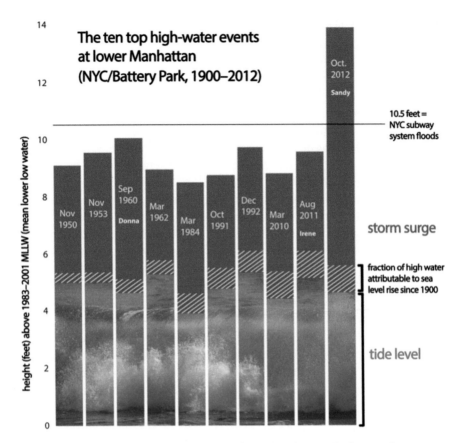

Fig. 6.11 Historic high water marks at Battery Park in lower Manhattan. Superstorm Sandy surpassed previous historical records by about four feet image by Carlye Calvin and Bob Henson, ©UCAR; data courtesy Chris Zervas, NOAA National Ocean Service

But did climate change cause—or contribute to—the size and intensity of Sandy? According to Kevin Trenberth of the National Center for Atmospheric Research, that is the wrong question. "All weather events are affected by climate change because the environment in which they occur is warmer and moister than it used to be."[32] Sea surface temperatures off the east coast were about 5 F higher than normal when Sandy hit, and it is warm seawater that is the energy source of hurricanes (Table 6.1).

Climate change in the form of sea-level rise and hurricanes will strongly impact the Gulf and Atlantic coasts. Meteorologists expect hurricanes to

---

[32] Russell, Muir, chair. *"The Independent Climate Change E-Mails Review"* July 2010. Available online: http://www.cce-review.org/pdf/FINAL%20REPORT.pdf

Table 6.1 Costliest U.S. Atlantic hurricanes. Cost refers to total estimated property damage

Rank	Hurricane	Season	Damage
1	Katrina	2005	$108 billion
2	Sandy	2012	$65 billion
3	Ike	2008	$29.5 billion
4	Andrew	1992	$26.5 billion
5	Wilma	2005	$21 billion
6	Ivan	2004	$18.8 billion
7	Irene	2011	$15.6 billion
8	Charley	2004	$15.1 billion
9	Rita	2005	$12 billion
10	Frances	2004	$9.51 billion

Source: National Hurricane Center https://en.wikipedia.org/wiki/List_of_costliest_Atlantic_hurricanes

track further north more often because of warming. Hurricane Emily, that caused such damage in New England, may be a harbinger of the future for the northeast. Every major city on the Gulf and Atlantic coasts has substantial areas near sea level. And given that hurricane surge can be more than 5 m, significant portions of densely populated areas are threatened. New Orleans is the worse case with 80 % of the city below sea level. New Orleans has over 300 miles of levees; no other city has anything approaching this. Miami is perhaps the second most threatened city. But many others are in danger including Galveston, Houston, Mobile, the densely populated east coast of Florida, Savannah, Charleston, Wilmington, Norfolk, Washington, Baltimore, Philadelphia, New York, and Boston. It is virtually unaffordable to build flood control systems like New Orleans for all these areas.

By late this century, models, on average, project an increase in the number of the strongest (Category 4 and 5) hurricanes. Models also project greater rainfall rates associated with hurricanes in a warmer climate, with increases of about 20 % averaged near the center of hurricanes.[33]

Over the past several years, stories have filled the news about extreme events. A drought and heat wave in Russia in 2010 led to greatly reduced grain production and wildfires that burned large areas of wheat fields. The country was forced to ban grain exports. In France, a heat wave and drought in 2003 lowered river levels so that there was not enough water to cool nuclear reactors. Electric power production was curtailed. Low precipitation in the U.S. southwest and mountain states has led to record low levels in reservoirs, including Lake Meade. Some speculated that the cur-

[33] Melillo et al. (eds) 2014, ibid.

rent dry conditions were similar to those documented in the middle twelfth and late thirteenth century that led to the collapse of Native American Ancestral Pueblo cultures of the area. In Pakistan, record floods in 2010 on the Indus River inundated a large part of the country. Six million people were left homeless.

In 2011, record flooding hit the Mississippi. The factors that led to the flood are all consistent with climate change projections. The intense storms that delivered so much precipitation are largely a result of the interaction of warm air masses from a warming Gulf of Mexico with colder continental air masses. This gave rise to the very intense storms that have also been documented in many parts of the world. Precipitation is generally expected to increase in higher latitudes, as is snowfall due to a wetter atmosphere.[34] According to a recent study, the combined impacts of changes in climate, CO_2 concentrations in the atmosphere, and land use may lead to increases in river discharge of 10–60 % by the end of the century.[35]

Given the trajectory of rainfall intensity in the temperate zone, floods like those in 2011 are likely to become more common. One possible indication is that the Bonnet Carré spillway, an overbank high water relief structure a few miles upstream of New Orleans on the Mississippi, has been opened 10 times in the past 80 years. It was opened three times during the first 40 years after it was built in the early 1930s, and seven times since 1973, during the second half of its operational life to date. Economic losses due to climate related disasters have increased over the past three decades, with losses exceeding $100 billion dollars in some years and $50 billion in most years since 1990.

According to Dr. Jeff Masters, cofounder of Weather Underground, weather in the U.S. during Spring 2011 reached new extremes, as a punishing series of billion-dollar disasters brought the greatest flood in recorded history to the Lower Mississippi River, an astonishingly deadly tornado season especially in Missouri and Alabama, the worst drought in Texas history, and the worst fire season in recorded history. There's never been a spring with such combined wet and dry extremes in the U.S. since record keeping began over a century ago.[36] The dramatic spring of 2011 was followed by an equally remarkable spring in 2012, when a record breaking heat wave in March smashed temperature records in 15 different states stretching from Canada down through the upper and lower Midwestern states. The warm weather continued through the spring 2012 months of March, April,

---

[34] IPCC, 2013, ibid.

[35] Bo Tao et al. 2014. Increasing Mississippi river discharge throughout the 21st century influenced by changes in climate, land use, and atmospheric CO2. Geophysical Research Letters, 21, 4978–4986. doi:10.1002/2014GL060361.

[36] http://www.cnn.com/2011/US/08/20/weather.disasters/index.html

and May, in the U.S. as the average temperature during these months was 5.2 °F above the long term average, hotter than any season on record.[37]

Thus, climate scientists are predicting that global climate change will lead to more dangerous and unprecedented extreme weather. This includes droughts, floods, extremely heavy precipitation, strong storms such as hurricanes, and heat waves. These conclusions were presented in a recent report by the IPCC on global warming and extreme weather. For example, the IPCC report (2013) predicted that heat waves that currently occur once in a generation will occur every 5 years on average by mid century, and every 2 years by the end of the century. In some areas, heat waves will be annual events. The heat waves will also become hotter. The report also predicted that storms with very heavy rainfall will happen much more frequently. If current trends of fossil fuel burning continue, such storms will occur about three times more frequently in the U.S. and four times more often in southeast Asia.

Orchestrated Denial

Despite the overwhelming scientific evidence of global climate change and its impacts, many in the public still do not believe this. Why is this the case? Before his untimely death in 2011, the noted climate scientist Stephen Schneider had agreed to write the climate chapter for this book. He had just begun this effort when he passed away. In discussions with him, he stressed that the science supporting climate change was robust but that there was an active campaign, especially in the U.S., to cast doubt on climate science. The well funded misinformation campaign regarding climate change that has continued for well over a decade is detailed in the 2010 book *Merchants of Doubt* by Naomi Oreskes and Erik Conway. James Powell followed in 2011 with *The Inquisition of Climate Science*. He shows how a small group of climate scientists with deep funding from big business has been able to create doubt in the public's mind. Some of this same group of scientists has been involved in anti-science efforts going back as far as Rachel Carson's *Silent Spring* and, most notably, the attack on the science showing the health effects of tobacco. The recent action by the North Carolina legislature fits into this tradition of dishonesty and deceit. As a result, the general public in the U.S. is more skeptical about climate change than in most other developed countries. But a fair and objective review of the science leaves little doubt that climate change is real and that society needs to be prepared for it.

In summary, we believe that the scientific evidence for human-induced climate change is sufficiently compelling to include in our regional sustain-

---

[37] http://www.reuters.com/article/2012/06/07/us-climate-warmth-usa-idUSBRE8561BK20120607

ability analyses, at least as a probable projection for the future. Since the projections and accumulating evidence indicate that the impacts of climate change will not occur evenly over the landscape, inevitably there will be winners and losers. The area that will be most negatively affected by drought is the Southwest including much of California and the southern Great Plains. These regions will dry and there will be less water available even as the population grows. There will be larger and more frequent wild-fires and loss of forests. Indeed, this is occurring now. And it is difficult to see how the region can adjust to these severe changes. Coastal regions will be impacted by accelerating sea-level rise and more frequent, stronger hurricanes. While the Mississippi delta and southern Florida will be most strongly affected, all of the Gulf and Atlantic coasts will experience stronger climate impacts. Unlike the Southwest, there are things that can be done to partially mitigate these impacts. People can move away from the coast and build higher and stronger. In the lower Mississippi, the resources of the river can be used to rebuild and restore the delta. Dr. Virginia Burkett, of the U.S. Geological Survey noted that, "Alaska is another hotspot of coastal vulnerability. Rising temperature is causing the ice to melt that binds coastal sediments. Sea ice, which protects the Alaskan coastline from erosion, is declining rapidly and sea level is rising. The land base of communities living on permafrost is calving like icebergs into the Arctic Ocean." She goes on to say that, "a bright spot in our report (the IPCC 2013 report) is evidence that people around the world are now starting to anticipate and adapt to climate change. Planning is more evident than action, however, and delaying action generally means higher cost and a greater risk of adverse and, potentially, irreversible outcomes. Our report also shows there are limits to our ability to adapt, some consequences are inevitable and some are already underway."

Chapter 7

Energy: The Master Resource

Introduction

Everyone has heard a lot about the topic of energy lately, and many people have an opinion on such issues as "U.S. energy independence," "renewable energy," "peak oil," "fracking," "nuclear energy," "coal," and so on. Often where one stands on these issues relates to where one sits politically, with politicians and a compliant media constituting the drivers of the public's knowledge about energy. Regardless of political ideology, an often thin veneer of "knowledge" is used to support the various opinions. Beneath this veneer lies a wasteland of ignorance about what energy really is, how important it is to modern society, how much is produced, and how much we currently consume—or might be able to consume—in the future. The spectrum of opinions held range from "the end of cheap fossil fuels will cause a death blow to civilization" to "we can easily wean ourselves off of fossil fuels," to "we have enough fossil fuels to last hundreds of years" to "wind and solar can provide all our energy needs." Among the least informed, it seems, are the very politicians who often drive the national conversation about energy. Few of them have scientific backgrounds, and all of them feel compelled to address the needs of particular constituencies. The press tends also to be notoriously uninformed, even misinformed, about energy. With all due respect to them and to former U.S. Vice President Al Gore, the state of America's energy security, as well as the enormous shortcomings of most visions of a sustainable future, as great or even greater "inconvenient truths" than even climate change.

© Springer Science+Business Media New York 2016
J.W. Day, C. Hall, *America's Most Sustainable Cities and Regions*,
DOI 10.1007/978-1-4939-3243-6_7

This chapter aims to provide a broad overview of critical topics in energy, buttressed by sound science and the best facts we have available to us. We ask readers to put aside their preconceptions about energy, energy policy and especially energy politics. This will be helpful in bringing a new level of sophistication to their thinking about energy. As former U.S. Sen. Daniel Patrick Moynihan so famously stated, "Everyone is entitled to his own opinion, but not his own facts." At the end of the chapter, we take a brief trip through western New York State to apply what we have learned about energy, showing how the city of Syracuse evolved in relation to energy—the master resource.

It's the Energy, Stupid!

As late as the 1990s, fossil fuels were cheap and abundant, and the U.S. economy thrived. Former U.S. Treasury Secretary, Robert Rubin, who is often credited with engineering this prosperity may be brilliant, but it is no coincidence that rapid economic growth and debt reduction during the 1990s took place against a backdrop of crude oil prices averaging around $20 per barrel. While the relationship between energy cost and economic performance seems simple once you think about it—and we will show the data if you have doubts—the subject of energy remains complex. Many people are averse to science and mathematics, and to be sure, one cannot really understand energy without embracing a bit of physics. So if you want to understand energy, you need to do some work (no pun intended, since one commonly repeated definition of energy is the ability to do work). Some understanding is essential as energy is sometimes referred to as "the master resource." In other words, there is no sustainability without completing a comprehensive energy balance sheet. In this chapter we outline what you need to know about energy, and what this balance sheet is likely to look like in the future.

The science of energy falls largely within the realm of thermodynamics, whose meaning is easy to determine by parsing out its Greek roots of "thermos" or "heat," and "dynamis" or "power." Thermodynamics explains how energy in one form can be transformed into another form and how much of it remains useable after the transformation. Any energy transformation results in the production of some heat either as the necessary by product or as indicator of inefficiencies in the process. Energy production and use are governed by the established and immutable scientific "laws of thermodynamics" that are never violated, and that we will come back to later in this chapter.

Energy vs. Power

An important distinction exists between energy and power that is often ignored or confused, even by scientists. Energy is a thermodynamic[1] quantity equivalent to the capacity of a physical system to do work or produce heat, independent of the rate at which that work might be done. Units of energy thus provide a way of measuring how much total work can potentially be done with a given quantity of energy. Power is defined as an amount of energy delivered or used per unit of time. It thus represents a rate of energy production or use. Thus, a watt is a unit of power and is, therefore, a rate of energy production or consumption. A 100-W bulb in your home burning for 1 h consumes 100 W hours of energy. A typical home in the U.S. consumes 909 kilowatt-hours (kWh) per month, or 909,000 watt-hours (Wh), according to the U.S. Energy Information Administration.[2] We use a lot of energy. But why is this important?

The answer is simple, even if the subject itself is not. In today's society virtually everything we do requires high-density fuels. By high density, we mean a large amount of energy contained in a given volume or weight. For example, a gallon of gasoline has a much higher amount of energy than a gallon of saw dust, even though both can be burned. We use fossil fuel energy more than any other source of energy, and for a variety of purposes; 80–85 % of world energy use is fossil fuels, about the same percentage as when we (the elder authors of this book) started to study energy back in the late 1960s. The major fossil fuels used by modern society are coal, often used for generating electricity and industrial heating; oil, mainly used for transportation; and natural gas, used for heating and, increasingly, for generating electricity. Industry uses all of these as a feedstock to make things like plastics or fertilizer. They are called fossil fuels because they were formed from ancient plants that died, became buried, and were pressure cooked for tens to hundreds of millions of years. From the point of view of human lifetimes, fossil fuels are a one-time storage of immense quantities of solar energy stored in the chemical bonds of once-living matter.

While our hunter-gatherer ancestors subsisted off the land, modern societies rely to a great extent on highly mechanized, fossil energy-intensive agriculture. The net result is that modern humans have achieved greater population numbers and have lived longer than ever before. But it is more than just agriculture. Think about it. Everything we do is

---

[1] Thermodynamics is defined below.

[2] Energy Information Administration (EIA). 2013. *2013 Average Monthly Bill-Residential.* http://www.eia.gov/electricity/sales_revenue_price/pdf/table5_a.pdf.

dependent on energy from fossil fuels and a few other sources. Our homes and places of work typically are heated and cooled directly or indirectly, primarily by fossil fuels. We fill our homes with stuff that is made from and with fossil fuels and that use fossil energy in one form or another. Fossil fuels are used as an energy source for manufacturing, as a petrochemical feed stock to make many products (virtually all plastics), to transport almost everything to stores, and to generate electricity to run a myriad of electronic gadgetry. Our transportation system, from cars to the Airbus A380, is highly dependent on petroleum derivatives such as gasoline, diesel, and jet fuels (not to mention that many of the materials in a car, truck, or plane are made from petroleum derivatives).

Think about the use of energy in your own life and consider our societal reliance on it. We use energy to be warm or cool, well fed and comfortable, to get to and from a job, and to do that job, as well as to avoid and recover from disasters. The average person in the developed world, or even one in the developing world, lives a far more comfortable life, with a diet, health, longevity, leisure activities and transportation options far more abundant and diverse than almost any king or the wealthiest person in past civilizations. Consider a typical day. The cup of coffee we drink in the morning depends on fossil fuels for growing, processing, transporting, selling the beans, and preparing the beverage. We use energy for hot water in showers, lights in our house, phone chargers, a non stick pan and a high density plastic spatula to make eggs. We store left overs in a plastic bag or container in an electrically-powered refrigerator. Many of us drive to work, grab a coffee or a soda in Styrofoam or plastic containers, and sit in a plastic office chair in an air conditioned or heated office working on a computer. When was the last day you did not use fossil fuel energy or a derivative? Probably never. How would your life change without these fuels?

And so on…there is really nothing in modern life or the modern global economy that does not depend primarily on the unique energy density, controlled reactivity, relative safety, and transportability of fossil fuels. Modern society would crumble without these fuels. So, too, would the economic base that provides the livelihood of almost all people. Without abundant and affordable supplies of energy, there would be chaos. Food would be in short supply, and the future would be grim indeed. For this reason, we are compelled to understand the importance of energy in society, and if we know what is good for us, we will use the energy we have wisely; much more so than we do now.

The Laws of Thermodynamics

If you are hoping to develop a perpetual motion machine to solve our energy problems, we have bad news for you. Because of the Second Law of thermodynamics it cannot be done. Any energy transformation, such as from stored energy in gasoline to vehicle movement, or anything with moving parts, will have losses to heat. This is a direct expression of the Second Law. We are reminded of an old cartoon of a car with a forward facing windmill on top, with the car being powered by the wind it encounters from its own motion (Fig. 7.1). The Laws of Thermodynamics tell us this cannot work.

If you want to understand energy, you will need a familiarity with these laws. To begin with, the Laws of Thermodynamics are the fundamental, inviolable energy laws of the universe. Albert Einstein said that thermodynamics "is the only physical theory of universal content…that will never be overthrown."[3] They apply to everything from why you heat up when you ride a bike to why an airplane can fly. Although humans, unlike airplanes, cannot consume and metabolize fossil fuels directly, modern agriculture transforms fossil fuel based fertilizers and mechanization

Fig. 7.1 Example of a violation of the second law of thermodynamics, a hybrid car as a perpetual motion machine (Cartoon by Mike Baldwin, permission obtained from cartoonstock.com)

"Do you have any hybrids that weren't designed by men?"

---

[3] Marvin Klein. 1967. Thermodynamics in Einstein's thought. *Science*, 157, 509–516.

(with an assist from fossil-fuel based pesticides, irrigation systems, and the sun) into the food we eat. In all cases, these laws impose very strict limitations on what kinds of work can be done, as well as limitations on how efficient work processes can be. The First and Second Laws of Thermodynamics constrain everything we do with no exceptions or plea-bargaining.[4] There is no use in going to the North Carolina legislature and asking them to repeal these laws.

The First Law of Thermodynamics is the "Law of Conservation of Energy." The First Law stipulates that the total quantity of energy in the universe is constant. Energy cannot materialize from thin air, nor can it be created *de novo*. Another way of saying this is that energy cannot be created or destroyed, only changed in form. Thus, what you start out with is what you finish with, although the form in which the energy is found can vary widely from storage in chemical bonds ("potential energy") that release energy when these bonds are broken—as by combustion—to falling water ("kinetic energy") hitting a paddle wheel or turbine. In other words the *quantity* of energy in a system is constant.

This sets up the Second Law, which states that the *quality* of energy in a system is not constant, but degrades each time the energy is converted from one form to another in the process of doing work. Other implications of the second law are that energy always must dissipate from a warmer body to a colder body, and when used some portion goes to a less organized and less useful state in the form of waste heat that can no longer do useful work. A second and related concept is entropy, which is a measure of the degree of disorder or randomness in a system. The reason that this is associated with the second law is that energy must be expended to create and maintain the order required for function, i.e. the non-random structure of molecules in things of value (such as a sandwich, an automobile, or a person). In the process some energy is lost as waste heat that can no longer do useful work.

The implications of the Second Law are profound. Processes that involve energy transformations are unidirectional and irreversible, leading to dissipation of heat and increases in "entropy" (or spontaneous dispersal of energy) with time. Another way of stating this is that if left to themselves, things become more disordered over time (think of a young child playing

---

[4] In contrast to laws, theories are less established principles that are believed to be true by most scientists, but are less easily demonstrated true than laws. Finally, hypotheses are in the realm of conjecture. They are falsifiable concepts that can be tested by observations or experimentation. Remember, scientific laws are not hypothesized. They are established truths, at least within the limits of what we know at a given time. No legislature, human desires, or other human institution can change them. This is hard for many humans to understand because they are used to believing that whatever they want will come to pass. Unfortunately, thermodynamics puts constraints on what we actually can do.

with blocks, or your closet). Or as one of our spouses said when entropy was explained to her, "eggs cannot be unscrambled." A corollary to this is that the "entropy" of the universe increases over time. Every energy transformation comes at a cost and cannot be 100 % efficient. Ecologists like us want all people to understand the importance of the Second Law because the Law, without exception, applies to every living organism and ecosystem alike, including humans, of course. While the energy of sunlight is immense, photosynthesis is a very inefficient process that converts less than 4 % of sunlight into plant biomass.[5] Similarly, herbivores typically convert only 10–20 % of the plant biomass they eat into their own flesh. The transfer efficiencies, which we ecologists call "ecological efficiency," are roughly similar for each succeeding "trophic (or food) level." By the time a top carnivore such as a wolf, lion, or tuna fish eats its prey, only a very small fraction of the original energy from the Sun remains in the food chain. The rest has been dissipated as heat. In a home, when an air conditioner is used to cool a room down, the outside must be heated up even more. If the power goes off and your refrigerator stops working, it heats up and the food inside spoils. Fortunately for both wild organisms and society, new sources of energy are always flowing in from the sun and, at least in the past 200 years, from fossil fuels. This sort of analysis helps us to understand the fundamental similarities in how ecosystems, economic systems, and other human systems function.

The more transformations in producing or using a given energy source, the more of that energy will be lost as waste heat without doing any useful work. Some transformations are more efficient than others. For example, internal combustion engines are notoriously inefficient, at best up to 40 %, at converting their fuel's chemical energy to mechanical energy. Engine friction and waste heat are major factors accounting for low efficiency, both of which relate to the Second Law. On the other hand, electric motors are highly efficient, approaching 90 % efficiency in converting electricity to mechanical energy. However, if the electricity itself is made from fossil fuels (the usual case), that process is done at only a 30 or 40 % efficiency. Thus converting oil into vehicle miles traveled would be roughly the same efficiency by the time you were done. The usual railroad locomotive has a diesel engine attached to a generator that runs electric motors that move the train. This is a little more efficient than using the diesel directly, but not much. Thus a general corollary to the Laws of Thermodynamics is that as energy is transformed, the quantity of energy may be less, but often the

---

[5] Zhu, X. G., Long, S., and Ort, D. R. 2008. What is the maximum efficiency with which photosynthesis can convert solar energy into biomass? *Current Opinion in Biotechnology*, 19(2), 153–159.

quality (ability to do work per Joule) is increased. Producing protein from grass or electricity from fossil energy are good examples.

Thus, although for some it would be a no-brainer to say that we should switch over completely to electric vehicles, you have to ask where the electricity comes from and what are the transformation costs at each step, including the capital costs. If the electricity comes from the grid (which in most countries is typically supplied by fossil fuels), the answer is not clear. And there is another big problem: the "energy density" or "specific energy" of a fuel, i.e. the amount of energy contained in a given weight or volume of that fuel, must be factored in. For gasoline or diesel fuels, the energy density is many times higher than for a charged battery. An automobile with an internal combustion engine needs to haul around only the weight of the fuel itself (in a light weight tank). An automobile with an electric motor must haul around a battery often weighing nearly 100 times its energy equivalent in gasoline (not to mention the metal that must be mined and processed into batteries). Because electric motors are themselves lighter and have a higher energy conversion efficiency than combustion engines, they can sometimes provide a higher deliverable mechanical energy density. Additionally, electric cars are expensive, implying a lot of energy being used in the "food chain" generating the car. Thus, we must take into consideration the energy requirements of getting the additional materials for batteries to store the electricity. And given that there are over a billion vehicles worldwide, this is a tall order. Finally, for heavy duty, long-haul trucking "18-wheelers" or airplanes or ocean going ships, electrically driven propulsion is simply infeasible.[6] If the advantages and disadvantages of electric vehicles more or less cancel out, there is really no point of going to these vehicles.

Another issue is that fossil fuels come out of the ground almost ready to use. Coal and natural gas can generally be burned as is. Oil needs to be refined or "cracked" a bit (broken into smaller pieces, such as eight carbon octane) before it can be used. Nature spent millions of years "cracking" and "refining" ancient plant material to produce fossil fuels. This is not the case for electricity or hydrogen. One cannot drill an "electricity well" or a "hydrogen mine" and pump or dig that energy out of the ground. Rather, another energy source has to be used to produce the electricity that drives the electric motor. And for the most part, electricity must be used at the time it is produced; it cannot be readily stored in large quantities. Fossil fuels are typically the main source of primary fuel to produce electricity. However,

---

[6] Chapman, L. 2007. Transport and climate change: a review. *Journal of transport geography*, 15(5), 354–367.; Friedemann, Alice. 2016. *When trucks stop running*. Springer Briefs, New York

electricity also can be produced from hydropower, nuclear fuel, wind, and solar. A nuclear plant or wind turbine produces high quality electric energy but requires a vast, fossil-fueled infrastructure to allow that to happen. Additionally each of these requires expensive infrastructure to produce the electricity and deliver it to its point of use.

Thus to say whether the electric or gasoline automobile is more efficient requires a much more comprehensive analysis than simply the efficiency of the respective motors. Unfortunately such *systems analyses*, often termed "life cycle analyses," are rarely undertaken, so that we rarely understand which approach might be more efficient in an overall sense. This has not stopped advocates of various technologies from making a case based on only part of the system in question.

For advocates of fuels from biomass (like ethanol from corn), the same thermodynamic realities apply as for ecosystems. The capture of sunlight by plants is very inefficient, as we said above, typically much less than 4 %.[7] Converting the plant biomass into a useable fuel takes a number of steps, and every step in the conversion process involves energy loss because of the Second Law. Thus, almost all fuels from biomass have very little net energy yield. This helps to explain why the vast reservoirs of energy in fossil fuels are not renewable, even though their original source was photosynthesis. The efficiency of producing these fuels was so low that it required tens of millions of years to create them.

History of Human Energy Use

For millennia, humans lived in some kind of fundamental balance with nature, although humans as hunters were often too effective and did in many other species, probably including mastodons in North America and rhinoceros in Europe. Human population is estimated to have reached less than 10 million some 10,000 years ago.[8] It grew very slowly and did not reach 1 billion until about 1800 near the beginnings of the Industrial Revolution. Death at an early age was commonplace, and while family sizes may have been large, few children survived to adulthood. People used almost no fossil fuels. As the human population spread across the globe, animal skins were more often worn, reducing exposure to the sun and limiting the production of "sunshine" vitamin D, leading to rickets and the

---

[7] Zhu et al. 2008, ibid.

[8] http://en.wikipedia.org/wiki/10th_millennium_BChttp://en.wikipedia.org/wiki/10th_millennium_BC.

breaking of bones. The dark skin pigment melanin, originally protecting all of our human ancestors from the direct tropical sunlight, was selected against in more northerly environments. The need to use fire to stay warm and for cooking was met by burning wood, peat, and dung. This was essentially a sustainable, but perhaps unpleasant from today's perspective, existence—"nasty, brutish and short." The vast majority of people enjoyed few luxuries. Life was mainly about survival. With a nomadic existence, humans lived at a subsistence level, although recent hunter gatherers, such as many North American natives and the !Kung tribesmen of Africa, appeared to be very satisfied with their life. Settled agriculture redirected the energy flow of the sun from a diverse suite of natural plants, most of no direct use to humans, to species that were most directly useful to people. As society developed and agriculture and urban settings became important, additional sources of energy were needed to sustain these "advanced" activities. Rivers and streams were dammed to harvest water power. Starting in the late eighteenth century, the steam engine caused an abrupt shift in human ability to do all kinds of economic work, as well as a massive increase in the need to burn more fuel. Bringing water to boil takes a lot of energy (this is called the heat of vaporization), specifically 2260 J/g, which is referred to as the latent heat of vaporization (think of the time it takes to bring a pot of water to boil on the stove). Forests were cut in Europe as the need to bring water to a boil to make steam spread almost literally like wild fire.[9]

The whaling industry prospered as the population used whale oil to provide fuel for illumination, but whale populations were nearly eradicated. In the nineteenth century, coal rapidly supplanted wood as a source of energy to run trains and power mills and factories. One of the first uses of coal was to pump water out of mines so that more coal could be mined (Fig. 7.2). In 1859, crude oil was discovered at a shallow depth in Titusville Pennsylvania, and "Colonel" Edwin Drake promoted its use as a substitute for whale oil in illumination. The Industrial Revolution had begun its course, and the Fossil Fuel Age was off and running, enormously increasing the economic work possible and allowing the very large increase in human populations. There is no turning back, at least until the fuels run out or become too expensive. Figure 7.3 shows the increase in the use of energy from "primitive" humans to the present.

Oil Baron John D. Rockefeller helped set in motion the twentieth century, the Century of Oil. His monopoly, the Standard Oil Trust, the ancestor of present day Exxon, helped standardize (hence "Standard" Oil) refining,

---

[9] For an overview of human development from an energy perspective see Chap. 2 of Hall, C.A.S., Klitgaard, K.A., 2012. *Energy and the Wealth of Nations: Understanding the Biophysical Economy.* Springer Publishing, New York, 407 p.

Fig. 7.2 One of the first coal driven engines used to drain mines so that more coal could be obtained. This is an early example of using fossil fuel to get more fossil fuel, a process that continues today. Fossil fuels are heavily involved in obtaining all forms of energy (building dams, nuclear plants, windmills, solar panels). What this tells us is that fossil fuels will never actually run out. Instead, we will stop producing them when the ratio between energy in and energy out reaches one (no net energy), or probably much sooner (Found in lobby of the Universidad Politécnica de Madrid, Photograph By Nicolás Pérez (GNU Free Liscence Agreement) https://en.wikipedia.org/wiki/File:Maquina_vapor_Watt_ETSIIM.jpg)

making petroleum products safer—and widely relied upon—for many uses from home heating to transportation. The development of the internal combustion engine enabled transportation entrepreneurs like Henry Ford to develop huge production capacity for gasoline powered vehicles. The American love affair with the automobile was consummated when the average wage earner could afford a mass-produced vehicle. So, too, was America's commitment to needing large quantities of cheap oil. Public transit in the U.S. never really had a widespread enduring presence as in many other countries, partly due to the influence of the automobile companies and partly due to the relatively dispersed U.S. population. When the Eisenhower Interstate Highway System was conceived and begun in the mid 1950s, it constituted a virtual death knell for most passenger rail transportation. The United States was as dependent on petroleum as a junkie on

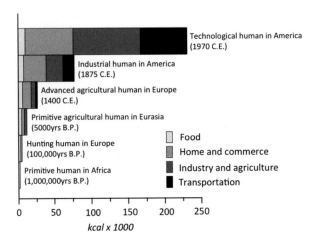

Fig. 7.3 Estimated energy utilization (e.g. food, wood, cattle, electricity) per person per day (in thousands of kcal/day) for one human at different levels of societal development. A kcal is equal to the amount of energy required to heat 1 kg of water 1 °C. A kcal (or kilocalorie) is also called a Large Calorie, a kilogram-calorie, 1 Nutritional Calorie or, confusingly, sometimes just Calorie with a capital C. The average adult eats about 2500 kcal per day. One kcal also equals 4.184 kJ or 3.96 BTUs (Redrawn From: Ngô, C., & Natowitz, J. 2012. *Our energy future: resources, alternatives and the environment* (Vol. 16). John Wiley & Sons. Data from: Cook, E. L. 1971. The flow of energy in an industrial society. *Scientific American*, (225), 135–142)

heroin. This has not changed since, and increasingly a similar phenomenon has been observed in virtually all developed and most developing nations. Oil, and more importantly shortages of oil, have since been drivers of foreign policy and often a *casus belli*. Understanding the factors that govern the production of oil leads naturally to the next topics, world energy use and "Peak Oil."

While a detailed history of the oil industry is beyond the scope of this chapter, it is worth pointing out some evolutionary changes that have occurred during the course of well drilling history. The first successful use of a rotatory drill was in 1901 for the Spindletop well in Beaumont, Texas. By today's standards, this was a very shallow well and a very productive one at that, yielding over 100,000 barrels per day. Its net energy yield was 100:1 or probably much greater (e.g., for each unit of energy used in getting the oil, 100 units of energy were produced), many times greater than is now possible in the industry, at least in the U.S. which has a yield of roughly 10:1.[10] For

---

[10] Guilford, M. C., Hall, C.A.S, O'Connor, P. and Cleveland, C.J. 2011. "A new long term assessment of energy return on investment (EROI) for US oil and gas discovery and production." *Sustainability*, 3(10), 1866–1887.

example, the notorious BP exploratory well, Macondo, had its well head roughly 1600 m (about 5000 feet) beneath the Gulf with the oil reservoir itself being approximately 5600 m (about 17,000 feet) below the sea bed. The capital and operating resources, technology, and personnel required to drill that well far exceeded that required for Spindletop, yet not a drop of oil from Macondo ever made it to a refinery. Even deeper wells ("ultra deep wells") are being drilled today, at considerable risk and great expense. The net energy yield of such wells is well below 10:1.[11] In general the mean EROI (Energy Return On Investment) has declined in the US and most places in the world from 30 to 50:1 in the middle of the last century to less than 10:1 today, as we must exploit deeper, smaller and more hostile oil fields. In addition EROI declines most rapidly when exploitation rates increase.

World and U.S. Patterns of Energy Production and Consumption

Global annual energy demand is about 500 quads (quadrillion BTU; roughly 500 ExaJoules or equal to about 80 billion barrels of oil) per year of energy, and is divided among different primary energy sources (Fig. 7.4). Of that, the U.S. demand is nearly one-quarter, despite the fact that the U.S. has only 5 % of the world's population. Total world energy use more than doubled between 1973 and 2011. As Fig. 7.4 demonstrates, the growth of energy demand globally has been substantial over the last two decades, although the growth rate is slowing, especially for oil, and the demand and use is projected to continue to grow over the next decade. But even if there is demand, this doesn't mean that there will be supply. It depends on the availability of oil at an affordable price.[12]

Several things are evident from Fig. 7.4. The first is that fossil fuels are overwhelmingly the most important sources of energy for the world, making up 80–85 % of total world energy use in the past and still today. This

---

[11] Moerschbaecher, M., & Day, J.W. 2011. Ultra-deepwater Gulf of Mexico oil and gas: Energy return on financial investment and a preliminary assessment of energy return on energy investment. *Sustainability*. 3: 2009–2026. doi:10.3390/su310200910.3390/su3102009.

[12] Where does one go for up-to-date energy statistics? The following sources provide up-to-date information about energy: For domestic information, U.S. Department of Energy (U.S.DOE) Energy Information Administration (EIA) produces reports available online (e.g., Annual Energy Outlook 2014; the Annual Energy Review 2011); For world information, BP publishes an annual review of world energy (2013 Statistical Review of World Energy); the International Energy Agency (IEA), and the annual World Energy Council survey (e.g., World Energy Resources: 2013) Survey provide information on global energy use.

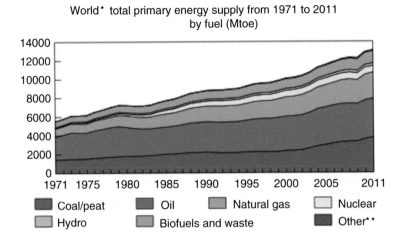

World* total primary energy supply from 1971 to 2011
by fuel (Mtoe)

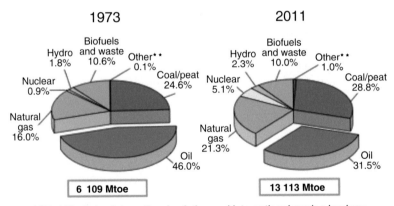

* World includes international aviation and international marine bunkers.
** Other includes geothermal, solar, wind, heat, etc.

Fig. 7.4 Total world primary energy supply (Mtoe or million tons of oil equivalent). The *upper panel* shows growth of different energy sources from 1971 to 2011. The *lower panel* shows the percent composition of energy sources and uses in 1971 and 2011. The quantity of different fuels is given in millions of tons of oil equivalents (Mtoe). The "Other" category includes solar, wind, and liquid biofuels (From: © OECD/IEA 2014 Key World Energy Statistics, IEA Publishing. Licence: www.iea. org/t&c/termsandcondition

proportion has hardly changed in the 50 years the authors have been thinking about energy. Oil is the most important of the fossil fuels, followed by coal and natural gas. As of 2015 about half of our oil is imported, even with the recent large (but now declining) increase in oil production

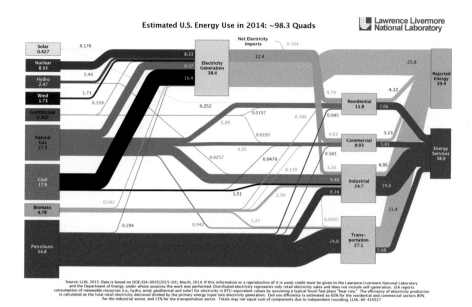

Fig 7.5 A detailed "Sankey diagram" produced by Lawrence Livermore National Laboratory, showing end use and efficiency of U.S. primary energy flow in 2014 (Lawrence Livermore National Laboratory and the U.S. Department of Energy, https://flowcharts.llnl.gov/content/assets/images/energy/us/Energy_US_2014.png)

from "horizontal drilling" and "fracking" technologies. Those who advocate US energy independence often are unaware of this important fact. All the other sources put together account for only a bit more than 15 %. Nuclear, hydropower, and biomass make up about 12–14 % of total energy use. Biomass refers to the so-called combustible renewables that include mainly wood, but also peat and dung, that have been used for millennia. Even under the most optimistic scenarios, our fossil fuel dependence is unlikely to change substantially in the next several decades. The reason is simple: alternative sources of renewable energy are not available in sufficient quantity or reliability to supplant traditional fuels. Additionally, the expense of replacing the existing energy infrastructure would be very large, both in financial and energy terms, to be able to change especially in a short time frame. Much of the fossil fuel infrastructure is aging and has to be replaced and/or maintained.

Consumption, or end use, of energy can be divided into four roughly equal categories: residential, commercial, industrial, and transportation. While a diversity of energy sources can be used for the first two, industrial energy use is primarily fossil fuels and transportation remains highly dependent on liquid fossil fuels, namely gasoline, diesel, and kerosene (Jet A fuel). This is an area of particular vulnerability for society, because any interruption in the supply of these petroleum products would have a considerable effect on society, and not just on transportation (Fig 7.5). Almost every segment of the economy and social society is dependent in one way or another on transportation. For example, agriculture is now highly dependent on fossil fuel for fertilizer production, tillage, weed and pest prevention, harvesting, processing, and transportation to market.

Peak Oil

This brings us to the issue of peak oil. In 1956, petroleum geologist M. King Hubbert, chief consultant to the Shell Development Company, shocked many of his colleagues when he introduced the concept of a peak or maximum production of oil in the United States, then the world's largest producer of oil (a concept often referred to now as "peak oil") at a meeting of the Southern District Division of Production, American Petroleum Institute. Using an extrapolation method based on previous production rates, the concept of finite exploitable stocks (called EUR, estimated Ultimate Recovery) and a normal or bell shaped curve of exploitation over time, Hubbert estimated that oil production in the U.S. would peak soon. "On the basis of the present estimates of the ultimate reserves of petroleum and natural gas, it appears that the culmination of world production of these products should occur within about half a century, while the culmination for petroleum and natural gas in both the United States and the state of Texas should occur within the next few decades." Specifically, Hubbert predicted that U.S. *conventional* oil production would peak in 1970 and that world production would peak around the first decade of the twenty-first century. He was derided for these predictions but the prediction came true for the US and nearly so for the world for conventional oil in the first decade of this century.[13] On a global basis, none of the alternatives to conventional oil (e.g., biofuels, synthetic fuels from tar sands, oil produced from "fracking") has anywhere near the production of conventional oil. We will come back to this later in this chapter. Meanwhile the majority of oil

[13] The details of Hubbert's predictions and methods and details on peak production are given a number of references. Campbell CJ, Laherrère JH. 1998. The end of cheap oil. *Scientific American* (March): 78–83. (29 January 2009; www.dieoff.org/page140.htm); Deffeyes KS. 2001. *Hubbert's Peak: The Impending World Oil Shortage*. Princeton (NJ): Princeton University Press.; Skrebowski C. 2004. Oil fields mega projects 2004. *Petroleum Review* (January): 18–20.; Hall and Klitgaard. 2012 ibid; Aleklett, K., 2012. *Peeking at Peak Oil*. Springer, New York, pp. 336.; Pascualli, R.C. and Hall, C.A.S. 2012. *The First Half of the Age of Oil*. Springer, NY.

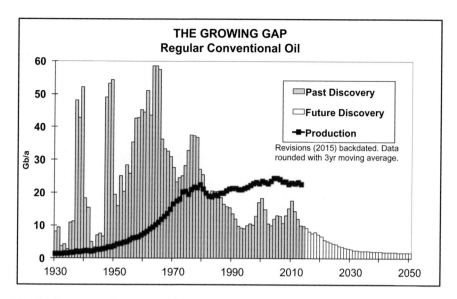

Fig. 7.6 Data on and estimates of finding oil (with revisions and extensions back dated to year of initial strike) and oil use. Obviously the total amount of oil used cannot be greater than the total amount of oil found (Courtesy of Colin Campbell 2015)

producing countries have clearly peaked and are on the downslope of their Hubbert curve.[14]

If we look at the history of discovery and production for world oil, the problem presented by peak oil becomes clear. Oil discoveries in an area generally precede peak production by 30–40 years. For example, U.S. oil discovery peaked about 1940 and production peaked in 1970. World conventional oil discoveries peaked by 1970 and have been falling since (at least to 2010), despite increased drilling efforts (see Fig. 7.6).[15] Most estimates of ultimately recoverable conventional oil for the world have been about two trillion barrels.[16] Production is now 2–4 times the discovery rate, and about 45 % of current production is from 45 of the largest oil fields, most of which are 30–40 years old. Four hundred or so giant fields, most discovered before 1960, provided about 80 % of the world's petroleum in the first decade of this

---

[14] Hallock, J. L., Tharakan, P. J., Hall, C. A., Jefferson, M., & Wu, W. 2004. Forecasting the limits to the availability and diversity of global conventional oil supply. *Energy*, 29(11), 1673–1696.

[15] Association for the Study of Peak Oil (ASPO), 2008. *Oil and Gas Liquids*. 2004 Scenario, www.peakoil.net/uhdsg, 5 March, 2009.

[16] Hall CAS, Tharakan PJ, Hallock J, Cleveland C, Jefferson M. 2003. Hydro- carbons and the evolution of human culture. *Nature*, 425, 18–322.

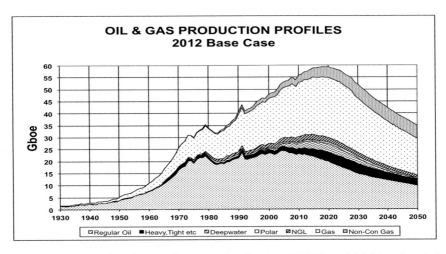

Fig. 7.7 Data on oil and natural gas production (until 2010) and highly educated extrapolations of production of various types of petroleum. "Heavy Tight etc." includes tight oil and extra heavy oil and bitumen produced in Canada and Venezuela. "NGL" stands for Natural Gas Liquids. The surge in 1992 was Kuwait war losses, this is treated as production because it reduced reserves by a like amount (Courtesy of Colin Campbell, 2015) (From Colin Campbell (Campbell.))

century.[17] About a quarter of these are declining by about 4 % annually. This comes at a time when world oil demand is increasing, especially in rapidly developing countries, particularly China and India. The problem is clear. The world is now using more oil than it is finding, and much of the oil that it is using is coming from giant fields discovered three to four decades ago. It's like living off your savings, rather than your income. If the amount of money put into savings (known as producible oil reserves) is less than what is taken out (oil use), then savings go down. If this continues, at some point you run out of money (or oil).

If the information on historical energy consumption patterns is put together with the concept of peak oil and projections for future oil availability, and the availability of energy in general, then a disturbing picture emerges. Perhaps our energy future is less rosy than we are often led to believe. Consider the following: Colin Campbell, one of the pioneers of research on future energy availability,[18] has presented the historical consumption patterns for the different kinds and sources of oil, and projections for their availability until 2050 (Fig. 7.7). Campbell's data show total world oil and natural gas production peaking around 2020 (in other words, soon)

---

[17] Skrebowski C. 2004. Oil fields mega projects 2004. *Petroleum Review*, (January), 18–20.
[18] Campbell, C. J., & Wöstmann, A. 2013. *Campbell's Atlas of Oil and Gas Depletion*. Springer.; Pascualli and Hall, 2012, Ibid.

and then declining. It is important to think about these trends in terms of several decades rather than short-term variability.

Energy Return on Investment: Why "Drill, Baby, Drill" Is Not a Sustainable Strategy

We noted that net energy is the difference between the amount of energy *inputs* used to produce a certain type of energy and the energy *output* of what is produced. Obviously, any energy source needs to have a positive value if it is to be useful to society. However, a better measure is one used by ecological economists, ecologists, and some engineers, "Energy Return on Investment" or EROI. EROI is the ratio between energy outputs of a given technology or fuel and the energy input required to produce the output. The obvious implication is that the higher the ratio, the better the energy source, other things being equal.

One of the authors of this book, Charles Hall, derived the concept when studying fish migration for his PhD project in ecology.[19] He and his then-student Cutler Cleveland applied the concept to oil in the early 1980s. They and their colleagues have been conducting EROI analyses on many energy sources for years. They have reached some very interesting and sobering conclusions. First, they conclude that an energy technology really needs to have an EROI greater than about five or even ten to one to be considered viable as a significant energy source for society.[20] In the first half of the "Century of Petroleum," i.e., the twentieth century, the EROI of oil and gas production was 30:1 or higher. This is a tremendous net energy yield that was responsible in large part for the spectacular economic, technological, and population growth of the twentieth century. That means that it took the energy equivalent of one barrel of oil invested to gain a yield of 30 barrels of oil. This ratio has declined over time, and is now near or below 10:1 for the U.S. and many other areas. This is because most of the easy-to-get oil with high EROI has already been found (much of which has been produced) and new oil that is being found has a low EROI. A case in point is ultra-deep water oil drilling that not only has low EROI but also very high costs, making an acceptable financial return on investment (ROI) of capital low as well. Unfortunately a decline in the quality of U.S. official statistics on energy costs within industry, and a failure to maintain good input–output inter-industry analyses, has made it increasingly difficult to update EROI analyses.

---

[19] Hall, Charles AS. 1972. Migration and metabolism in a temperate stream ecosystem. *Ecology* 585–604.

[20] Hall, C.A.S, Balogh, S. and Murphy, D.J. 2009. What is the minimum EROI that a sustainable society must have? *Energies* 2(1), 25–47.

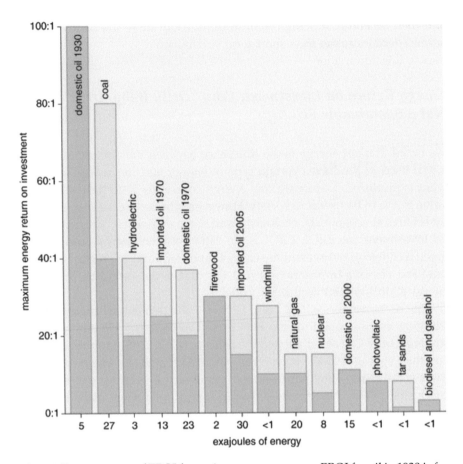

Fig. 7.8 One summary of EROI for various energy sources. EROI for oil in 1930 is for finding oil. Producing that oil is closer to 20–30:1, still high by today's values. The different colors represent the range of estimates (Source: Hall, C. A., & Day, J. W. 2009. Revisiting the Limits to Growth After Peak Oil. *American Scientist*, **97**(3), 230–237. See also: Hall, C.A.S., Lambert, J.G., Balogh, S.B. 2014. EROI of different fuels and the implications for society. *Energy Policy Energy Policy*, **64**, 141–152)

There are a number of important points about EROI, or net yield of different energy sources that must be understood (as shown in Fig. 7.8). First, oil and coal had very high EROI in the first half of the twentieth century, ranging from 80:1 for coal to 100:1 for oil for the best oil wells. The EROI for both domestic and imported oil in the US has declined significantly over the past half century. Imported oil had an EROI of about 30:1 in 2005 while the EROI of domestic oil was around 10:1 in 2006. Clearly the net energy yield of oil has declined. The interesting thing is that compared to oil and coal at their highest net energy yield, most other energy sources have considerably lower EROI values. Hydroelectric power can be as high as 40:1, and wind—

at its best—can be 20:1 to as high as nearly 40:1. In most cases, solar photo-voltaics, tar sands, and biofuels are considerably less than 10:1. We will come back to this information on EROI as we discuss each of these energy sources, because there are other issues that effect their utility and value.

Will Unconventional Oil and Gas Replace Conventional Supplies?

Even among those who agree with the idea of peak oil and gas, many argue that non-conventional oil and gas will be a long-term replacement for conventional supplies. They further argue that the ability to get better yields out of known reserves will delay the effects of peak oil by essentially constituting new discoveries. While there may be some truth to this, we urge caution against over-optimism. There have been numerous claims in the last several years, and even in the last decades, that shale oil and gas, tar sands, ultra-deep and Arctic reserves, and other unconventional sources are game changers, that the U.S. will soon become or remain the number one oil producer in the world, with North America becoming energy independent or perhaps even a significant exporter with a 100-year supply of oil and gas. Indeed, recent success in recovering shale gas and oil in the Bakken geological formation in North Dakota and elsewhere has transformed local economies and even reduced the overall U.S. dependency on imported oil and gas. This has led many to conclude that a "steady-as-she-goes" course of action is what is needed, and despite considerable evidence to the contrary, they conclude that greenhouse gas emissions such as carbon dioxide from fossil fuel burning is not a problem. But is the optimism justified or wishful thinking, climate issues notwithstanding? Let's take a closer look.

First, it is important to remember what is meant by conventional oil and gas. These are supplies that occur in relatively compact underground reservoirs that allow oil and gas to flow relatively freely to the surface through a pipe when a well is drilled. Normally a conventional field is made up of geologically porous material where oil and gas accumulate. These "trap rocks" are areas where oil and gas, generated in lower "source rocks," have migrated and been trapped in dense concentrations. These conventional reservoirs can be reached by conventional drilling technology with vertical bores. After an initial push from associated gas deposits, these usually need pumping and pressurization to force the oil through the formation and to the surface. This means that these reservoirs are generally relatively shallow and in easy to reach areas. Typically about 35 % of the oil in place can be recovered. Sometimes, with "enhanced oil recovery" (processes that use additional energy), another few percent can be squeezed out. Oil produced from conventional fields typically comes to the surface as a liquid that normally has a low-sulfur content.

Now, compare these fields to unconventional sources of oil and gas, for example "tight" oil and gas from shale. Here, drillers exploit source rocks, the original places where the oil and gas were formed. The fields tend to be of low porosity (hole space) and permeability (ability for oil to flow through), with the oil and gas trapped in tiny bubbles in shale rock. These supplies cannot be produced by simply drilling a well into a shale formation, because the oil and gas cannot flow out of the trapped bubbles. To make the supplies flow, the rock must be broken up. This is done by pumping a fluid into the formation under very high pressure. This breaks up or fractures the rock in a process called hydraulic fracturing or in the common parlance, "fracking." To do this, wells are first drilled vertically to the desired depth and then horizontally for a mile or so. Fracking fluid is then pumped into the well bore under very high pressure, breaking up the rock. A small percent of the oil or gas in the total formation can then flow freely out of the formation.

Because the rock is fractured over a relatively small area and does not collect over time in large reservoirs (even with 1–2 mile laterals), many more wells must be drilled compared to conventional technology to obtain similar amounts of oil and gas. Figure 7.9 shows wells drilled into a shale oil formation in the Bakken region of North Dakota. The close spacing of the wells shows over how small an area oil is extracted from a typical fracked well. Shale formations cover large areas of the U.S. (see Fig. 7.10), but productive areas are restricted to relatively small locations called "sweet spots." The left half of the image in Fig. 7.9 above is from one of the few sweet spots in the Bakken formation. This area is running out of available drilling locations. Recent government reports made when oil prices were above $100 per barrel predict that the Bakken area will likely peak by 2020 and then decline.[21] However, lower oil prices since mid 2014 have already led to declines in most fields, and the nation as a whole. Thus future production will depend both on the physical characteristics of the field and the price of the oil produced. This will also depend somewhat on the demand for this expensive-to-produce form of oil and gas as well as technological developments. One measure of an oil well's productivity is the number of barrels of oil per day produced per foot drilled (bpd per foot). The Saudis get 3.8 bpd per foot drilled, while the US shale producers get only about 0.039 bpd per foot drilled—almost a hundred time less.[22]

[21] EIA. 2014. *Annual Energy Outlook*. Energy Information Agency, U.S. Department of Energy.
[22] A new peak in conventional crude oil production, post and comments. http://euanmearns.com/a-new-peak-in-conventional-crude-oil-production/

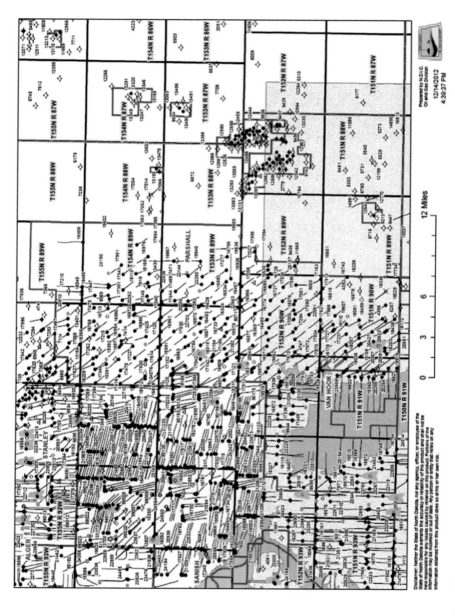

Fig. 7.9 Map of well bore and laterals in Bakken Region of North Dakota shows that production comes from concentrated 'sweet spots' source: North Dakota Industrial Commission, Oil and Gas Division, 2012

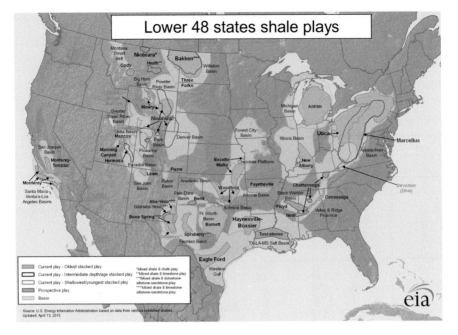

Fig. 7.10 EIA map of shale plays in the lower 48 states. Only very small portions of these plays have reserves that can be produced economically (Source: U.S. Energy Information Administration. 2015. More maps are available on: http://www.eia.gov/pub/oil_gas/natural_gas/analysis_publications/maps/maps.htm)

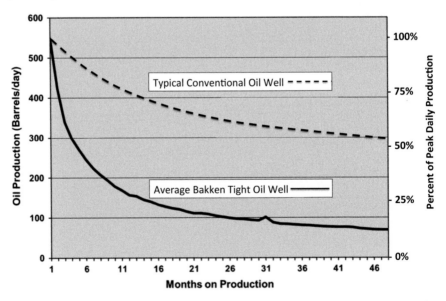

Fig. 7.11 Average decline curve for a Bakken shale oil well from Hughes 2014 and a conventional oil well with the same peak production. Decline curves for shale gas wells are very similar (Figure adapted from Hughes, J.D., 2014. *Drilling Deeper: A reality check on US government forecasts for a lasting tight oil and shale gas boom.* Post Carbon Institute, Santa Rosa, CA, http://www.Shalebubble.org, retrieved July 2015. Decline curve for conventional oil production from: oilprice.com, http://oilprice.com/Energy/Energy-General/The-Fracking-Revolution-Promise-and-Perils.html, retrieved July 2015)

Fracked wells deplete much more rapidly than conventional ones. Whereas a conventional well may produce for 30–40 years, a typical fracking well may decline by 50 % or more in the first year and play out completely in 4–5 years (Fig. 7.11). This means that there must be continual new drilling just to maintain flow from a field. Producers call this a production treadmill, and for Alice in Wonderland fans, "The red queen phenomenon." Both reflect the high financial and energy costs involved. If drilling is stopped, production from a shale play declines rapidly. In addition, oil prices must stay relatively high for shale oil to be economically feasible. This was made very clear by the fall in oil prices in 2014–2015 when many shale producers shut down. Oil analyst Arthur Berman estimates that the majority of investors in shale oil made no profit.

Because productive areas in a shale play occur in sweet spots, which are generally the ones that are exploited first, initial extrapolations of shale reserves may have been significantly overstated, leading to the early 100-year projections of shale reserves. Moreover, previous EIA estimates were that 60 % of shale reserves in the U.S., some 15.4 billion barrels, were in California. But because of the complicated geology, the EIA reduced initial estimates of total California reserves by a whopping 96 % to a mere 0.6 billion barrels.[23] Because of all these factors, estimates of total reserves have fallen and some estimates project that shale oil and gas production will peak out in or before the early 2020s.[24]

Other considerations must be taken into account concerning fracking. Fracked wells require an enormous amount of water and sand. Richard Heinberg reports that a single well might use 60 million gallons of water, equivalent to that carried by 600–1000 large semi-trucks.[25] This is another example of the "use the best first" principle at work. In some formations, such as the Haynesville/Bossier in Louisiana, sufficient water is generally available to meet the need, although there is still concern about what to do with the wastewater produced by fracking there. In formations such as the Barnett in Texas, water availability is another matter. There, due to a drier climate and burgeoning urban and suburban development, water problems existed even before fracking. In essence, fracking has now become "another mouth to feed." The right kind of sand is also a critical commodity needed for fracking, and it may be that sand shortages will be pressing concerns as well.

---

[23] Hughes, D. (2013). *Drilling California: a reality check on the Monterey Shale*. Post-Carbon Institute.

[24] Art Berman. http://www.artberman.com/blog/; Hughes, D., 2013. *Drill, Baby, Drill. Can Unconventional Fuels Usher in a New Era of Energy Abundance?* Post Carbon Institute, Santa Rosa, CA, 166 p. Available online: www.postcarbon.org

[25] Richard Heinberg. *Snake Oil – How Fracking's False Promise of Plenty Imperils Our Future*. Post Carbon Institute, Santa Rosa, CA. 251 p. 2012. www.postcarbon.org

The 100-Year Supply and Other Myths

Given all the recent talk about potential U.S. energy independence and projections that the U.S.'s energy future is nothing but bright, you might ask, is this too good to be true? Are there any skeptics out there? In his provocative book, *"Cold, Hungry and in the Dark,"* energy analyst Bill Powers asserts that "the impact shale gas will have on America's energy future is vastly overstated."[26] He emphasizes that estimates implying that the U.S. has a 100-year supply of shale gas are based on a failure to understand the difference between "proven" (i.e. realistically recoverable) and potential (i.e. possibly recoverable) resources. In other words, it is not possible to predict resources without taking into consideration their quality. In addition, Powers lists what he terms three myths that contribute to this over optimism.

The Manufacturing Myth. This relates to a perception that all acreage in a shale play is homogeneous and that a manufacturing approach should be used in developing it. In contrast, Powers cites the work of petroleum geologist Arthur Berman, who contends that a shale play has geological features that create "sweet spots" that greatly enhance productivity in certain, but not all, localities. For that reason, production projections based on assumptions of homogeneity are incorrect.

The Decline Curve Myth. Again, derived from the work of Arthur Berman, shale gas operators often make overly optimistic projections of wells' estimated ultimate recoveries (EUR) based on the first few months of "flush production." Typically, as Fig. 7.11 shows, there is an exponential decline in production rate over a very few years. Failing to take into account this decline will result in an overestimate of production capacity and EUR.

The 40 to 65 Year Reserve Life Myth. Powers identified a third myth, i.e. that shale gas wells have reserve lives of between 40 and 65 years. Again, one respected expert on this is Arthur Berman, whose research on the Barnett has shown that a significant fraction of the wells drilled in the last decade are already plugged and abandoned or produce less than 1 million cubic feet of natural gas per month. Analyses by Canadian Geologist David Hughes give very similar results.[27]

Taken together, the above three myths raise important questions concerning the optimism about the U.S.'s oil and natural gas future. Clearly, while it is possible that new technologies may be developed that enable the

[26] Bill Powers. *Cold, Hungry and in the Dark. Exploding the Natural Gas Supply Myth.* New Society Publishers. Gabriola, Canada. 312 p.

[27] Hughes, D. 2014. *Drilling Deeper: A Reality Check on the U.S. Government Forecasts for a Lasting Tight Oil and Gas Boom.* Post Carbon Institute. October 2014.

extraction of more oil and gas from fracked wells, it seems improbable that sufficient new geological formations will be found in the U.S. and elsewhere to greatly increase our estimates of proven reserves. Consequently, this is certainly not what we should rely on. Suppose Powers, Berman, Hughes, and others are right? What then? Can we count on yet higher prices to make the next downward, lower EROI, resource grades available at the level we need to run our modern industrial societies indefinitely? Neither biophysical nor financial analysts have addressed this question.

Tar Sands

The Alberta tar sand formation in Canada is another example of a reserve whose impact has probably been over stated. Estimates of oil reserves in the tar sands are as high as two trillion barrels (almost twice the amount of oil already burned) but only 167 billion barrels are listed as recoverable reserves. The majority of tars sands are not profitable produce at oil prices below $50 per barrel.[28] If there were indeed two trillion barrels of recoverable oil in the tar sands, the Canadian inentory would double the remaining oil reserves for the world, although even the best grades at present—the ones being exploited now—have a very unfavorable EROI, estimated to be less than 5:1.[29] Most of the reason is that tar sands (or the more lovable "oil sands") are not liquid oil. They are bitumen, a tar-like substance mixed with sand and clay, in the Athabasca tar sands of Alberta, Canada, that requires a considerable energy investment to recover and refine.

The reason that tar sands have such a low EROI is that bitumen does not flow to the surface like conventional oil. It must be mined, either on the surface by large machines (followed by massive heating and hydrogen injection), or liquefied in place with water superheated with natural gas (Fig. 7.12). For surface-mined bitumen, about 4 tons of soil must be moved for each barrel of "oil" produced. Because tar sands consist of a sticky solid substance, which is in fact what bitumen is, producers and the government of Canada, in a public relations slight of hand, have taken to calling it oil

---

[28] Canadian Energy Research Institute. 2014. Canadian Economic Impacts of New and Existing Oil Sands Development (2014–2038).; According to recent estimates the average tar sands producers using the "SADG" method, about half of total production, need oil prices at least $80 per barrel "to recover shipping and other costs plus a 10 % return on investment." Chester Dawson. "Canadian Oil-Sands Producers Struggle" *Wall Street Journal*, August 2015. Online at: http://www.wsj.com/articles/oil-sands-producers-struggle-1440017716.

[29] Brandt, Adam R., Jacob Englander, and Sharad Bharadwaj. 2013. "The energy efficiency of oil sands extraction: Energy return ratios from 1970 to 2010." *Energy*, 55, 693–702.

Fig. 7.12 Picture of a tar sands operation in Alberta Canada ("Alberta Tar Sands NWF Flight Oct 2010 #1" by NWFBlog, Creative Commons, CC by 2.0, https://creativecommons.org/licenses/by-nc-nd/2.0/legalcode)

sands. When bitumen is extracted from the sand and clay, it must be highly processed to make a liquid synthetic oil. Production and processing of tar sands uses tremendous amounts of natural gas that must be imported from the Mackenzie River Delta many miles away.[30] Since gas has been relatively cheap for the last decade, production of tar sands is economically feasible, albeit not by much. One Canadian environmentalist said, "we are using natural gas, which is the cleanest fossil fuel, to wash sand and make a dirtier fuel. It's like using caviar to make fake crabmeat."[31] Mining tar sands is also extremely environmentally damaging. Large areas of the Peace-Athabasca delta where mining takes place have been so disturbed that the impacted area is clearly visible from space. In his book, *The Peace-Athabasca Delta*, Kevin Timoney gives a detailed description of the area and the impacts of mining.[32] Oil from tar sands is usually referred to as a rate limited resource, not a stock-limited resource, because the resource is huge but it can be produced at only a certain rate due to water, natural gas and

---

[30] Brandt, et al. 2013, ibid.

[31] Marlo Raynolds, executive director of the Pembina Institute, a Canadian environmental group. http://money.cnn.com/2006/10/04/news/economy/oil_sands/?postversion=2006100700

[32] Kevin Timoney. 2013. *The Peace-Athabasca Delta – Portrait of a Dynamic Ecosystem*. The University of Alberta Press, Edmonton. 596 p.

environmental limitations. As the best near-surface resources are depleted, extraction will have to move to deeper, presumably more energy-intensive resources.

Ultra-deep reserves are another source of unconventional oil and gas. Most oil and gas produced from ultra-deep wells is not different from conventional supplies. What is unconventional is the depth of the reservoirs and the technologies used to get it. Most ultra-deep oil and gas is currently produced in the Gulf of Mexico off Louisiana and Texas. Deepwater oil is produced in water depths of 300–1500 m while ultra-deep comes from greater than 1500 m.[33] In addition to the very deep water, the oil and gas reservoirs are generally located deep beneath the seafloor. For example, the Macondo Prospect into which the Deepwater Horizon was drilling when it blew out, was in 1500 m (nearly 5000 feet) of water and 5400 m (almost 18000 feet) beneath the ocean floor.[34]

To produce oil from such depths is extremely challenging and pushes the limits of drilling technology. The Deepwater Horizon rig was valued at over a half billion dollars when delivered to Transocean Limited in February 2001.[35] Such a rig is not only expensive in money and energy terms to construct but is extremely expensive to operate. Total daily operational costs run in the range of one million dollars.[36] Because of these high costs, oil companies are under pressure to complete drilling at any particular site as quickly as possible. This has been suggested as part of the reason for the oil spill.[37] Back when oil was easy to produce from shallow fields, a drilling rig cost $10–20 million dollars, almost two orders of magnitude less than the Deepwater Horizon rig. With such escalating costs, the days of high return on investment appear over, and the economic feasibility of such massive investments can only be supported if oil prices are much higher than at present.

Accidents such as the Deepwater Horizon spill are probably not highly unusual events that are unlikely to be repeated. Michael Klare states that the "challenges posed by drilling in deep-offshore locations are partly a product of the physical environment itself: the immense pressures

[33] Moerschbaecher, M. and Day J.W. 2011. Ultra-deepwater Gulf of Mexico oil and gas: energy return on financial investment and a preliminary assessment of energy return on energy investment. *Sustainability*, 3(10), 2009–2026.

[34] Moerschbaecher and Day, 2011, ibid; Tainter and Patzek. 2012. *Drilling Down – The Gulf Oil Debacle and Our Energy Dilemma*. Springer, New York. 242 p. 2012.

[35] Offshore-technology.com. *Deepwater Horizon: A Timeline of Events*. Available online: http://www.offshore-technology.com/features/feature84446/ (accessed on 1 February 2011).

[36] Rigzone Inc. *Offshore Rig Day Rates*. http://www.rigzone.com/data/dayrates/(accessed on 31 May 2011).; Leimkuhler, J. *Shell Oil: How Do We Drill For Oil?* Presented at the 2nd Annual Louisiana Oil & Gas Symposium, Baton Rouge, Louisiana, USA, August 2010.

[37] Moerschbaecher and Day, 2011, ibid; Tainter and Patzek, 2012, ibid.

encountered at these depths produce unique and often unforeseen stresses on personnel and equipment. Offshore rigs are particularly vulnerable to the elements, including hurricanes, typhoons, and, in far northern latitudes, floating ice. In addition, many offshore fields lie in politically contested waters."[38] Tainter and Patzek argue that the escalating complexity of the technology leads to unforeseen problems.[39] As the search for oil and gas moves to more and more hostile environments, these authors argue that not only will such accidents as Deepwater Horizon become more probable, they will be inevitable.

All these factors influence net energy yield or EROI. The obvious implication is that just to stay even in coming years, we will have to produce more and more gross oil, because the net will be proportionally less. Almost no projections of future energy needs by any entity, government or private, mention the concept of EROI, much less take it into account in their energy forecasts. Estimates of the EROI of ultra deep wells are considerably less than 10:1.[40]

The increasing complexity and difficulty of exploring for—and producing—oil and gas in these challenging environments inevitably means that energy is becoming much more expensive to produce and that net energy yields are dropping. In other words, you have to put more energy in (which comes at a financial cost as well) to get energy out. Ultimately, if obtaining energy from certain sources (e.g., deep water or poor quality tar sands) takes more energy than is produced, it will not be worth the effort. For example, in the first half of the twentieth century, oil was easy to get and producers could make a profit on oil that cost $10 a barrel. By investing the equivalent of a barrel's worth of energy, a hundred barrels could be produced. Now oil produced by fracking needs to sell for as much as $100 a barrel to make money for the operators. Canadian tar sands need cheap natural gas to make production profitable, and the net energy yield is very low. For many tar sands, crude prices may have to be as high as $150 a barrel. Oil production in the deep waters off the coasts of Africa and Brazil may require oil prices between $115 and $127/barrel to be profitable, according to Carbon Tracker Initiative.[41]

---

[38] Michael Klare. 2012. *The Race for What's Left*. Holt and Company, New York. 306.

[39] Tainter and Patzek, 2012, ibid.

[40] Hall, C. A., Powers, R., & Schoenberg, W. (2008). Peak oil, EROI, investments and the economy in an uncertain future. In *Biofuels, solar and wind as renewable energy systems* (pp. 109–132). Springer Netherlands.

[41] Carbon Tracker Initiative. http://www.carbontracker.org/

Coal

Coal is still an important source of primary energy for electrical generation in the U.S. providing in 2015 about a third of the energy used to make electricity. However, U.S. coal use is declining sharply, being supplanted with presently cheap and abundant U.S. natural gas.

Coal has several environmental downsides: First, we have gone through most of the high grade, low Sulfur (S) anthracite coal that burns hottest and most cleanly. Coal burning has the attendant problem that it produces sulfur and nitrogen oxides that contribute to atmospheric pollution such as acid rain. Scrubbing technology used to remove sulfur and nitrogen is expensive and lowers EROI. Second, compared to natural gas, the burning of coal produces twice as much CO_2 per unit of energy. Third, coal burning releases other atmospheric pollutants, such as mercury and other heavy metals. The U.S. is one of the top coal exporting nations in the world, and a great irony from a climate change perspective is that the coal that the U.S. does not burn may be burned elsewhere anyway.

Coal does have some positives as well. Most importantly, the U.S. has abundant supplies of coal, albeit low grade, that could last for more than 100 years. One underappreciated fact is that coal has a high EROI, which may mitigate the negatives to some extent. For example, the amount of usable energy from coal combustion per molecule of CO_2 released may actually be higher than for an equivalent quantity of natural gas.

Renewable Energy to the Rescue?

Many governments, environmental groups, and industries are encouraging the migration from fossil fuels to renewable energy as the only viable solution to meet the world's future energy needs while controlling CO_2 emissions. Typically, people assume that this is an easy thing to do if only the political will is there, and that different energy sources are extremely fungible. We think that instead, the markets are in control and are likely to continue to be in control for some time because of the high EROI of fossil fuels. But we could be wrong, as the political will and the technologies are changing. But first consider the following quote, rather typical of many from renewable advocates:

> Within 50 years, the world should be able to achieve a 100 percent clean energy economy. Within the next couple of decades, every time you turn on a light or power up your computer, every bit of that electricity will come from

clean, renewable, carbon-free sources. Soon after that, solar and wind will displace nuclear as well, at which point we'll be getting 100% of our electricity from renewables. By 2030 we should be able to cut transportation oil use in half and then cut it in half again a decade later. Once we're finally fossil-fuel free, we'll not only see our climate stabilize but we'll also rest secure knowing that we can get all our power from sources that are safe, secure, and sustainable. It's already within our grasp.[42]

Unfortunately, the same thermodynamic constraints that are confronting oil executives also confront renewable energy producers. Let's consider some of the issues involved.

First, what renewable sources of energy are being considered? Renewables include hydropower, wind, solar, and biofuels. Two sources of renewable energy are already widely used and have been for centuries, hydropower and combustible renewables (mostly wood). Hydropower generally refers to electrical energy produced as water stored behind a dam flows past a turbine. Hydropower currently provides about 3–4 % of total world energy use. Hydropower and wind power have been used since antiquity to power machines that did work such as grinding grain, sawing logs, and pumping water and generally has a high EROI, as much as 40:1 or better, for good sites. Prior to the development of steam driven pumps, the Dutch used about 10,000 windmills to keep their country dry.

Combustible renewables, such as wood, obviously refers to burning organic matter to produce heat. Wood is far and away the most important combustible, but peat and dried animal dung are also used. These renewables have been used for hundreds of thousands of years when our ancestors first warmed themselves by fires and began to cook. John Perlin gives an especially good account of the importance of wood fuel in past civilizations.[43] Burning wood and other combustibles has been used throughout history for a wide variety of activities, including cooking, heating, melting metal ores, and in the early industrial revolution, boiling water for steam engines. Using firewood has a relatively high EROI, especially when burned close to its source. Combustible renewables are estimated to supply from 4 to 9 % of world energy use. In our view, it is a great leap of faith to assume that the energy intensive global economy, as we know it now, can obtain much more energy than at present from such renewables, especially when issues surrounding deforestation are so important.

Much of the discussion about renewable energy nowadays refers to the so-called new renewables that include solar, wind, waves, tides, and high

---

[42] Michael Brune, Executive director of the Sierra Club. *National Geographic Magazine.* January 2015.
[43] Perlin, J. 2005. *A forest journey: The story of wood and civilization.* The Countryman Press.

technology biofuels. Let's consider each of these. Solar energy produces electricity in two ways. Sunlight falling on a photovoltaic (PV) panel produces electricity directly when the energy of sunlight causes electrons to flow in the matrix of the panel. Electricity can also be produced in what is called a concentrating solar array in a solar thermal process. Numerous mirrors focus sunlight on a boiler that turns a turbine that generates electricity. A major problem with solar energy is that the sun does not shine all the time. Also, some areas such as the U.S. Southwest have lots of sunshine while others have much less. So solar energy is intermittent and is often not abundant in areas of the greatest demand.[44]

For wind energy, moving air turns blades of some sort that turn turbines to generate electricity. Under optimum conditions, wind energy can produce electricity very efficiently and with a high EROI. But as in the case of solar energy, the wind doesn't blow all the time (about 30 % on average) and the places where the wind blows most consistently (like the Great Plains) are not where people live (dense urban areas). So both solar and wind suffer from intermittency and spatial distribution. Winds may be more consistent in coastal areas, but the cost of anchoring turbines in often shifting sands or in the coastal ocean, and the high requirement for zinc, chromium and other elements to make steel less corrosive, can greatly increase the dollar and energy costs of the turbines placed there.[45] If solar and wind were to completely replace fossil fuels, it would require an enormous amount of resources, so much so "that the growth of the renewable energy sector may impact investment in other areas of the economy and stymie economic growth."[46]

Tidal and wave energy can produce electricity from moving water. In some high tide range coastal areas, water is trapped behind a barrier or dam at high tide and then let out rapidly at low tide to produce electricity as the falling water flows through a turbine. Wave energy uses the up and down motion of waves to move a lever that turns a generator to produce electricity. Wave and tide energy currently account for much less than 1 % of total energy use. It is important to note that wind, solar, tide, and wave

---

[44] For detailed analysis of solar PV see: Palmer, G. 2014. *Energy in Australia: Peak Oil, Solar Power, and Asia's Economic Growth*. New York, NY: Springer. And; Prieto, P. A., & Hall, C. 2013. *Spain's photovoltaic revolution: the energy return on investment*. Springer Science & Business Media.

[45] For a detailed discussion on the sustainability of wind energy see: Davidsson, S., Grandell, L., Wachtmeister, H., & Höök, M. 2014. Growth curves and sustained commissioning modelling of renewable energy: Investigating resource constraints for wind energy. *Energy Policy*.

[46] M. Dale, S. Krumdieck, and P. Broder. 2012. Global energy modeling – A biophysical approach (GEMBA) Part 2: Methodology. *Ecological Economics*. 73: 158–167.

energy all produce electricity. A problem is that storms may destroy the devices long before their intended use is done.

Biofuel refers to energy produced when biochemical processes convert plant material into liquid fuels. While the concepts are straightforward and ancient, the technology for doing this has been advancing steadily. The most notable biofuel produced in the U.S. is ethanol produced from corn, and in Brazil from sugar cane. Other plants have been suggested for biofuel production such as switch grass. Palm oil is derived from palm trees, whose plantations cover vast areas in the tropics and supplant natural habitat or land area that could be used for growing food crops.

Unfortunately, biofuels suffer from severe limitations. One is that they have a notoriously low net energy as they require nearly as much energy to produce as they yield. A second issue is that biofuel production competes with the production of food and the preservation of natural forests. Many consider it immoral to use agriculture for the production of ethanol or other liquid fuels for use by people who are well off enough to own a car or SUV, rather than producing food for a growing population. Finally, there is a limit to the amount of plants that the earth can grow. Even if all plant growth on earth were turned into energy, it could not supply world energy.[47] The EROI of most biofuels is barely greater than one showing that these fuels are not a viable source of energy.

The net energy yield for wind and solar energy is generally better than for biofuels but still relatively low, on average. The EROI of wind can be as high as 20:1 or 40:1 in the best cases and solar energy can be, at best, a little higher than 10:1, but more usually in the vicinity of 5:1 or less. Three recent studies which looked at the energy costs in depth found EROI values of about 3:1 of the devices lasted 25 years, although as noted there inputs were lower quality fossil fuels and outputs higher quality electricity.[48] But as we stated earlier, these are intermittent sources. Wind and solar produce kinetic energy in the form of electricity, not potential (stored) energy such as in biomass fuel. This is fortuitous if immediate demand exists for this electricity at the time of its generation, but is extremely problematic if it must be stored for later use. Since photovoltaic cells now can achieve 20 % efficiency or better at high cost, or 10–15 % at prices that are becoming

---

[47] Dukes, J.S. 2003. Burning buried sunshine: human consumption of ancient solar energy. *Climatic Change*, 61(1–2): 31–44.

[48] Weißbach et al., 2013. Energy intensities, EROIs, and energy payback times of electricity generating power plants. *Energy*, 52(210).; Dale, M., & Benson S.M. 2013. Household Solar Photovoltaics: Supplier of Marginal Abatement, or Primary Source of Low-Emission Power? *Energy Environ. Sci.* doi:10.1039/c3ee42125b; Palmer 2014, ibid.; Prieto and Hall 2013, ibid.

economically viable, they seem to have the advantage of efficiencies that are 3–6 times higher than photosynthesis.

The energy costs of storage have yet to be taken into consideration when evaluating the EROI of renewables (or for that matter any energy source) but are probably very high. All electrical storage devices can store only 2.6 percent of instantaneous world electrical generation (and 2.4 percent for US). Batteries are a small portion of this and hence can store only much less than one percent of output, almost useless in smoothing out load at the national level. Storage of water and quick response gas turbines offer better alternatives. Smart grids can help, but what do we do when there is no wind for 2 weeks, which happens in many places? This is a tough problem being resolved only very slowly. If the wind blows only 30 % of the time, does that mean we need two units of storage or backup for each unit of wind power? Would that cut the EROI to one third? These are questions that must be addressed.

How Much Energy Does Society Need?

How much energy is necessary to provide for societal needs is a complex issue that can provoke endless debate. We will not add to speculation here, but we think it would be worthwhile to pose some questions. Could society get along with 15 % less energy? Certainly! Western societies, in particular, are very wasteful of energy; a small amount of belt tightening is clearly possible, and may even be beneficial. Could society get along with 30 % less energy? Probably. Homes, offices, and vehicles can be made much more energy efficient than they currently are. We can also eat lower on the food chain (less meat). We can travel less and use more efficient transportation. We can generally consume less. So, by instituting reasonable conservation measures, there is little doubt that we could get along just fine by using 15 % less energy and probably even 30 % less. Beyond that it is much more problematic, especially as populations and economic aspirations grow. As recently as the mid twentieth century, Americans, on average, used half of the energy on a per capita basis that we use today. That was not a time of deprivation, at least not for the great majority of people in the U.S.So yes, we could get along using 15 % less energy without much impact on our lifestyle, certainly in first world countries. Beyond that, it is a different story.

Modern society could function perfectly well only with fossil fuels. After all, fossil fuels constitute 80–85 % of the total energy used. So as we indicated above, a 15 % reduction of total energy use would not be that difficult. But fossil fuels are finite and will be mostly exhausted within the life span of many people alive today, and burning fossil fuels appears to be the main cause of climate change. But without fossil fuels and the 85 % of energy use they provide, Western society would be very different than it is now. Most people would lose easy access to affordable food, clothing, housing, health care, transportation, and the other amenities of modern life.

Of course, people could turn off almost all heating and cooling systems, get rid of cars, consume a low calorie diet that is much less diverse, provide for our own health care, and buy many, many fewer things. After all, this is what most people did for more than 99 % of human history, and what many people in the world currently experience. But this is not what is meant when it is suggested that society can move to a green future powered by renewable energy. Those who make those suggestions mean that energy from solar, wind, biofuels, and other renewables can sustain an energy future that will allow reasonable access to affordable food, clothing, housing, health care, transportation, and so forth. In other words, a life style that is not too dissimilar from what the majority of the population currently experiences in developed countries.

An important point to consider about all other non-fossil fuel energy sources is that fossil fuels are necessary to produce and use them.[49] Mining, processing, and transport of ores, construction and operation of a nuclear plant, and dealing with wastes are largely done with fossil fuels. The same can be said of "alternative" energy sources. The construction of windmills, solar panels, and dams to impound water for hydropower all require fossil-fuel powered machinery. It is estimated that for solar and wind to replace fossil fuels for electricity generation alone, it would require metals equivalent to what is used by the global automobile industry.[50] Even most biomass feedstocks like wood are obtained with fossil fuel-powered machinery.

Another problem is the inevitable consequence of the Second Law of Thermodynamics. Things wear out and break down. What happens when a windmill has to be replaced and affordable fossil fuel is not available? Accidents happen. When a powerful storm supercharged by a heated atmosphere destroys an offshore wind farm, what will be the consequences? What happens when an earthquake-generated tsunami destroys a nuclear power plant? Before the Fukushima disaster, most would have said that this was impossible. What happens when hail the size of golf balls or even baseballs destroys a 5 year-old solar array 50 years from now when sufficient fossil fuels are not available to replace it? These events will happen.

Fossil fuels can do everything in the energy economy. They can mine the energy, refine it, and pay for getting the energy to the consumer—even

---

[49] Palmer, 2014, ibid.; Dale et al., 2012, ibid.
[50] Dale et al., 2012, Ibid.

while supporting all the other work the economy does. They can provide liquid fuels and generate electricity. They can be used to construct wind turbines, dams and PV systems. They can be burned to provide space heating and to power many industrial processes. This is not the case for all renewables. Solar, wind, and nuclear energy produce electricity. While electricity can be used for many different kinds of work in the economy that is presently done directly by fossil fuels (small electric vehicles, electric heating), it is improbable, if not impossible, that electricity can be used to power large ships, airplanes, 18-wheelers, and tractors. Mining and many energy-intensive industrial processes cannot run on electricity alone. Agriculture cannot run on electricity, nor can dredging harbors or building large levee systems to protect against hurricanes. Biofuels can be used for liquid fuels. But the quantity of such fuels required by industrial society could not be supplied by biofuels even if most plant growth on earth were turned into liquid fuels. All alternatives to fossil fuels are very diffuse, as they occur in nature. Humans must pay the costs to concentrate these energy sources so that they are useful in the industrial economy. In this respect, alternative energy is dependent on daily flows of solar energy, whereas fossil fuels are storehouses of sunlight that was captured by plants over millions of years of earth history.

An illuminating way to look at the issue of net energy and its relation to society is to consider what we call the EROI pyramid (Fig. 7.13). In effect, the degree of economic well-being, educational achievement, cultural sophistication, and health of society is related to the total amount of energy that flows into society as well as the EROI of that energy.[51] This is not to say that pre-industrial revolution societies did not attain great cultural and technological achievements. They certainly did. But in those societies, most people worked in basic economic activities, especially farming. It took the work of thousands and thousands of such workers to support one person at an affluent life style that many take for granted today. University education or great art was available mainly to a select few of the nobility and wealthy Bourgeois, but again dependent upon many thousands of workers exploiting solar energy through agriculture, forestry, and mining. But as large quantities of high EROI fossil energy flowed into society during the industrial revolution, most people moved away from farming and other basic economic activities into more technologically advanced production and service occupations. Energy intensive technology replaced human labor in farming, mining, forestry and fishing. As more and more energy flowed into the economy, economic activity, technological progress, and

---

[51] Lambert, J.G., Hall, C.A.S., Balogh, S.B., Gupta, A., Arnold, M. 2014. Energy, EROI and quality of life. *Energy Policy*, 64, 53–167.

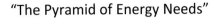

"The Pyramid of Energy Needs"

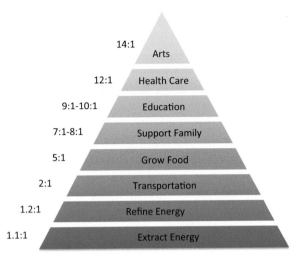

14:1 Arts

12:1 Health Care

9:1-10:1 Education

7:1-8:1 Support Family

5:1 Grow Food

2:1 Transportation

1.2:1 Refine Energy

1.1:1 Extract Energy

Fig. 7.13 "The Pyramid of Energetic Needs" representing the minimum EROI required for conventional oil, at the well-head, to be able to perform various tasks required for civilization. If the EROI of a fuel (say oil) is 1.1:1 then all you can do is pump it out of the ground and look at it. Each increment in EROI allows more and more work to be done and more and more of the trappings of civilization. EROI values are given on the left of the pyramid. The *blue* values are published values, the *yellow* values are increasingly speculative (Adapted from: Lambert, J.G., Hall, C.A.S., Balogh, S.B., Gupta, A., Arnold, M. 2014. Energy, EROI and quality of life. *Energy Policy*, **64**, 53–167.; with EROI data from: Hall, C. A., Balogh, S., & Murphy, D. J. 2009. What is the minimum EROI that a sustainable society must have? *Energies*, **2**(1), 25–47)

population grew rapidly, although often the resource base (such as soils and wild fish) was increasingly seriously overused.

There are energy losses up the economic "food chain" just as occurs in the biological food chains in nature. The more energy, and the higher its EROI at the bottom of the pyramid, the more is available to be spent higher up the pyramid on such areas as universities, art museums, music festivals, and theme parks. But as energy moves up the pyramid, there is less and less energy available to do work that is discretionary but that enriches society. As the EROI of available energy sources entering the bottom of the pyramid decreases because we must frack, drill deeper offshore, process tar sands, or capture diffuse daily flows of solar energy, the amount of net energy available at higher levels also likely decreases. The EROI of solar PV is generally less than 10:1. Wind energy's EROI can be higher but on average is less than 20:1, and even less when the additional expenses of dealing with intermittency are included (the wind isn't always blowing and the sun doesn't always shine). Most liquid biofuels have an EROI of not much greater than 1:1. Hydropower EROI can be very high for the best sites, but most of these were developed long ago, especially for first world countries, and there can be high environmental costs. By comparison, fossil fuels have provided the world with an average EROI greater than 30:1 for much of the previous century. So it will be very difficult, and perhaps impossible, for renewable fuels to power society as we know it. For remaining stocks of fossil fuels such as the Canadian tar sands, it is not just how big the resource is, it is how much energy must be used to get the energy to the point of use. A barrel of oil from any source may look the same after exploration, production and processing, but in fact the energy inputs mean that the oil from unconventional sources will be effectively, on a net energy basis, some fraction of the oil from the conventional fields on which our global economy was built.

Let's summarize the importance of EROI in this way. If you live on an island with one energy source, say an oil well, and the EROI is 1.1:1 then all that can be done is to pump the oil out of the ground and look at it. If it is about 3:1, then it can be refined and made into gasoline or diesel and shipped to where it is used. But in this case all of the net energy has been used, and there is none left to do other things besides drive the tanker truck. If you want to drive a truck to do other work, then you need an EROI of greater than 3:1 at the well head to account for the depreciation (making and servicing) of the truck and the roads and bridges it drives on, but there would be no energy left over to transport something in the truck.[52] The EROI numbers get more speculative the higher we proceed up the pyramid. If we want to put something in the truck, say food, it would require an EROI at the well of perhaps 5:1. If the food needs to be transported long distances, as in the current food system, then the EROI has to be still higher. Society also needs many different types of workers, including truck drivers, farmers, sales people, carpenters, teachers, and doctors to provide all the needs of society, including education and health care. All of this takes energy. In order for society to function as it did during the twentieth century, the EROI probably needs to be at least 10:1 at the energy source, i.e. the bottom of the pyramid. To maintain our current economic system, the EROI at the bottom of the pyramid probably needs to be fairly large, perhaps 15:1 at the bottom. But the average EROI of most renewables is less than 10:1. So this is one of the critical problems with powering the whole society on renewables.

Another factor to consider when discussing the potential role of renewables is the time it will take to develop the industry and bring on line enough energy production to replace declining non-renewable fuels, espe-

---

[52] Hall, Murphy and Balogh, 2009 ibid.

cially fossil fuels. Many point out that wind and solar have been growing so rapidly that they will easily be able to replace other fuels. Vaclav Smil, who has written extensively on energy, disagrees.[53] He points out that over the past 150 years, it took decades for new energy technologies to grow into dominant proportions of the energy market. For example, it took oil 40 years to grow from 5 to 25 % of the global primary energy supply, and it will likely take natural gas 60 years to do the same. Wind and solar rose from about 0.1 % of total U.S. primary energy consumption in 2000 to about 1 % in 2010 and 2.2 % in 2014. Even if this rapid rate of growth were to continue, and the materials and manufacturing capacity were not limiting factors, fossil fuels would still be supplying 78 % of U.S. primary energy in 2030, and 75 % in 2040. It will clearly take decades, if ever, for renewables to become the dominant source of energy for the U.S. and world.

We end this section by comparing projected future fossil fuel production with solar and wind energy produced to date. Drs. Gaetano Maggio and Gaetano Cacciola of the Instituto di Tecnologie Avanzate per l'Energia in Messina, Italy, forecasted future trends in world fossil fuel production.[54] They used a sophisticated analysis based on the approach used by M. King Hubbert, discussed earlier. Their results are shown in Fig. 7.14. The light green, yellow, and black lines are the forecasts for oil, natural gas, and coal, respectively. The symbols on the lines show actual data up to 2011 for the three fossil fuels. The grey line is the total past and projected future production of all fossil fuels. All forms of energy are expressed as billions of barrels of oil equivalents per year. We also included in the figure the total production of wind and solar energy (dark green line) as of 2014 also in barrels of oil equivalents. *We did not include in the graph any data for biofuels because as we discussed earlier, their production is very low and they yield almost no net energy, so are not a viable future fuel for society.*

What is clear from the graph is that it is extremely unlikely, if not impossible, for wind and solar to produce anything like the energy levels that fossil fuels now provide. Beyond mid century, oil and natural gas production will be in steep decline. Coal will be the only fossil fuel to provide significant amounts of energy for the second half of the twenty-first century and beyond. Renewables can provide significant amounts of energy to society, but they do not seem poised to provide the necessary energy to maintain a modern society as we now know it.

---

[53] Vaclav Smil. *2015 Revolution? More like a crawl.* http://www.politico.com/agenda/story/2015/05/energy-visionary-vaclav-smil-quick-transformations-wrong-000017

[54] G. Maggio and G. Cacciola. 2012. When will oil, natural gas, and coal peak? Fuel, 98, 111–123. http://dx.doi.org/10.1016/j.fuel.2012.03.021

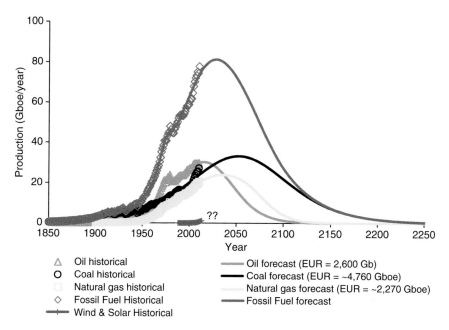

Fig. 7.14 Past energy production for fossil fuels and wind and solar and future projections for fossil fuels. The *grey line* is the total for all fossil fuels. The width of the figure can be considered as a rough approximation of the age of the industrial revolution (Fossil fuel information is from Maggio and Cacciola, 2012, ibid. Wind and energy data from: British Petroleum. World Energy Outlook 2014)

It is obvious from this graph that for many people reading this book, and certainly their children and grandchildren, the future is going to require some very difficult choices about energy and lifestyle, something that very few people are doing today. For example, we took a very informal survey of 17 people on the streets of London in June 2015. We asked them if they could define peak oil or tell us what the oil production of England was now compared to 20 years ago. Of the 17, 16 had no clue as to what peak oil was and only one guessed that UK production was less than 20 years ago. That person guessed a little less, whereas the actual figure is that in 2013 UK produced only about 20 % of the oil that was produced at its peak about 20 years ago.

Thus we think there is a great deal of education required to get people ready for whatever will transpire. Since most people will see a decrease in energy availability (from peaking production and declining EROI) in terms of their own economic situation, and given the propensity of all of us to blame other people for their economic misfortunes, the greatest concern is civil unrest, as is already occurring on a relatively small scale in some nations today.

Socio-economic trends

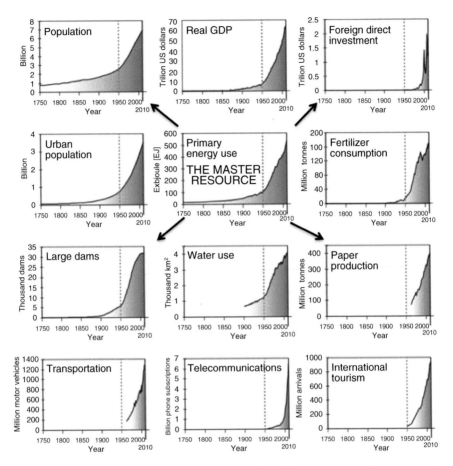

Fig. 7.15 Trends from 1750 to 2010 at the global level for a number of measures of social, economic, and energy development. *Arrows* indicate that energy, the master resource, affects all other measures (Adapted from: Steffen, W., Broadgate, W., Deutsch, L., Gaffney, O., & Ludwig, C., The Anthropocene Review 2(1), pp. 81–98, copyright © 2013 by (Copyright Holder) Reprinted by Permission of SAGE Publications, Ltd.

Let's consider the implications of energy for urban living by studying another city, Syracuse, New York. If you look up synonyms for "economy" or "economics" online or in a thesaurus, you will not find the word "energy." But you should, as the two concepts are so tightly coupled that money can be viewed as a "lien" on energy—energy enables all things economic. To illustrate this point, Fig. 7.15 shows graphs of a number of socio-economic indicators of human development from 1750 to 2010. All of these curves show geometric growth that is especially apparent in the twentieth, or so-called exponential, century. Because the impacts of human activity on the earth have been so pervasive during this period, many scientists are calling this era the Anthropocene. This means that for the first time in human history, humans are the dominant force controlling a geological era.

It is important to understand that these trends are interdependent. All are connected, but the central trend that enabled all the rest was energy. Cheap fossil fuels provided the energy that underwrote the dramatic growth in population, technology, and the work that enabled the economy over the past century and a half. Without fossil fuels almost none of this would have happened. For this reason, we call energy the "master resource."

Now let's look at these relationships through the lens of one moderate-sized city, Syracuse, New York. As we have seen in Chap. 4, the distribution of population centers across the landscape is not random. There are a variety of resource-driven reasons for cities and towns to be located where they are, and most of those reasons can be linked to energy and other resources. Energy is necessary to exploit other resources, or in some cases to substitute technology for ecosystem services. Increasing energy availability became the basis for everything that we now take for granted, from routine airline flights to supermarkets full of food from around the globe. Although not one of our 12 model cities, Syracuse provides a particularly vivid picture of the energy-development nexus.

Syracuse was named after the city of that name in Sicily. A bureaucrat, probably colonial state legislator Robert Harpur, was given the task of naming towns in what was then known as the "Military Tract" because homesteads in this region were given to European–American soldiers as a reward for military service in the Revolutionary War. Harpur clearly had a taste for classical literature as so many of these towns were given names from the Greek and Latin classics. The obvious ones are the larger settlements of Syracuse, Rome, Utica and Troy, but we also find Camillus, Cicero, Cincinnatus, Corinth, Fabius, Ilion, Ithaca, Homer, Macedon, Marcellus, Pompey, Romulus, Scipio, Sparta, Ulysses, Virgil, and many others.[55] Harpur was obviously not impressed that many of these regions already had Native Americans living there with names of their own.

Syracuse, only 300 feet/100 m above sea level, is located at the intersection of the major North–south and (especially) East–west transportation

---

[55] William R. Farrell. 2002. *Classical Place Names in New York State: Origins, Histories and Meanings*. Pine Grove Press.

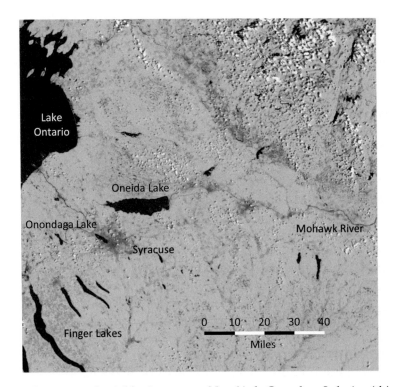

Fig. 7.16 Syracuse and neighboring upstate New York. Onondaga Lake is within the urban area and Oneida Lake is just to the northeast of the city; the Mohawk River runs east from the lake. The Finger Lakes are south and west of Syracuse. *Pink* areas, usually in fertile valleys or other flat areas, are agriculture and the *green* background is mostly forest. Fine wines are produced near deep lakes where the deep water, with its large heat energy storage, protects grape vines from frosts. Interstate highway 90 runs north of the Finger Lakes through Syracuse and follows the Mohawk River and the route of the once-important Erie Canal. Interstate 81 runs north south through the city. Lake Ontario is in the *upper left. Grey spots* represent some of the cities, including Syracuse, smaller Rome and Utica to the East, all along the East West transportation corridor, Watertown to the North and Cortland to the South. This region was, and still is, an important agricultural region. This distribution of cities in upstate New York is similar to distribution patterns of other organisms in nature. (Image from NASA)

corridors of upstate New York (Fig. 7.16). These corridors lie along low-lying and flat lands that were relatively energy-efficient for transporting people and freight around the many steep hills elsewhere in upstate New York. Syracuse's Onondaga Lake also connects readily with waterways important for transportation, including the Great Lakes, the Seneca River and the Mohawk River. Long before the time when European–

Americans settled upstate New York, the location of present Syracuse was an important transportation nexus, dwelling place, and spiritual focus for the Onondaga and other Native Americans. The Onondaga nation today is located just south of the city of Syracuse.

Settlement during the nineteenth century also made use of these geographic advantages for transport of people and goods, starting with wagon trails built along the native trails and culminating in the Erie Canal and the Mohawk Valley railroad line. Later, Syracuse grew into the fourth largest city in New York, after Buffalo, Rochester and of course New York City.

The East–west transportation corridor stretches all the way across upstate New York, connecting the towns of Albany, Utica, Syracuse, Rochester and Buffalo at 50–90 mile intervals (Fig. 7.16). This corridor evolved through time from Native American trails to early wagon trails to the Erie Canal to railroad main lines to today's Interstate 90. All of them have used the low-lying flat land to decrease energy expenditures for moving people and goods, although the source of energy changed from the human body and draft animals to flowing water to coal to gasoline and truck diesel. Agricultural areas also save energy in valleys or between the Finger Lakes where good soil and flatter land yield high agricultural energy return on energy invested.

In his classic book, "Environment, Power and Society," ecologist Howard Odum showed a general pattern in nature where large areas of plant growth are sprinkled with smaller areas of intense consumption (Fig. 7.17).[56] Plants use solar energy to convert soil and water containing essential elements such as carbon, nitrogen, and phosphorus into the stuff of the plant. Odum gives examples ranging from a glass slide immersed in a stream on which organisms grow, to tree trunks surrounded by the forest canopy, to houses dotting croplands in Kansas farm country.

The top graphic shows a glass laboratory slide that has been submerged in a fertile stream for 6 weeks. Tiny one-celled plants called diatoms are spread evenly over the slide while the worms that feed on them are unevenly spaced. This pattern is also observed from an airplane flying over Kansas farm country. Farm fields cover the landscape and roads lead to farm houses and barns. The same pattern is also evident in a tropical rain forest with limbs holding photosynthesizing leaves and roots spreading out from the trunk. The roots collect nutrient resources from large soil areas while the leaves use sunlight to convert water and soil resources to plant material. Cities likewise are concentrated centers spread over the landscape that are dependent on collecting resources from an area much larger than the city to generate specialized economic products.

---

[56] Howard T. Odum. 1971. *Environment, Power, and Society*. Wiley-Interscience, New York.

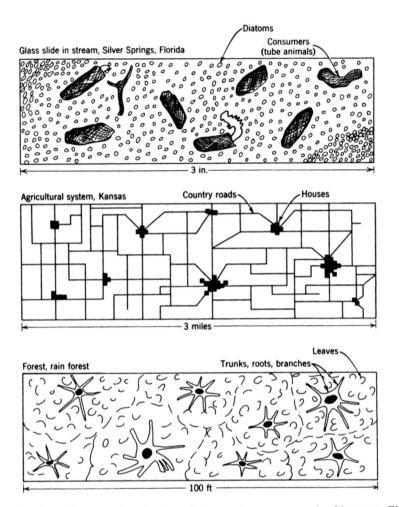

Fig. 7.17 Three diagrams showing how living systems are organized in space. Plant growth or photosynthesis is spread more or less evenly over the surface while consumption (*darker areas*) is concentrated in centers that are linked to plant growth by converging pathways (Reprinted from: Howard Odum. 1971. *Environment, Power, and Society*. Wiley-Interscience, New York.)

In a similar manner, the cities along this New York transportation corridor are centers of consumption—of food, energy, economic products and so on—surrounded by larger areas of plant production, including forests and relatively rich agricultural land. In this respect, early Native American settlements were similar to cities today, except that today's human settlements are generally much larger and often more closely spaced because of the high energy density available from fossil fuels (Fig. 7.18). Of course, this

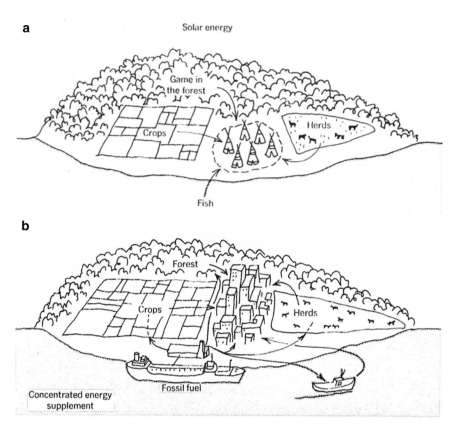

Fig. 7.18 Comparison of an agrarian (**a**) with an industrial system (**b**). Both have the same general structure with large areas of productive resources flowing into a dense human community with high metabolism. The difference is that fossil fuels allow a much higher concentration of humans in cities (Odum, ibid.). These diagrams certainly represent the region we call Syracuse, the first in 1820 and the latter today (Reprinted from: Odum, 1971, ibid.)

means that current human settlement patterns with large dense cities are also *dependent* on cheap and abundant supplies of fossil fuels.

Oyster reefs are natural analogs to cities. Oyster reefs develop dense concentrations of animals that require a continual input of resources (food and oxygen in the water flowing over the reef) and flushing of wastes in order to survive (Fig. 7.19). If the reef were enclosed in a large plastic dome, most of the organisms would die in a matter of hours. A dense city is also dependent on continuous inputs of energy, food, materials, and money, with outputs of wastes carried in water, air currents, and in trucks, train cars, and barges. Like the oyster reef, if these flows were stopped by

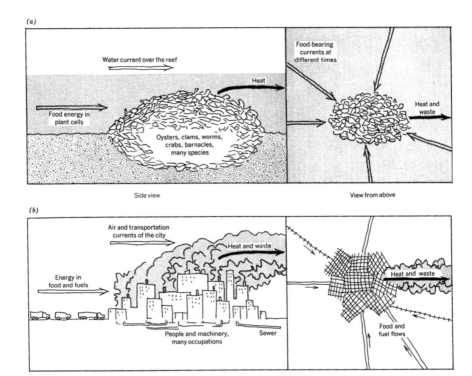

Fig. 7.19 Comparison of two systems of concentrated consumers whose survival is dependent on large energy flows that bring in resources and carry away wastes. Oyster reefs (**a**) with dense populations of oysters and many other organisms are found in estuaries around the world. Modern cities (**b**) are similarly dependent on large areas to provide resources and accept wastes. Neither of these systems could survive without the large flows in and out (Reprinted from: Odum, ibid.)

covering them with a large dome as in some futuristic movies, the city would not survive for very long at all.

Other natural resources contributed to the original citing of Syracuse, notably salt and limestone, both remnants of ancient seas. Salt was especially valuable as a preservative for meat before the advent of refrigeration. Salt water was taken from springs and later from wells near Syracuse and evaporated, first using the sun, then wood, and then coal brought in from Pennsylvania. Accordingly, Syracuse was often called "Salt City." Limestone was valuable as the main raw material for concrete. But more important for Syracuse, glass in Colonial times could be made only by importing expensive sodium carbonate (Na_2CO_3), or soda ash. Some Americans went to Germany as apprentices to the glass makers and came back with the recipe for making Na_2CO_3. Syracuse, or more specifically the small town next to

Syracuse named Solvay (after the Scotch originator of the glass-making process used), turned out to be the perfect place for producing this valuable raw material. Sodium was extracted from the salt (NaCl), and carbonate from the limestone ($CaCO_3$), to make sodium carbonate and huge amounts of waste calcium chloride. In the process, the adjacent Onondaga Lake, once a beautiful and productive fishing lake, sacred to the Onondaga people and a great recreational location a century ago, became one of the most polluted lakes in North America.

In the first half of the twentieth century Syracuse became a relatively wealthy blue collar industrial town based on the chemical industries and the manufacturing of china, firearms, automobiles, automobile parts, air conditioners and many other industrial commodities. The manufacturers were originally fueled by firewood from the abundant local forests, followed by cheap coal from Pennsylvania, and then oil, gas and the cheap hydroelectricity from water routed around Niagara Falls. The last of these also powered the great industrial development of the Buffalo region, usually considered the origin of large scale industrialization in the United States. These abundant and cheap energy resources were critical to Syracuse becoming a successful and relatively wealthy regional city.

Unfortunately, the subsequent loss of many of the city's critical industries followed the "oil supply shocks" of the 1970s. Many companies relocated overseas, leaving Syracuse to become a classic "rust belt" city. Today there are relatively few heavy industries, with their high paying blue collar jobs, but the city continues to function, although at a lower population level, mostly on "Eds and Meds" (the world class educational and medical establishments that are concentrated there). Syracuse remains an important transportation hub, and some limestone is still quarried from remnant resources. Interestingly, one of the larger industrial activities is cleaning up the pollutants left from a half century before.

The history of Syracuse over the past century and a half reflects what has happened to much of America—the finding and exploitation of valuable resources, the rapid development of industries based on those resources and cheap, readily available energy, the generation of massive amounts of wastes, the gradual exhaustion of the resource base (sometimes combined with other losses of revenues) and then a long period of decline marked by remnant pollution. Fortunately there was a base of less energy-intensive "Eds and Meds" that have helped to maintain Syracuse as a good place to live and work. An important question is whether such high levels of Eds and Meds can be maintained in an energy—constrained future. As we saw in the previous section, Eds and Meds are at the top of the EROI pyramid. As energy input to the bottom of the pyramid decreases, less economic activities such as these can be maintained. These patterns and issues related to Syracuse are common to the other cities and regions that we discuss.

In this and other chapters, we have seen how energy and other resources controlled the course of development, and sometimes decline, across the American landscape. But if the oil and gas industry, along with technologies to capture "renewable" energy from the sun, promise a secure energy future as we are often told, why do any of these issues matter to the sustainability of cities and regions? As we have shown in this chapter and throughout the book, it is very unlikely, probably impossible in the foreseeable future, that renewables can come anywhere near fully replacing fossil fuels, which will be largely gone by the end of this century. In addition, new fossil fuel resources as well as renewables (based on solar flows) are likely to have a much lower net energy or EROI.

Landscape Implications

It is important to understand that energy scarcity will profoundly impact all sectors of society. The age of cheap energy allowed dramatic changes in society and dramatic growth of energy use, population, and the economy. Energy scarcity will lead to changes just as profound, because the wealth accumulation of the past century and a half cannot be sustained. Cheap energy allowed tremendous growth in arid regions with the development of huge water projects. Development in coastal areas such as south Louisiana and south Florida was possible due to energy intensive flood control and hurricane protection. It will be difficult to maintain this extensive and expensive infrastructure in an energy-scarce, climate-challenged world. Drought, sea level rise and stronger storms will make living in the southern Great Plains, the Southwest, and many coastal areas more difficult. Cheap energy supercharged the industrial food production system and allowed it to expand into arid areas of the west. The mechanization of the basic economic activities of farming, fishing, forestry, and mining underwrote the migration to growing cities. Maintaining large urban populations will become more difficult as the age of cheap energy wanes. We will come back to these questions in Chap. 10 as we consider the relative sustainability of our 12 cities and 10 regions.

Chapter 8

Feeding America's Cities: Putting Food on the Table in the Twenty-First Century

Food is a major preoccupation of all animals, and people are no exception. The number of magazines, whole sections of newspapers, and shows on television attest to our delight in food and cooking. An incredible diversity of restaurants, with many celebrity chefs, offer their services in the U.S. today. There are probably more people who know who Emeril or Rachel Ray are than could name a majority of the justices on the U.S. Supreme Court or a single Nobel Prize winner.

But when the senior author (J.D.) was coming of age in Baton Rouge in the early 1960s, food options were limited. There were restaurants of standard American food, a few locally-based fast food restaurants serving mainly hamburgers and other sandwiches, and Louisiana food (fish, shrimp, crabs, crawfish, gumbo, jambalaya, etouffee, etc.). There were also a few Italian, Chinese and Mexican places reflecting segments of the local population. To the south, New Orleans was well known for its cuisine, but it was still largely Louisiana-based. In 1960, iced tea and soda were the non-alcoholic drinks of choice. It was impossible to find a latte or any of the other coffee concoctions commonly sold nowadays off most Interstate exits. This was largely the case for almost all middle-sized American cities. Wider varieties of food in large U.S. cities like New York were products of greater cultural diversity.

But this has changed. Baton Rouge now has numerous restaurants serving Thai, Sushi, Indian, Nepali, etc., as well as blended or fusion cuisines. This is true for practically all moderate-sized and larger American cities (and many smaller ones) no matter where you are in the country.

Many Americans are concerned about the healthfulness of what they eat. Much of the U.S. diet is made up of high sugar (or high fructose corn

© Springer Science+Business Media New York 2016
J.W. Day, C. Hall, *America's Most Sustainable Cities and Regions*,
DOI 10.1007/978-1-4939-3243-6_8

syrup), high fat, and high salt items, all catering to the vestiges of needs conditioned by evolution when food was less abundant and relatively hard to come by. Consumption of junk and fast foods with these characteristics in the U.S. is at an all-time high. We face an epidemic of obesity and diabetes. Many recent books have addressed the issues of unhealthy diet and alternatives, including *Fast Food Nation* and *In Defense of Food*.[1] Clearly Americans take a great interest in food, diet, and health.

But few Americans are similarly aware of the details of how food actually gets to their plates, whether at home or in a restaurant. They know relatively little about how and where food is produced and by whom, how it is handled, and what goes into it other than natural ingredients. Even fewer people are aware of how much energy it takes to produce food, and the impact that climate change may have on food production and vice versa. In-depth knowledge of these issues raises many troubling questions. In this chapter, we sink our teeth into the food system.

First, let's consider a generalized typical meal. You and your family sit down to a meal of bread, green salad, a baked potato or rice, vegetables, meat or seafood, apple pie, and a nice bottle of wine, with soda or milk for the kids. What does it take to get this meal to your plates?

A Brief History of Food Production

Humans have been farmers for about 10,000 years. But for a much longer period, hundreds of thousands of years, our ancestors were hunter-gatherers. According to different estimates, our species, *Homo sapiens*, came into existence somewhere around 200,000 years ago, but our hominoid ancestors go back several million years. Most of us have heard of Lucy (*Australopithecus afarensis*) whose skeletal remains were discovered in the Afar Depression of Ethiopia in 1974. Archeologists estimate that Lucy lived about 3.2 million years ago, and she shared many characteristics with us today, or at least those key differences that distinguish us as humans in relation to the great apes.

Several lines of evidence indicate that the original way to make a living for modern humans (*Homo sapiens*, but probably earlier species on our lineage as well) involved hunter-gathering and living in small, mostly nomadic groups.[2] So to settle down and become farmers was a significant departure for the human lifestyle ten millennia ago. For almost all of the past 10,000 years, practically all food that humans consumed was exploited

---

[1] E. Schlosser. 2001. *Fast food nation*. Mariner Books, Boston, U.S.A.; M. Pollan. 2009. *In defense of food: an eater's manifesto*. Penguin Books, London, U.K.

[2] Jared Diamond. 1997. *Guns, germs, and steel: the fates of human societies*. W.W. Norton and Company, New York, USA.

Table 8.1 Six independent centers of crop origin identified and ages of earliest C-14 dated crop remains[a]

Location	Crops	Earliest C-14 dated crops (years before present)
Mesoamerica (Southern Mexico & Northern Central America) & Central America	Maize, Phaseolus beans, Sweet potato, tomato	Squash (10,000); Cassava, Dioscorea yam, arrowroot, maize (7000–5000)
The Andes of South America	Potato, cassava (manioc), pineapple	
Southwest Asia including the "Fertile Crescent"	Wheat, barley, pea, lentil	Einkorn wheat (9400–9000); Lentil (9500–9000); Flax (9200–8500)
The Sahel region & Ethiopian highlands of Africa	Sorghum, coffee, melon, watermelon	
China	Asian rice, soybean, adzuki bean, orange, apricot, peach, tea	Rice (9000–8000)
Southeast Asia	Cucumber, banana, plantain	

[a]Crop Origins and Evolution. http://en.citizendium.org/wiki/Crop_origins_and_evolution. *After Paul Gepts, 2003, in Chapter 13 of Plants, Genes and Crop Biotechnology, Chrispeels and Sadava*

or produced locally, and these local food sources were what made up the human diet (Table 8.1). Where one lived determined the staple food crop; e.g., wheat in Mediterranean lands, rice in Southeast Asia, maize or corn in the Americas, and sorghum and root crops in Africa. All of these crops were domesticated from relatively productive (i.e., high energy yielding) wild species between 6 and 10,000 years ago. This condition persisted for the most part until the great age of colonization, beginning in the sixteenth century, when wheat, rice, corn, and many other crops spread to almost all suitable areas of the world. Until early in the twentieth century, however, most farming work was still done by humans and draft animals such as horses, water buffalo, and cattle (Fig. 8.1a, c).

The next great leap forward in agriculture came in the twentieth century. Increasing use of cheap fossil fuels and technological development brought about a dramatic transformation of agriculture. During this period, fossil fuel-driven tractors and other machines replaced the labor of humans and draft animals (Fig. 8.1a, c). During the nineteenth century, up to 72 % of the U.S. labor force worked on farms. By the end of the twentieth century, that work force had dropped to less than 3 %, with muscle power having been largely replaced by fossil fuel-powered machines. Astonishingly, fewer Americans were working on farms in 2000 to support the diets of burgeoning urban populations than the number of farm workers in 1840 (Fig. 8.1a). This drastic decline in human labor on farms, and the amazing increase in agricultural productivity that occurred simultaneously, were possible because of the use of oil and gas to make and transport fertilizers and

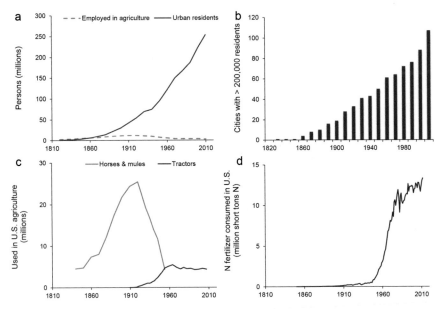

Fig. 8.1 (**a**) U.S. persons employed in agricultural pursuits and residing in urban areas. (**b**) Number of U.S. cities with >200,000 residents. (**c**) Horses/mules and tractors used in U.S. agriculture. (**d**) Nitrogen fertilizer consumption in the U.S (All data for this figure was obtained from reports issued by the United States Census Bureau with the exceptions of: % employed on farms 1980–2010 (The World Bank. Employment in agriculture—% of total employment. Available online: http://data.worldbank.org/indicator/ SL.AGR.EMPL.ZS), % residing in urban areas 1950–2010 (United Nations. 2011. World urbanization prospects, the 2011 revision. Available online: http://esa.un.org/unup/ CD-ROM/Urban–rural-Population.htm), and nitrogen fertilizer consumed in 1943–1960 (United States Geological Survey. 2014. Historical statistics for mineral and material commodities in the United States. Available online: http://minerals.usgs.gov/minerals/ pubs/historical-statistics/) and 1961–2012 (http://faostat.fao.org/). Means are shown where multiple overlapping estimates existed from census data and these sources.)

pesticides, mechanized farming using fossil fuels, along with the development of new crop varieties. The invention of the Haber–Bosch process, which converts atmospheric nitrogen gas into ammonia, was critical to making fertilizers abundant (Fig. 8.1d). The same can be said for technological advances that enabled the mining of phosphate rock deposits. Irrigation became widespread, allowing crops to be grown in arid and semi-arid areas, such as the Central Valley of California. Worldwide, crops became more homogenized,[3] and food processing became widespread. All of these changes allowed the development of a globalized industrial food system.

---

[3] Khoury, C.K. et al. 2014. *Increasing food homogeneity in global food supplies and the implications for food security.* Proceedings of the National Academy of Sciences of the United States of America. www.pnas.org/cgi/doi/10.1073/pnas.1313490111.

Crop Production in the U.S.

Agriculture in the U.S. occupies an immense area. In 2007, over 922 million acres of farmland spread across the country, down from about 1.16 billion acres in 1950, a decrease of more than 20 %.[4] Of this farmland, roughly half comprised pasture, while crops covered much of the rest. A good place to examine the spatial distribution of U.S. agriculture is the website for *The 2009 Cropland Data Layer* by David Johnson and Richard Mueller of the U.S. Department of Agriculture (see Fig. 8.2).

The Cropland Data Layer website allows one to view the distribution of crops produced in different parts of the U.S. down to the county level. The

AGRICULTURE

Pasture/Grass
Corn
Soybeans
All Wheat
Other Hay
Fallow/Idle Cropland   NON-AGRICULTURE
Alfalfa   Woodland
Cotton   Shrubland
Other Crops   Urban/Developed
Sorghum   Wetlands
Vegetables/Fruits/Nuts   Water
Other Small Grains   Barren
Rice   Perennial Ice/Snow

Fig. 8.2 The cropland data layer mapped at the national level (United States Department of Agriculture. National Agricultural Statistical Service. Available Online: http://nassgeodata.gmu.edu/CropScape/)

---

[4] United States Department of Agriculture. 2007. *Census of Agriculture*. Available online: http://www2.census.gov/prod2/decennial/documents/21895591v2ch02.pdfhttp://www2. census.gov/prod2/decennial/documents/21895591v2ch02.pdf.

image of the whole country provides an excellent overview of what is produced where. A number of things immediately stand out. One of the most distinctive regions is the so-called American breadbasket of the Midwest. The greens and yellows in Fig. 8.2 show where corn and soybeans, respectively, were grown in 2009. Corn production for grain is a staple of America's agro-industrial system (28 % of harvested cropland in 2007),[5] with the majority of acreage confined to the central portion of the country bounded by eastern Colorado to the west, the Dakotas to the north, Ohio to the east, and the junction of Kentucky, Illinois, and Missouri to the south. Soybean production, another staple of American agriculture (21 % of harvested cropland in 2007), grows in rotation with corn, and therefore overlaps with the major corn producing areas. America's third most widespread crop is wheat (16 % of harvested cropland in 2007, brown in the Cropland Data Layer), which is primarily raised in the central portion of the country on the Great Plains extending from North Dakota to northern Texas. The Midwestern region where these three crops are grown is one of the most important farming areas in the world. Other U.S. regions contain smaller areas of corn, soy, and wheat production (Fig. 8.2).

Another feature apparent from the map is that most important food producing regions of the country are west of the Mississippi River, with the exception of the eastern side of the corn and soybean production area. This includes the aforementioned wheat production stretching down through the Great Plains. Further west lies the Central Valley of California, which is easily identified on The Cropland Data Layer. The Central Valley stands as the granddaddy of fruit, vegetable, and nut production in the U.S. California has approximately 25 million acres of farmland,[6] much of which is in the Central Valley, where more than 50 % of the vegetables, fruits, and nuts of the country are grown.[7] Unsurprisingly given this region's dry climate, this is also the greatest concentration of irrigated vegetable production land in the country. California is a predominant source of the asparagus, broccoli, cantaloupes, lettuce, onions, tomatoes, almonds, walnuts, avocados, lemons, strawberries, plums/prunes, pears, cherries, grapes, and peaches that we eat. Other concentrated agricultural production zones in the west include the Willamette Valley in Oregon (vegetables, fruit), southeastern Washington extending into Oregon (vegetables, fruit, wheat, potatoes), and southern Idaho (potatoes).[8]

---

[5] Estimates of % of harvested cropland are based on harvested cropland area of 309,607,601 acres and acres harvested by crop reported in the 2007 Census of Agriculture.

[6] USDA, 2007, ibid. http://www.agcensus.usda.gov/Publications/2007/Full_Report/Volume_1,_Chapter_1_State_Level/California/st06_1_008_008.pdf.

[7] U.S. Committee on Natural Resources. 2014.

[8] The information in this paragraph and the next comes from the 2007 US Census of Agriculture.

Notable crop production zones east of the Mississippi River in addition to the eastern portion of the Midwestern corn and soy area include Florida (oranges, sugarcane, and strawberries), western/central New York (vegetables, fruit, corn), the intersection of Pennsylvania, Maryland, Virginia, Delaware, and New Jersey (vegetables, fruit, corn, soybeans, wheat), and a ribbon of land extending northeast from Georgia to southeastern Virginia (vegetables, fruit, peanuts, corn, soybeans, wheat). Terrain, soil fertility, and climate limit crop production in many locales across the eastern U.S. in comparison to the relatively ideal conditions found in the Midwest. Crop production in the U.S. is, therefore, spatially heterogeneous, with certain areas (the Midwest, Great Plains, and Central Valley of California) providing the bulk of our domestic food supply. A surprising amount of land produces little or no food crops at all.

Thomas Jefferson once remarked that, "civilization itself rests upon the soil." Currently, soil degradation, especially by erosion and salinization, threatens the finite, unequally distributed supply of soil resources on Earth.[9] The dominant soil order as defined by the USDA in much of the Midwestern Great Plains heart of U.S. agriculture is Mollisol, typically found where grassland ecosystems have flourished. The long-term addition of organic materials derived from grassland plants creates a thick, dark, fertile surface soil. This in turn enables incredibly productive agriculture. Around the world, Mollisols are used extensively for growing food.[10] Much of the rest of the Midwest is characterized by Alfisols, highly fertile soils mainly formed under forest cover.[11] Although Mollisols and Alfisols are widespread in the U.S. (together they make up more than one third of U.S. land area[12]), other soil types found around the country are not as favorable for growing food unless significant resources are used to boost productivity.[13] Figure 8.3 shows the soil fertility index derived by Ralph Schaetzl,

---

[9] Lal, R. 2013. Food security in a changing climate. *Ecohydrology & Hydrobiology*, 13, 8–21.

[10] Brady, N.C., and R.R. Weil. 2007. *The nature and properties of soils*, 14th edition. Prentice Hall.

[11] Brady and Weil, 2007, ibid.

[12] Soil & Land Resources Division, College of Agricultural and Life Sciences, University of Idaho. *"The twelve soil orders."* http://www.cals.uidaho.edu/soilorders/.

[13] Taking a global perspective, it is important to note that much of the new agricultural land expansion is occurring in tropical environments having highly weathered soils. These soils formed under year-round hot, moist conditions and over time weathering (i.e., physical, chemical, and biological transformations) has resulted in conditions where essential nutrient forms (e.g., phosphate) are less available to plants. This has important implications for resource use. An example is the Cerrado of Brazil, where soil properties lead to the use of substantially greater amounts of phosphorus fertilizer in soy production relative to locations in Iowa: Riskin, S.H., S. Porder, M.E. Schipanski, E.M. Bennett, and C. Neill. 2013. Regional differences in phosphorus budgets in intensive soybean agriculture. *BioScience*, 63, 49–54.

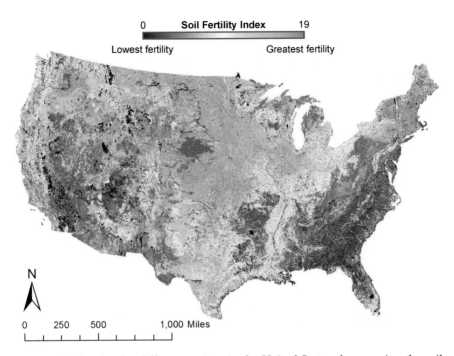

Fig. 8.3 Soil fertility for different regions in the United States shown using the soil fertility index developed by Schaetzl et al. (Map produced using data from: http://foresthealth.fs.usda.gov/soils, The Soil Fertility index is described in: Schaetzl, R.J., F.J. Krist, and B.A. Miller. 2012. A taxonomically based ordinal estimate of soil productivity for landscape-scale analysis. *Soil Science*, **177**, 288–299)

Frank Krist, and Bradley Miller,[14] illustrating the variability in soil quality for plant growth across the nation. Inevitably, utilizing soils for agriculture causes changes in soil properties and can enhance the risk of erosion.[15] Sustainable soil management must accompany sustainable production of crops. As Franklin D. Roosevelt wisely stated, "The nation that destroys its soil, destroys itself."

Local climate, particularly precipitation, is also critical for the success of agriculture. Low rainfall can exacerbate the challenge of overcoming poor soil fertility.[16] The Midwest is characterized by fairly ideal annual rainfall during most years. For example, Cedar Rapids, Iowa receives around 35 in.

---

[14] Schaetzl, R.J., F.J. Krist, and B.A. Miller. 2012. A taxonomically based ordinal estimate of soil productivity for landscape-scale analysis. *Soil Science*, 177, 288–299.

[15] Powlson, D.S., P.J. Gregory, W.R. Whalley, J.N. Quinton, D.W. Hopkins, A.P. Whitmore, P.R. Hirsch, and K.W.T. Goulding. 2011. Soil management in relation to sustainable agriculture and ecosystem services. *Food Policy*, 36 (Suppl. 1), S72-S87.

[16] Pimentel, D., et al. 1997. Water resources: agriculture, the environment, and society. *BioScience*, 47, 97–106.

of rain on average annually. Later in this chapter we will also explore how local precipitation characteristics can influence the feasibility of future food production in and around different U.S. cities.

Now, let's move up the food chain and explore the current production of meat, dairy, and eggs in the U.S.

Livestock, Dairy, and Seafood Production in the US, Where and How Much

Around the globe, in both developing and developed nations, people are consuming much more meat than they did 50 years ago (Fig. 8.4). This trend is expected to continue. Lester Brown refers to the growing global appetite for animal products as "moving up the food chain" and highlights the fact that the increase in consumption of livestock products (in addition to the conversion of some grain to biofuel) is responsible for doubling of global grain demand during the 2000s.[17] A high-meat, so-called "Western diet" (sometimes referred to as the "meat-sweet diet") continues to characterize the food intake of many in the U.S. today. Increasingly, residents of developing countries are

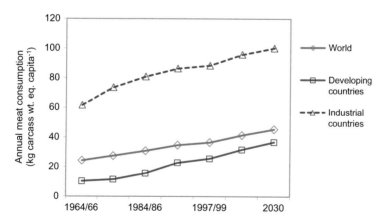

Fig. 8.4 Moving up the food chain. The global increase in meat consumption from the 1960s to 1990s and forecasted up to 2030 (World agriculture: Towards 2015/2030. An FAO perspective. http://www.fao.org/docrep/005/y4252e/y4252e05b.htm)

---

[17] L.R. Brown. 2012. *Full planet, empty plates: the new geopolitics of food scarcity.* W.W. Norton and Company, New York, USA.

moving up the food chain as well (Fig. 8.4). How does the U.S. agricultural system meet this unending craving for meat and other animal products?

First, let's examine the population numbers for various livestock in the U.S. from the 2007 U.S. Census of Agriculture. There were nearly 33 million beef cows in inventory that year and another 96 million cattle and calves. The population of hogs was no less impressive at nearly 68 million. Finally, the number of poultry inventoried in 2007 is astounding. There were more than 30 times more chickens (broilers and egg layers) than humans in the U.S. in 2007—over nine billion of them.[18] Data for total meat production by weight in the U.S. during 2011 provides more eye-opening figures: over 26 billion pounds of beef, nearly 23 billion pounds of pork, and a whopping 44 billion pounds of chicken.[19] These animals are spread across the U.S., with some confined more to certain regions than others.[20]

Terrestrial agriculture provides the large majority of the food we eat. However, fisheries still form a foundational food supply in several coastal communities, and the production and trade of sea catch and aquaculture is important in today's global economy. Globally, capture fisheries and aquaculture provide three billion people with almost 20 % of their average intake of animal protein, and another 1.3 billion with about 15 % of their per capita intake.[21] In 2012, U.S. commercial fisheries produced a total of 7.5 billion pounds of fish and shellfish for human consumption worth $5 billion.[22] The U.S. has rich fisheries along all three coasts and Alaska (Fig. 8.5).

The two top states for commercial fisheries landings are by far Alaska and Louisiana. These two states accounted for over 68 % of total U.S. landings by weight in 2012 (Alaska = 56 %, Louisiana = 13 %).[23] Major fishery resources in Alaska include Pacific salmon, groundfish, Pacific halibut, shellfish, and herring. Overall, catches in Alaska remain below the long-

---

[18] http://www.agcensus.usda.gov/Publications/2007/Online_Highlights/Ag_Atlas_Maps/Livestock_and_Animals/.

[19] USDA Economic Research Service (ERS).

[20] According to the 2007 USDA Census of Agriculture, Beef Cattle are predominantly raised in the Great Plains, Dairy Cows are in California, the Northern Great Lakes and Western New England. Poultry is spotted throughout the country, with broiler chickens raised in the southeast from Arkansas to Georgia, while egg layer hens are prevalent in the Ohio Missouri River Drainage from Pennsylvania to Iowa. Pork is primarily in North Carolina, and the Midwest, especially Minnesota, and Iowa. Most Goats are raised in Texas. Table 8.6 provides a listing of regions and states producing large amounts of animal products.

[21] HLPE. 2014. *Sustainable fisheries and aquaculture for food security and nutrition*. A report by the High Level Panel of Experts on Food Security and Nutrition of the Committee on World Food Security, Rome 2014.

[22] NOAA. 2012. U.S. Commercial Landings. 21pp.

[23] NOAA, 2012 ibid.

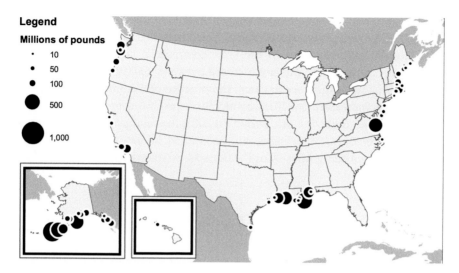

Fig. 8.5 Commercial fishery landings at major U.S. ports in 2012. Alaska, the fertile fisheries crescent around the Mississippi delta, the Pacific northwest, and Chesapeake Bay stand out (Source: NOAA. 2012. U.S. Commercial Landings. 21 pp.)

term potential. The most valuable fishery in the Southeast region of the U.S. including the Gulf of Mexico is shrimp. This shrimp fishery is fully exploited despite efforts to boost yields. In fact, the FAO reports that only half of the 2005 shrimping effort was actually needed to produce the yield obtained.[24] Other important Louisiana species include oyster, blue crab, and menhaden. Salmon, halibut, groundfish, pelagic fishes, and nearshore resources make up the available species on the Pacific coast of the Lower 48, with most—including all five salmon species—being fully or over utilized.[25] In the Northeast, overfishing has resulted in severe declines in the traditionally most valuable groundfish fishery, including cod, haddock, and yellowtail flounder. The region also hosts commercial fisheries for very valuable crustaceans and bivalve mollusks (American lobster, sea scallop, clams, blue crab, oyster, and mussel)—nearly all of which are fully exploited.[26]

As vast as the oceans of Earth are, there is a limit to the seafood bounty that humans can acquire. In 1950, global ocean catch was less than 20 million tons. In subsequent decades, we greatly increased our ability to obtain food

---

[24] Food and Agricultural Organization of the United Nations (FAO). 2005. Fishery country profile: The United States of America.

[25] FAO, 2005, ibid.

[26] FAO, 2005, ibid.

from the sea, and catch increased to over 80 million tons by 1990. However, things then slowed down considerably. Catch flattened out during the 1990s at around 90 million tons.[27] In the U.S., commercial landings of fish and shellfish for human food did not increase between 2003 and 2012.[28] During this period of stagnating ocean catch, the practice of fish farming—or aquaculture—has increased. By 2010, aquaculture produced around 60 million tons of fish and shellfish worldwide.[29] In 2005, the total production value of aquaculture in the United States was around $1 billion, with the top four producing states being Mississippi, Arkansas, Alabama, and Louisiana.[30]

International Trade of Food Products

If you pay attention to food labels while grocery shopping, you will have certainly noticed that food today is global. Plant products, animal products, and seafood enter the U.S. from all over the world each year. A traditional U.S. meal today may, in reality, contain food from far-flung places including Mexico, Chile, Italy, and Indonesia. So what comes from where?[31] Canada was the chief source of foreign grains and vegetable oils to the U.S. in 2011, with Mexico and multiple Asian nations also contributing to our supply of these diet staples. Mexico is by far our largest source of foreign vegetables and fruit. Canada and China were the next two top sources of vegetables, while Chile and China were numbers 2 and 3 for fruit. Massive amounts of meat and livestock move in both directions across U.S. borders during exchange with countries including Canada, Australia, New Zealand, Mexico, Nicaragua, Japan, South Korea, Hong Kong, Russia, and Angola.[32]

---

[27] FAO, 2005, ibid.; L.R. Brown, 2012 ibid.

[28] NOAA, 2012, ibid.

[29] FAO. 2010. *State of World Aquaculture 2010*. Available online: http://www.fao.org/fishery/aquaculture/en; L.R. Brown, 2012 ibid.

[30] United States Census of Aquaculture. http://www.agcensus.usda.gov/Publications/2002/Aquaculture/aquacen2005_01.pdf.

[31] Trade statistics and information for this chapter were obtained from the United States Department of Agriculture Economic Research Service.

[32] Over 2 billion pounds of beef and veal were imported to the U.S. in 2011. The top five sources of this beef and veal were Canada, Australia, New Zealand, Mexico, and Nicaragua. During the same year, the U.S. exported nearly 3 billion pounds of beef and veal to countries including Mexico, Japan, Canada, South Korea, and Hong Kong. For other meat products and livestock, the trade balance varies between net import and net export. The U.S. is a net importer of live cattle, live hogs, lamb, and mutton, while exporting enormous quantities in 2011 of pork (~5 billion pounds, top 3 destinations = Japan, Mexico, Canada) and chicken (over 7 billion pounds, top 3 destinations = Mexico, Russia, Angola). Source: USDA ERS.

Italy, New Zealand, and France export substantial amounts of dairy products to the U.S. And one can find Italian or French wine, as well as many other varieties, to have with their foreign cheese at most grocery stores. Substantial trade of seafood and freshwater fish across U.S. borders takes place with countries including China, Canada, Thailand, Chile, and Norway.[33]

The Family Meal

Now, let's come back to the family meal. Where did the food come from? For the average family, most, if not all of the food certainly did not come from close to home. Let's start with the bread. The wheat that went into making the bread—assuming it was domestic—most likely came from the fields of the northern Great Plains. If foreign, Canada was the likely source. But there is more than wheat in today's typical loaf of bread. Here are the ingredients in a loaf of Nature's Own 100 % whole wheat bread:

> Whole wheat flour, water, sugar, wheat gluten, cane refinery syrup, yeast, soybean oil, salt, raisin juice concentrate, cultured wheat flour, vinegar, dough conditioners (contains one or more of the following: sodium stearoyl lactylate, calcium stearoyl lactylate, monoglycerides and /or diglycerides, calcium peroxide, calcium iodate, datem, ethoxylated mono- and diglycerides, enzymes), calcium sulfate, soy lecithin, wheat starch. Topping: wheat cuts, wheat flakes, wheat bran, flaxseed.

The sugar and cane syrup could have come from sugar cane fields in Louisiana or Florida, or may have been imported from any number of tropical countries (El Salvador, for example). Soybean oil probably originated from the fields of the Midwest, but again could have been sourced from Canada. A lot of bread today has high fructose corn syrup that also comes mostly from the Midwest. Salt could have come from the mines in southern Louisiana or been imported from Canada, Chile, or Mexico. The raisin juice likely came from California, but the grapes could have been imported from Chile. The other ingredients were likely produced in a network of far-flung factories and producers in the U.S. and elsewhere. Finally, wherever you live, the loaf of bread was not likely baked nearby.

---

[33] The top 3 sources of total fish and shellfish imported into the U.S. during 2011 were China, Canada, and Thailand. With a few exceptions, the U.S. is largely a net importer of aquatic foods. For example, in 2011, 48 pounds of shrimp were imported for every pound of shrimp exported. The U.S. imported over 1 billion pounds of frozen or fresh shrimp worth more than $5 billion, primarily from several Asian nations. Foreign Atlantic salmon comes to the U.S. in greatest quantities from Chile (157 million pounds in 2011), Canada (120 million pounds in 2011), and Norway (42 million pounds in 2011). Source: USDA ERS.

Most of the vegetables of the family meal likely came from the Central Valley of California. But they could have also come from Mexico, Canada, China, or elsewhere. Very few vegetables are actually grown by the people who consume them (a tiny fraction of 1 %). The oil in the salad dressing could have come from the Midwest, the Great Plains, or Mexico. If the apple pie was bought ready-made, it likely has a list of ingredients longer than the loaf of bread. If domestic, the apples likely came from Washington State or upstate New York, if not Chile or New Zealand. The meal's potatoes most likely came from Idaho. If the rice is domestic, it could have been produced in Louisiana, California, or parts of the lower Mississippi Valley alluvial floodplain. If the rice was imported, the odds are that it came from Thailand.

Now let's consider the animal protein. Much of the domestic beef consumed in the U.S. comes from the Great Plains but there is production in many other parts of the country as well. If you are having pork, chances are it came from eastern North Carolina. The origin of any fish or seafood included in the meal will of course depend on the species. Catfish could very likely have come from Mississippi, Salmon from Alaska, shrimp from Louisiana, and lobster from New England. Of course, there is a good possibility that none actually originated in the U.S. There's a good chance that the shrimp actually came from Thailand, Indonesia, Ecuador, India, or Vietnam. And the tilapia fillets that have become more common in the seafood section of most U.S. grocery stores very likely originated in China.

The drinks also have far-flung sources. California is the most important wine producer in the U.S. but wine is also produced in the northwest, New York, Virginia, and a number of other states. However, the wine could have come from Europe, South America, Australia, or even China. The milk cows may have been in Wisconsin or New York. But milk is not necessarily processed where it is produced, and it is shipped far and wide. Soda is produced all over the country, although containing high levels of high fructose corn syrup produced in the Midwest.

So food is produced in many parts of the U.S. and much is imported and exported. A lot of it is highly processed, and it is shipped all over the country and the world. No matter where a consumer lives, he or she would be hard pressed to supply the majority of the ingredients of their diet from local sources. A typical U.S. meal today may in reality not be very "American" after all (except in the sense that foreign suppliers conform to our wants, i.e., their export economy becomes "Americanized").

What Does It Take to Produce Our Food?

Vast Expanses of Suitable Land

As indicated previously, it takes hundreds of millions of acres of farmland with good quality soil and adequate water to produce the food we eat. Cropland now occupies 17 % of the total land area in the U.S., but little additional land is available or even suitable for future agricultural expansion.[34] Roughly half of the original prime agricultural soil in Iowa and the Palouse region in the northwest has been eroded by previous agricultural activity, and is no longer available.[35] The growing global population will require either more arable land or an increase in yields on existing land. Don't believe that the latter happens only because of more and better technology. It happens largely because of higher inputs of fossil fuel powered fertilization and mechanization. Essentially, the intensive agriculture that we practice trades fossil fuel energy for food energy that we can ingest and digest. Global population increased from just over three billion to nearly seven billion between 1961 and 2010. During this time period, available arable land and permanent cropland globally decreased from 0.44 to 0.22 hectares per capita, while available permanent meadows and pastures decreased from approximately 1.00 to below 0.49 hectares per capita (note that 1 hectare = 2.47 acres).[36] However, the diverse and protein-intensive diet commonly enjoyed in the U.S. requires, on average, about 0.5 hectares per person of land in production, and up to nearly 0.9 hectares for the most meat-rich diets.[37] It is simply impossible for everyone in the world to eat like U.S. residents, who consume the most meat, using existing croplands and pastures, even if high yields were obtained on all land. Increasing world demand for meat will continue to put positive pressure on agricultural land expansion,[38]

---

[34] USDA. 2004. *Agricultural Statistics 2004*. U.S. Department of Agriculture, U.S. Government Printing Office, Washington, DC.

[35] Pimentel, D. 2006. Soil erosion: a food and environmental threat. *Environment, Development and Sustainability*, 8, 119–137.

[36] FAOSTAT. *Food and Agricultural Organization of the United Nations*. http://faostat.fao.org; Pimentel, D., and M. Pimentel. 2008. Corn and cellulosic ethanol cause major problems. *Energies*, 1, 35–37.

[37] Peters, C.J., J.L. Wilkens, and G.W. Fick. 2007. Testing a complete-diet model for estimating the land resource requirements of food consumption and agricultural carrying capacity: The New York State example. *Renewable Agriculture and Food Systems*, 22, 145–153.

[38] Alexandratos, N., and J. Bruinsma. 2012. *World agriculture towards 2030/2050*. The 2012 revision. ESA Working Paper No. 12–03. Agricultural Economics Division, Food and Agriculture Organization of the United Nations. http://www.fao.org/docrep/016/ap106e/ap106e.pdf.

which has recently resulted in the loss of vast expanses of tropical forests.[39] One of the most straightforward ways to reduce your environmental impact is to eat less meat. Researchers at the Environmental Sciences and Forestry School at the State University of New York at Syracuse (including one of us—C.H.) found that reducing meat intake to a few small portions a week would decrease the land area needed to feed the population in Onondaga County, New York by over 70 %.[40] Other researchers similarly report that a lacto-vegetarian diet (no meat or eggs) can require as little as 0.2 hectares of land per person per year vs. the 0.5 ha used for the average American diet.[41]

Energy

Agriculture requires much more than land. It takes huge amounts of energy, material inputs, water, machines, and factories to produce food today (Table 8.2). But what it doesn't take, at least not down on the farm, is large numbers of people (Fig. 8.6).

Modern farms are not your grandparents' or great grandparents' farms. In the US, tractors powered by fossil fuels were still relatively rare as late as 1940, when only 8 % of farms within the Mid-Atlantic region reported solely using tractors.[42] By 1954, 56 % of farms in the Mid-Atlantic region relied exclusively on tractors. Use of animal power declined as tractor use became increasingly standard practice (Fig. 8.1c). The capacity to allocate substantial amounts of energy to agriculture via fossil fuels in the mid-twentieth century supercharged food production, increasing food supply, reducing hunger, and improving nutrition for many (but certainly not all).[43] An industrial food system emerged, characterized by linkages among

---

[39] Gibbs et al. 2010. Tropical forests were the primary sources of new agricultural lands in the 1980s and 1990s. PNAS, 107, 16732–16737.

[40] Balogh, S.B., C.A.S. Hall, A.M. Guzman, D.E. Balcarce, and A. Hamilton. 2012. The potential of Onondada County to feed its own population and that of Syracuse, New York: Past, present, and future. In: Pimentel, D. (ed). *Global economic and environmental aspects of biofuels*. CRC Press.

[41] Peters et al. 2007, ibid.

[42] United States of America Census 1954; Balogh et al. 2012, ibid.

[43] Tilman, D., K.G. Cassman, P.A. Matson, R. Naylor, and S. Polasky. 2002. Agricultural sustainability and intensive production practices. *Nature*, 418, 671–677.

Table 8.2 Production expenses on U.S. farms in 2007[a]

Farm input	Total U.S. expenses in 2007 (million US$)
Total farm production expenses	$241,114
Fertilizer, lime, and soil conditioners	$18,107
Chemicals	$10,075
Seeds, plants, vines, and trees	$11,741
Livestock and poultry purchased or leased	$38,004
Feed	$49,095
Gasoline, fuels, and oils	$12,912
Utilities	$5918
Supplies, repairs, and maintenance	$15,897
Hired farm labor	$21,878
Contract labor	$4514
Customwork and custom hauling	$4091
Cash rent for land, buildings and grazing fees	$13,275
Rent and lease expenses for machinery, equipment, and farm share of vehicles	$1385
Interest expense	$10,881
Property taxes paid	$6223
All other production expenses	$17,119

[a]Data is from the 2007 United States Census of Agriculture

distant locations,[44] immense energy inputs,[45] and dramatic influence on the cycling of important chemical elements on Earth.[46]

The availability of cheap energy in the U.S. allowed the vast majority of the population (>97 %) to take part in non-agricultural economic activities by the end of the twentieth century (Fig. 8.7).[47] This transition has also been underway in developing nations in recent decades. Almost 70 % of China's population was employed in agriculture in 1980. By 2010, this percentage had declined to less than 40 % (Fig. 8.7). The critical role in a society played

[44] Erb, K.H., F. Krausmann, W. Lucht, and H. Haberl. 2009. Embodied HANPP: Mapping the spatial disconnect between global biomass production and consumption. *Ecological Economics*, 69, 328–334.

[45] Steinhart, J.S., and C.E. Steinhart. 1974. Energy use in the US food system. *Science*, 184, 307–316.

[46] Vitousek, P.M., J.D. Aber, R.W. Howarth, G.E. Likens, P.A. Matson, D.W. Schindler, W.H. Schlesinger, and D.G. Tilman. 1997. Human alteration of the global nitrogen cycle: sources and consequences. *Ecological Applications*, 7, 737–750.; Smil, V. 2000. Phosphorus in the environment: natural flows and human interferences. *Annual Review of Energy and the Environment*, 25, 53–88.; Seto, K.C., B. Guneralp, and L.R. Hutyra. 2012. Global forecasts of urban expansion to 2030 and direct impacts on biodiversity and carbon pools. *Proceedings of the National Academy of Sciences of the USA*, 109, 16083–16088.

[47] Note that, according to the USDA ERS, the total food system (farms, restaurants, grocery stores, transport systems, etc.) accounted for 9.2 % of U.S. employment in 2012.

Fig. 8.6 Farming on the American plains. (**a**) 33 horsepower (HP) controlled by five workers. (**b**) 200 HP combine controlled by one worker (Images courtesy of Mario Giampetro)

by the injection of fossil energy and mechanization into food production is captured in currency notes from various nations that feature images of tractors (Fig. 8.8).

The American food supply today is largely dependent on non-renewable energy sources. The energy efficiency of U.S. agriculture declined dramatically between 1910 and 1970 as inefficient use of cheap fuels increased.[48] Despite increased energy efficiency since 1970, the total food-related share of the national U.S. energy budget equaled approximately 16 % in 2007.[49]

---

[48] Cleveland, C.J. 1995. The direct and indirect use of fossil fuels and electricity in USA agriculture, 1910–1990. *Agriculture, Ecosystems and Environment*, 55, 111–121; Steinhart, J.S., and C.E. Steinhart. 1974. Energy use in the US food system. *Science*, 184, 307–316

[49] Canning, P., A. Charles, S. Huang, K.R. Polenske, and A. Waters. 2010. *Energy use in the U.S. food system*. Economic Research Report no. 94. United States Department of Agriculture Economic Research Service; Hamilton, A., S. Balogh, A. Maxwell, and C. Hall. 2013. Efficiency of edible agriculture in Canada and the U.S. over the past three to four decades. *Energies* 6, 1764–1793.

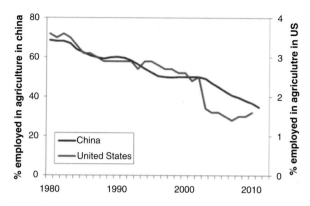

Fig. 8.7 Declines in the agricultural labor force as a percentage of total employment in the United States (*right y-axis*) and China (*left y-axis*) (The World Bank. Employment in agriculture (% of total employment). Available online: http://data.worldbank.org/indicator/SL.AGR.EMPL.ZS). The United States' fossil fuel-intensive industrial food system now operates with <2 % of the population working in agriculture. The more recent rural to urban transition and industrialization of agriculture in China has resulted in dramatic declines in the proportion of the population working in the agricultural sector (Cao, S., G. Xie, and L. Zhen. 2010. Total embodied energy requirements and its decomposition in China's agricultural sector. Ecological Economics 69: 1396–1404.)

The total U.S. food system (i.e., farms, transportation, processing, storage, preparation, etc.) consumes more than 7 units of energy to deliver 1 unit of edible food energy (Fig. 8.9).[50]

Fertilizers, Other Chemicals, and Irrigated Water

In 2007, growers irrigated[51] 56.6 million acres of farmland, while 266 million acres were treated with commercial fertilizer, lime, and soil conditioners, 90.9 million acres were treated with chemicals to control insects, and 226 million acres were treated with chemicals to control weeds, grass, or brush.[52] For comparison, organic production systems, which eliminate chemical

---

[50] Heller, M., and G. Keoleian. 2000. *Life cycle-based sustainability indicators for assessment of the U.S. food system*. The University of Michigan Center for Sustainable Systems, pub. No. CSS00-04.

[51] USDA, 2007, Census of Agriculture, ibid.

[52] USDA, 2007, ibid.

Fig. 8.8 Currency notes from India, Cambodia, and Vietnam featuring tractors. Images courtesy of Robert Lane

inputs, decrease soil erosion, conserve water, and improve soil organic matter and biodiversity, only covered 2.6 million acres.[53] The industrial food system in the US and elsewhere has relied on "feeding fossil fuels to soil" in the form of fertilizers containing primarily high concentrations of the nutrients nitrogen, phosphorus, and potassium.[54] Annual applications of nitrogen and phosphorus fertilizers to US farmland increased 254 % and

---

[53] Pimentel, D., P. Hepperly, J. Hanson, D. Douds, and R. Seidel. 2005. Environmental, energetic, and economic comparisons of organic and conventional farming systems. *BioScience*, 55, 573–582.

[54] Ramírez, C.A., and E. Worrell. 2006. Feeding fossil fuels to the soil: an analysis of energy embedded and technological learning in the fertilizer industry. *Resources, Conservation and Recycling*, 46, 75–93.

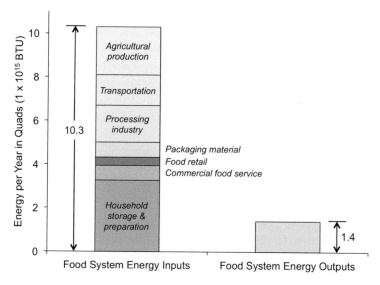

Fig. 8.9 Energy flow in the US food system, as estimated by Heller and Keoleian (2000) and CSS (2011). Similar findings were reported by Hamilton et al. (2013). Figure redrawn from CSS (2011) (CSS (Center for Sustainable Systems). 2011. U.S. Food System Factsheet, pub. No. CSS01-06. University of Michigan. Hamilton, A., S. Balogh, A. Maxwell, and C. Hall. 2013. Efficiency of edible agriculture in Canada and the U.S. over the past three to four decades. Energies 6: 1764–1793.)

95 %, respectively, between 1961 and 1980. While fertilizer consumption has plateaued in many developed nations, including the US, it is still on the rise in countries such as China, Brazil, and India.[55] Application of lime is also commonly required to adjust soil pH to a level suitable for crop production. Collectively, U.S. farmers spent over $18 billion on fertilizers, lime, and other soil conditioners in 2007. Chemicals, including pesticides, required an additional $10 billion in investment.[56]

Crops require large amounts of water for photosynthesis, growth, and reproduction.[57] Various crops use water at rates between 265 and 530 gallons of water per kg dry matter of crops produced. The water required for corn per hectare is about 2.6 million gallons. Even with an annual rainfall of 1000 mm (about 39 in.) in the U.S. Corn Belt, corn frequently suffers from insufficient water during the critical summer growing period. Wheat, which produces

---

[55] FAOSTAT. Food and Agricultural Organization of the United Nations. http://faostat. fao.org.

[56] USDA, 2007, ibid.

[57] Unless otherwise noted, information in this section is from: Pimentel, D., B. Berger, D. Filiberto, M. Newton, B. Wolfe, et al. 2004. Water resources: agricultural and environmental issues. *BioScience*, 54, 909–918.

less plant biomass than corn, requires only about 634,000 gallons per hectare for a yield near 3 metric tons. On average, soybeans require about 1.6 million gallons per hectare for a yield of 3 metric tons. Under semiarid conditions, yields of non-irrigated crops, such as corn, are low (1.0–2.5 metric tons per hectare) even when ample amounts of fertilizer are applied. Irrigated land in the U.S. covered about 0.18 acres per capita in 2008.[58] Most of the irrigated farmland lies in the west. Food production in the dry climate of the important Central Valley of California requires an expansive and intensive water distribution system, currently threatened by drought.[59] In the Great Plains, the Ogallala aquifer is a critical source of irrigation water that has been subjected to a severe imbalance between recharge and output.[60] Irrigation for agriculture is the primary use of consumed water globally, and pumping water from aquifers at rates greater than the rate of recharge (as with the Ogallala aquifer) is common, leading to falling water tables. Approximately 31 % of water withdrawn in the U.S. in 2005 was used for irrigation, with livestock and aquaculture operations accounting for an additional 3 %.[61] Transportation of irrigated water across long distances also consumes considerable energy (e.g., 10 % of energy expended in crop and livestock production annually in the U.S.).[62] Reducing irrigation dependence in the U.S. would save significant amounts of energy, but would probably require that crop production shift from the dry and arid western and southern regions to the more favorable climate of the eastern U.S.

The Capacity of Different U.S. Cities and Regions to Produce Food

By reducing the number of people required for food production and other primary economic activities (forestry, fishing, mining), cheap fuels were a key factor in the growth of large urban areas. The urban transition is gener-

[58] USDA, 2007, ibid.

[59] Joyce, B.A., V.K. Mehta, D.R. Purkey, L.L. Dale, and M. Hanemann. 2011. Modifying agricultural water management to adapt to climate change in California's central valley. *Climatic Change* 109 (Suppl 1): S299–S316.

[60] William Ashworth. 2007. *Ogallala blue: Water and life on the Great Plains*. Countryman Press, pp. 330; Sophocleous, M. 2012. Conserving and extending the useful life of the largest aquifer in North America: the future of the High Plains/Ogallala aquifer. *Ground Water*, 50, 831–839.

[61] Barber, N.L. 2009. *Summary of estimated water use in the United States in 2005*. U.S. Geological Survey Fact Sheet 2009–3098, 2 pp.

[62] Pimentel, D., et al. 1997. Water resources: agriculture, the environment, and society. *BioScience*, 47, 97–106.

ally more advanced in developed nations. For example, over 80 % of the US population now resides in urban areas.

A growing scientific and popular literature extols the virtues of cities in the context of sustainability.[63] However, urban populations in the U.S. and elsewhere are nearly entirely dependent on the industrial food system and could feasibly face price shocks and supply interruptions within the next few decades due to global megatrends. The recent 2008 Food Crisis provides insight into the causes and consequences of rapid increases in food prices, including the disproportionate impact on the world's urban poor.[64] Food insecurity is not solely a problem in developing countries. Over 14 % of U.S. residents are currently food insecure, and one in four relies on some form of government assistance to obtain food.[65] With these issues in mind, is it possible for urban areas to enhance their self-sufficiency in food production, and thus their resiliency to future industrial food crises?

It is exciting for those of us concerned about the future of food to imagine cities teeming with edible vegetation, chicken coops integrated into urban residential lots, and efficient urban recycling of nutrients and water. This vision becomes more seductive if we consider the social benefits of food localization, such as connection to place and economic opportunity.[66] Designers and urban planners have put forth ideas for breaking down the boundary between the "agrarian" and the "urban," and continue to produce visions of an "agrarian urbanism."[67] Is a truly agrarian urbanism feasible? If so, can U.S. cities actually feed themselves sustainably from local sources?

Urban agriculture is not new; there are examples from several Asian cities, along with Moscow and the African cities of Dar es Salaam, Kinshasa,

---

[63] Edward Glaeser. 2011. *Triumph of the city: how our greatest invention makes us richer, smarter, greener, healthier, and happier*. Penguin, New York, NY.; D. Despommier. 2011. *The vertical farm: feeding the world in the 21st century*. Picador, New York, NY.

[64] Heady, D., and S. Fan. 2008. Anatomy of a crisis: the causes and consequences of surging food prices. *Agricultural Economics*, 39(Suppl. s1), 375–391; World Bank (2008).

[65] United States Department of Agriculture. 2012. *Food security in the U.S.* USDA Economic Research Service; Oliveira, V. 2013. The food assistance landscape: FY 2012 annual report. United States Department of Agriculture, Economic Research Service, Economic Information Bulletin No. 109.

[66] Ackerman-Leist, P. 2013. *Rebuilding the foodshed: How to create local, sustainable, and secure food systems*. Post Carbon Institute, Chelsea Green Publishing, VT, USA.

[67] C. Waldheim. 2010. Notes towards a history of agrarian urbanism. In: M. White and M. Przybylski (eds.), *Bracket 1: [on farming]*. Actar, Barcelona.

Kampala, and Maputo.[68] Following the economic crisis in Cuba during the late 1990s, critical food shortages along with decreased availability of fossil fuels and industrial agrochemicals led to a dramatic rise in food cultivation within Havana that was characterized by agroecological principles (no synthetic chemical pesticides or fertilizers), recycling, use of local resources, and human and draft animal labor to offset the lack of fossil fuels. These systems reportedly produced substantial amounts of fruits and vegetables.[69] However, first-hand observations by one of us (D.P.) suggested that the overall impact of Cuba's urban agriculture on satisfying caloric demand was limited. Historically in Europe, allotment gardens served as important safeguards against urban food supply interruptions.[70] During the "war garden" or "victory garden" movement of the World War II period in the US, approximately 20 million gardens were planted largely in non-rural areas to boost food supplies and save resources (Fig. 8.10). There is no reason to believe that a contemporary, innovative agrarian urbanism is impossible in the US, and urban agriculture can provide benefits ranging from income generation to nutrition improvement.[71]

Benefits created by the proximity of food production and consumption are typically at the center of arguments in favor of urban agriculture. Production–consumption proximity enables short market chains leading to lower prices in some cases,[72] as well as creating opportunities for the recycling of food waste, nutrients, and water. However, maintaining large-scale centralized recycling infrastructure in high-density cities (e.g., a central wastewater treatment plant that harvests nutrients in municipal wastewater) would likely be costly and energy-intensive.[73] While recycling greywater

---

[68] Hamilton, A.J., K. Burry, H.F. Mok, F. Barker, J.R. Grove, and V.G. Williamson. 2014. Give peas a chance? Urban agriculture in developing countries. A review. *Agronomy for Sustainable Development*, 34, 45–73; Mok, H.F., V.G. Williamson, J.R. Grove, K. Burry, S.F. Barker, A.J. Hamilton. 2014. Strawberry fields forever? Urban agriculture in developed countries: a review. *Agronomy for Sustainable Development*, 34, 21–43.; Smit, J., A. Ratta, and J. Nasr. 1996. *Urban agriculture. Food, jobs, and sustainable cities*. United Nations Development Programme, Publication series for Habitat II, Volume 1. New York, USA.

[69] Palma, I.P., J.N. Toral, M.R.P. Vasquez, N.F. Fuentes, F.G. Hernandez. 2014. Historical changes in the process of agricultural development in Cuba. *Journal of Cleaner Production*. doi: 10.1016/j.jclepro.2013.11.078.

[70] Barthel, S., C. Folke, and J. Colding. 2010. Social-ecological memory in urban gardens – retaining capacity for management of ecosystem services. *Global Environmental Change*, 20, 255–265.

[71] Smit, J., A. Ratta, and J. Nasr. 1996. *Urban agriculture. Food, jobs, and sustainable cities*. United Nations Development Programme, Publication series for Habitat II, Volume 1. New York, USA.; Ackerman-Leist, P. 2013. *Rebuilding the foodshed: how to create local, sustainable, and secure food systems*. White River Junction, VT: Chelsea Green Publishing.

[72] De Bon, H., L. Parrot, and P. Moustier. 2010. Sustainable urban agriculture in developing countries. A review. *Agronomy for Sustainable Development* 30: 21–32.

[73] Cordell, D., and S. White. 2013. Sustainable phosphorus measures: strategies and technologies for achieving phosphorus security. *Agronomy*, 3, 86–116.

Secretary Plowing Boston Common 4/11/44 - victory Garden Program

Fig. 8.10 Plowing a "victory garden" in Boston Common in 1944 during World War II (Franklin D. Roosevelt Library [Public Domain] Photographs, U.S. National Archives and Records Administration)

for irrigation is reportedly not a significant public health risk, reuse of wastewater and biosolids requires monitoring and regulation of metals, pathogens, vectors, and other contaminants to ensure public safety.[74]

Despite the advantages that production–consumption proximity creates for urban food production, there are several significant challenges limiting both self-sufficiency and sustainability. For megacities (e.g., Los Angeles and New York), it is more accurate to define many of these challenges as outright barriers. Insufficient available arable land is a significant hurdle to feeding large urban centers from within their boundaries. Assuming an average diet land requirement of 0.5 hectare capita^{-1} (about 1.25 acres per person), urban agriculture could satisfy only a small proportion of food demand from populations in the New York–Newark and Syracuse urban areas (5 % and 12 % of food demand, respectively) in extremely optimistic scenarios where 50 % of urban land area is converted to agriculture and yields match those

---

[74] Wortman, S.E., and S.T. Lovell. 2013. Environmental challenges threatening the growth of urban agriculture in the United States. *Journal of Environmental Quality*, 42, 1283–1294.

currently achieved in rural settings.[75] Substantial portions of U.S. urban areas are paved for transportation uses (roads, sidewalks, parking lots) and thereby not available for farming.[76] Widespread adoption of meat-free diets requiring 0.2 ha capita^{-1} could increase the capacity of urban agriculture to support resident calorie requirements up to 12 % and 31 % of demand in New York–Newark and Syracuse, respectively. Changing dietary norms to vegetarianism, however, is a major challenge; the 2012 Gallup "Consumption Habits Survey" indicated that only 5 % of U.S. residents are vegetarian, the same percentage as in 1999.

What about local and regional agriculture occurring outside of cities? By local, we mean agriculture occurring outside of the city limits, but within a relatively short distance (e.g., 100 miles). Recent research has examined "foodsheds"—the geographic areas from which populations derive their food supply.[77] These studies reveal six key insights.

First, meeting the food demand of megacities with local production is not feasible. For example, allocating 100 % of New York State's entire current food production to feeding residents of New York City would satisfy only 55 % of the city's demand.[78]

Second, high regional population density, along with competition between cities for foodshed resources, limits capacity to meet demand with local or

---

[75] Loss-adjusted per capita daily availability of meat (red meat + poultry + eggs) in the U.S. in 2012 was 180 g, with 90 % attributed to red meat and poultry (USDA ERS 2014). This quantity of meat consumption was compared to land requirement results for New York State presented by Peters et al. (2007) (see their Figure 1). The diet documented for Peters et al. (2007) characterized by the following was used: low fat (52 g fat capita^{-1} d^{-1}), 2,308 kcal capita^{-1} d^{-1}, 190 g meat and eggs capita–1 d–1. This diet requires approximately 0.5 ha capita^{-1} y^{-1}, again accounting for food waste losses during processing, distribution and after purchase by the consumer, as well as inedible portions of harvested crop weight. See Peters et al. (2007) for further details. Previous authors have reported an identical land requirement estimate for a meat-based U.S. diet (Pimentel and Pimentel 2003). Urban area populations and land area were estimated using the 2010 US Census. References: Peters CJ, Wilkins JL, and Fick GW. 2007. Testing a complete-diet model for estimating the land resource requirements of food consumption and agricultural carrying capacity: *The New York State example*. Renew. Agr. Food Syst. 22, 145–153. Pimentel D and Pimentel M. 2003. Sustainability of meat-based and plant-based diets and the environment. *Am. J. Clin. Nutr.*, 78(suppl), 660S-663S. USDA ERS (United States Department of Agriculture Economic Research Service). 2014. Food availability (per capita) data system. http://www.ers.usda.gov/data-products/. US Census (United States Census Bureau). 2010. 2010 Census urban and rural classification and urban area criteria. https://www.census.gov/geo/reference/ua/urban–rural-2010.html.

[76] Rose, S.L., H. Akbari, and H. Taha. 2003. *Characterizing the fabric of the urban environment: a case study of Metropolitan Houston, Texas*. Lawrence Berkeley National Laboratory. Report no. LBNL-51448.

[77] Peters, C.J., N.L. Bills, A.J. Lembo, J.L. Wilkins, and G.W. Fick. 2009. Mapping potential foodsheds in New York State: a spatial model for evaluating the capacity to localize food production. *Renewable Agriculture and Food Systems*, 24, 72–84.

[78] Peters et al., 2009, ibid.

regional production. Current production in New England can support only 16 % and 36 % of the region's plant-based and animal-based food demands, respectively.[79] Philadelphia could feasibly be fed by agriculture occurring on lands within a 100-mile radius if the demand of other cities were ignored. However, this land area includes 10 % of the U.S. population, and the total population's food demand is far greater than the local supply.[80]

Third, diet exerts a tremendous influence on foodshed requirements. Low-meat or vegetarian diets can reduce the arable land area requirement to <30–40 % of what would be required for a heavy meat diet.[81]

Fourth, current yields can meet food demand for small to mid-size cities using local or regional agriculture in many cases, but not all. Zumkehr and Campbell (2015) report that local foodshed potential could satisfy the majority of US food demand, including that of most small to mid-size cities (with the assumption that existing local crop production practices could be changed to meet the needs of a well-rounded diet).[82] Sufficient capacity of local or regional production to meet city or regional demand has been reported in additional studies for small cities in upstate New York, southeastern Minnesota, and the Midwestern U.S. region.[83] However, even the relatively productive Willamette Valley in Oregon currently fails to produce adequate food for local consumption demand. Current agriculture in the Willamette Valley can meet only 67 % of grains required annually by the resident population, 10 % of vegetable needs, 24 % of fruit, 59 % of dairy, 58 % of meat and bean needs, and 0 % of dietary oil requirements.[84]

Fifth, substantial challenges exist related to local environmental and economic factors, such as crop failure due to interannual variability in cli-

---

[79] Griffin, T., Z. Conrad, C. Peters, R. Ridberg, and E.P. Tyler. 2014. Regional self-reliance of the Northeast food system. Renewable Agriculture and Food Systems. doi:10.1017/S1742170514000027.

[80] Kremer, P., and Y. Schreuder. 2012. The feasibility of regional food systems in metropolitan areas: an investigation of Philadelphia's foodshed. *Journal of Agriculture, Food Systems, and Community Development*. doi: 10.5304/jafscd.2012.022.005.

[81] Balogh et al., 2012, ibid.

[82] Zumkehr, A., and J.E. Campbell. 2015. The potential for local croplands to meet US food demand. *Frontiers in Ecology and the Environment*, 13, 244–248.

[83] Peters et al., 2009, ibid.; Hu, G., L. Wang, S. Arendt, and R. Boeckenstedt. 2011. An optimization approach to assessing the self-sustainability potential of food demand in the Midwestern United States. Journal of Agriculture, Food Systems, and Community Development, 2, 195–207.; Galzki, J.C., D.J. Mulla, and C.J. Peters. 2014. Mapping the potential of local food capacity in Southeastern Minnesota. Renewable Agriculture and Food Systems. doi:10.1017/S1742170514000039.

[84] Giombolini, K.J., K.J. Chambers, S.A. Schlegel, and J.B. Dunne. 2011. Testing the local reality: does the Willamette Valley region produce enough to meet the needs of the local population? A comparison of agriculture production and recommended dietary requirements. *Agriculture and Human Values*, 28, 247–262.

mate, the common absence of local sources of dietary oils and grains that make up a significant portion of food demand, competition with non-edible crops or export crops for agricultural land, and mismatch between local supply and demand.[85]

Finally and significantly, existing foodshed analyses largely miss considerations of energy and material input requirements needed to sustain local food systems.

Installing urban agricultural systems and associated infrastructure does not necessarily result in sustainable production. Local food systems are often as reliant on nonrenewable energy resources for production, transportation, and consumption as long-distance systems, and can be far-from-sustainable.[86] Trucking food relatively long distances may actually save energy, as then food can be produced in optimal climates for that crop. Additionally large semi trucks are vastly more efficient in ton-miles per gallon than are pickup trucks used for local food production.[87] Thus while "eat local" may make sense in many ways it is not clear that it is always the most efficient way to turn energy into food. As usual what sounds "green" usually has little good systems research behind it. To date, energy associated with food transportation has received considerable attention[88] and is often cited in promotions of urban agriculture. Transportation, however, only accounts for 11–14 % of energy invested in the U.S. food system (Fig. 8.9), so solely accounting for "food miles" is insufficient for quantifying sustainability.[89] Energy and material consumption associated with food production, processing, storage and waste handling must be factored in.[90] Accessing agricultural inputs can be a significant obstacle to urban food

[85] Giombolini et al., 2011, ibid.; Desjardins, E., R. MacRae, and T. Schumilas. 2010. Linking future population food requirements for health with local production in Waterloo Region, Canada. *Agriculture and Human Values*, 27, 129–140.

[86] Beck, T.B., M.F. Quigley, and J.F. Martin. 2001. Emergy evaluation of food production in urban residential landscapes. Urban Ecosystems 5: 187–207; Mariola, M.J. 2008. The local industrial complex? Questioning the link between local foods and energy use. *Agriculture and Human Values*, 25, 193–196.

[87] Balogh et al., 2012, ibid.

[88] For an example of a study focused on food miles, see: Weber, C.L., and H.S. Matthews. 2008. Food-miles and the relative climate impacts of food choices in the United States. *Environmental Science & Technology*, 42, 3508–3513.

[89] Edwards-Jones, G., L.M. Canals, N. Hounsome, M. Truninger, G. Koerber, B. Hounsome, P. Cross, E.H. York, A. Hospido, K. Plassmann, I.M. Harris, R.T. Edwards, G.A.S. Day, A.D. Tomos, S.J. Cowell, and D.L. Jones. 2008. Testing the assertion that 'local food is best': the challenges of an evidence-based approach. *Trends in Food Science & Technology*, 19, 265–274.

[90] Heller, M.C., G.A. Keoleian, and W.C. Willett. 2013. Toward a life cycle-based, diet-level framework for food environmental impact and nutritional quality assessment: a critical review. *Environmental Science and Technology*, 47, 12632–12647.

production.[91] And producing a lot of food in a city without large amounts of fossil energy and chemicals would require that a significant number of city dwellers become farmers! This is something that hardly exists at all— even potentially—in most U.S. cities.

We define "sustainable urban agriculture" as food production occurring within urban area boundaries that does not require excessive throughput of materials (e.g., irrigated water, chemical fertilizer) and energy and can therefore be expected to be more viable in a resource-constrained future. Whether or not urban agriculture can be "substantial" depends on the size of the local population relative to the land available and suitable for production. Of the urban areas considered in this book, Cedar Rapids and Flint have the greatest *relative* potential for substantial, sustainable urban agriculture due to a combination of lower population densities, suitable climate, and fertile soils. Large, densely populated urban areas (e.g., New York–Newark, Los Angeles–Long Beach–Anaheim), and arid urban areas such as Las Vegas and Phoenix, have the least potential. Location-specific constraints will affect efforts in cities like Asheville (poor soil fertility) and Amarillo (low water availability). Even in relatively high potential areas, urban agriculture should not be expected to meet total resident caloric demand, especially with current high-meat diets (Tables 8.3 and 8.4).

Pollution effects of human activity on urban soils must also be considered.[92] Rooftop runoff can contaminate soils with heavy metals (Pb, Al, Zn, Cu) depending on roofing material, and may lead to plant toxicity over time. Lead (Pb), arsenic, mercury, cadmium, and polycyclic aromatic hydrocarbons (PAHs), often derived from paints, are all soil contaminants of concern in urban soils. Of these, Pb is most commonly observed at elevated concentrations in the U.S. as documented, for example, in the cities of Chicago, Syracuse, and Boston.[93] PAHs are also of concern to urban farmers, particularly because many are classified as carcinogens.[94] Urban soils

---

[91] De Bon, H., L. Parrot, and P. Moustier. 2010. Sustainable urban agriculture in developing countries. A review. *Agronomy for Sustainable Development*, 30, 21–32.

[92] ormation contained in this paragraph is reviewed in: Wortman, S.E., and S.T. Lovell. 2013. Environmental challenges threatening the growth of urban agriculture in the United States. *Journal of Environmental Quality*, 42, 1283–1294.

[93] Clark, H.F., D.J. Brabander, and R.M. Erdil. 2006. Sources, sinks, and exposure pathways of lead in urban garden soil. *Journal of Environmental Quality*, 35, 2066–2074.; Kay, R.T., T.L. Arnold, W.F. Cannon, and D. Graham. 2008. Concentrations of polycyclic aromatic hydrocarbons and inorganic constituents in ambient surface soils, Chicago, Illinois: 2001–2002. *Soil and Sediment Contamination*, 17, 221–236.; Johnson, D.L., and J.K. Bretsch. 2002. Soil lead and children's blood levels in Syracuse, NY, USA. *Environmental Geochemistry and Health*, 24, 375–385.

[94] Srogi, K. 2007. Monitoring of environmental exposure to polycyclic aromatic hydrocarbons: a review. *Environmental Chemistry Letters*, 5, 169–195.

Table 8.3 Climate and soils information for 11 urban areas discussed in this book

Urban area	Climate	Annual precipitation (inches, NOAA 1981–2010)	Average soil fertility index[a] (0 = lowest fertility, 16 = highest fertility)
New York–Newark, NY-NJ-CT	Humid continental (warm summer)	44	6
Los Angeles–Long Beach–Anaheim, CA	Mediterranean	13	10
Houston, TX	Humid subtropical	50	11
Portland, OR-WA	Marine westcoast	36	11
Orlando, FL	Humid subtropical	51	4
Las Vegas–Henderson, NV	Midlatitude desert	4	6
Baton Rouge, LA	Humid subtropical	61	11
Flint, MI	Humid continental (warm summer)	32	11
Asheville, NC	Humid subtropical	46	5
Cedar Rapids, IA	Humid continental (warm summer)	35	12
Amarillo, TX	Semiarid steppe	20	12

Precipitation values are averages from local airport monitoring sites recorded by NOAA. Generally, agricultural productivity is limited when annual precipitation is less than 20 in.
[a]Index developed and mapped by: Schaetzl, R.J., F.J. Krist, and B.A. Miller. 2012. A taxonomically based ordinal estimate of soil productivity for landscape-scale analysis. Soil Science 177: 288–299. Mean values shown here were calculated using Zonal Statistics for US Census 2010 urban areas in ArcMap 10.2.

have been found to have substantially higher concentrations of PAHs than rural soils in Switzerland.[95] Both Pb and PAHs can come from atmospheric deposition in cities, and the primary risk of exposure to humans appears to be through direct ingestion or inhalation of contaminated soils and aerosols, not plant uptake and consumption.[96] Soil remediation options such as soil removal, washing, or capping are cost-prohibitive and energy-intensive.

"Vertical farming" has been proposed as a way to circumvent constraints in cities imposed by land availability, climate, and soil quality. This idea has been championed by Dickson Despommier of Columbia University and included as part of a 2011 feature article in *Scientific American* on sustainable cities.[97] While there has yet to be a full-scale sustainability assessment of

[95] Bucheli, T.D., F. Blum, A. Desaules, and O. Gustafsson. 2004. Polycyclic aromatic hydrocarbons, black carbon, and molecular markers in soils of Switzerland. *Chemosphere*, 56, 1061–1076.

[96] Wortman, S.E., and S.T. Lovell. 2013. Environmental challenges threatening the growth of urban agriculture in the United States. *Journal of Environmental Quality*, 42, 1283–1294.

[97] Despommier, D. 2011. *The vertical farm: feeding the world in the 21st century*. Picador; Fischetti, M. 2011. "*The efficient city.*" Scientific American, 305, 74–75.

Table 8.4 Characteristics and relative potential for substantial, sustainable urban agriculture for 11 urban areas in the United States

Urban area	Population density (persons per km²)	Relative potential for substantial, sustainable urban agriculture	Key limitation
Cedar Rapids, IA	822	High	
Baton Rouge, LA	626	Moderate	Water-logging
Flint, MI	583	High	
Asheville, NC	409	Moderate	Poor soil fertility
Houston, TX	1150	Low-moderate	Insufficient land
Portland, OR-WA	1362	Low-moderate	Insufficient land
Orlando, FL	976	Low	Poor soil fertility, Water-logging
Amarillo, TX	935	Low-moderate	Low water availability
New York–Newark, NY-NJ-CT	2054	Very low	Insufficient land, poor soil fertility
Los Angeles–Long Beach–Anaheim, CA	2702	Very low	Insufficient land, low water availability
Las Vegas–Henderson, NV	1747	Very low	Insufficient land, low water availability, poor soil fertility

Population density data are from the US Census Bureau

vertical farming, there is a simple way to begin to assess the potential. Any engineered process or system that replaces an ecosystem service provided by nature will require energy investment and compromise long-term sustainability. The bottom line: control comes at a cost. Artificial lighting costs could be immense for multi-story greenhouse operations as discussed in an article in *The Economist* in 2010.[98] The same would be true for climate control systems. Hydroponic systems that continuously supply water and nutrients may also require significant investment of materials and energy. The costs for land in the New York metropolitan area and rural Iowa vary dramatically. On average, one could buy more than 2000 hectares of Iowa farmland, some of the richest in the world, for the mean price of one hectare in the New York metro area.[99] This suggests that low-cost, vertically-farmed

[98] The Economist. December 2010. "Vertical farming: does it really stack up?" Available online: http://www.economist.com/node/17647627.

[99] In 2006, the average cost of one hectare of land in the New York metropolitan area was about $45.6 million in 2013 constant dollars (Haughwout et al. 2008). In contrast, record high farmland prices in Iowa in 2013 were about $21,500/hectare (Duffy et al. 2013). References: Haughwout A, Orr J, and Bedoll D. 2008. The price of land in the New York Metropolitan Area. Federal Reserve Bank of New York. *Current Issues in Economics and Finance* 14(3): April/May 2008. Duffy M, Johanns A, Klein W. 2013. *Farmland value reaches historic $8,716 staewide average.* Iowa State University Extension and Outreach. Available online: http://www.extension.iastate.edu/article/2013-land-value-survey, Accessed February 2015.

food in central locations of megacities is likely not on the horizon, and claims of sustainable vertical farming need to be backed up by rigorous life-cycle, energy, and economic analyses. Again, such comprehensive, systems-oriented studies are extremely rare in the "green" literature, which is probably, on average, misleading.

In summary, urban food production may play a role in future food security (assuming cities remain economically viable otherwise), but this role should not be overestimated. There are opportunities to establish more virtuous recycling systems that better couple production and consumption. However, biophysical limits must be considered carefully and not cast off as worrisome topics for pessimists. Attempts at agrarian urbanism may be ill-advised, far-from-sustainable flops if energy and material flows, along with considerations of climate and soil, are not part of the accounting system.

Impacts of Megatrends on Food Production

The twenty-first century megatrends related to energy and climate present critical challenges for food production. First, U.S. agriculture is extremely energy intensive, and therefore vulnerable to declines in energy availability and affordability. The cost of primary inputs to agriculture including fertilizers, other chemicals, and machinery operation are all tied directly to energy availability and prices. For example, a 2009 report from the Economic Research Service of the US Department of Agriculture concludes that fossil energy prices will likely be the dominant driver of fertilizer prices in the future.[100] Furthermore, a recent research article published in the journal *Global Environmental Change* reports that the fertilizer production sector is especially vulnerable to high energy prices.[101] The fertilizer price shock of 2008 provided a taste of what the future likely holds, greatly influencing, for example, production practices on small farms in Louisiana.[102] Resource-intensive diets including large amounts of meat and seafood will be increasingly more difficult and expensive to produce on a large scale. Governments may react to fuel price changes by providing fuel subsidies to assist farmers

---

[100] Huang, W.Y. 2009. *Factors contributing to the recent increases in U.S. fertilizer prices, 2002–08.* Economic Research Service Report AR-33, United States Department of Agriculture.

[101] Kerschner, C., C. Prell, K. Feng, and K. Hubacek. 2013. Economic vulnerability to peak oil. *Global Environmental Change*, 23, 1424–1433.

[102] Roy, E.D., J.R. White, and M. Seibert. 2014. Societal phosphorus metabolism in future coastal environments: insights from recent trends in Louisiana, USA. *Global Environmental Change*, 28, 1–13.

and fishers in the short term. However, decreasing energy return on investment (EROI) is not a short-term problem, and subsidies that support existing energy-intensive production practices do not encourage the adoption of sustainable practices.[103] Keep in mind that oil and gas prices are subject to short-term volatility, but the overall trend to lower EROI is clear and demonstrable, even for shale oil and Canadian tar (oil) sands. Achieving global food security will be more difficult in a future with lower EROI than it is today.

Second, climate change greatly threatens agriculture in the drier production regions of the Western US. Increases in atmospheric CO_2 concentrations can stimulate plant productivity, and this may potentially offset the negative effects of changes in temperature and precipitation on cereal yields in the U.S. overall.[104] However, this is not the case for more arid regions of the U.S. agricultural system where irrigation and fertilization currently support large amounts of food production. California is a prime example. The megatrends of the twenty-first century will come down hard on California. Climate change will place additional stress on the extensive water management system to the extent that some water users will face decreased water supply reliability.[105] Increasing temperatures, low ecosystem services, and economic disruption will also likely severely stress the state.[106] But the coup de grace may be what will happen to the wine industry. Wine is an iconic product of California and is deeply embedded in the psyche and lifestyle of the state. But a recent analysis of climate change impacts paints a dim future for the California wine industry. In a 2013 study appearing in the prestigious *Proceedings of the National Academy of Sciences*, Lee Hannah of Conservation International and colleagues investigated the impact of climate change on

---

[103] Pelletier, N., et al. 2014. Energy prices and seafood security. *Global Environmental Change* 24: 30–41; Sumaila, U., L. The, R. Watson, P. Tyedmers, and D. Pauly. 2008. Fuel price increase, subsidies, overcapacity, and resource sustainability. *ICES Journal of Marine Science* 65: 832–840; Grafton, Q. 2010. Adaptation to climate change in marine capture fisheries. *Marine Policy*, 34, 606–615.

[104] Parry, M.L., C. Rosenzweig, A. Iglesias, M. Livermore, and G. Fischer. 2004. Effects of climate change on global food production under SRES emissions and socio-economic scenarios. *Global Environmental Change*, 14, 53–67.

[105] Joyce, B.A., V.K. Mehta, D.R. Purkey, L.L. Dale, and M. Hanemann. 2011. Modifying agricultural water management to adapt to climate change in California's central valley. *Climatic Change* 109 (Suppl 1): S299-S316.

[106] For example, see the following publications: Hayhoe, K., et al. 2004. Emissions pathways, climate change, and impacts on California. *Proceedings of the National Academy of Sciences of the United States of America*, 101, 12422–12427; Lobell, D.B., C.B. Field, K.N. Cahill, C. Bonfils. 2006. Impacts of future climate change on California perennial crop yields: model projections with climate and crop uncertainties. *Agricultural and Forest Meteorology*, 141, 208–218.

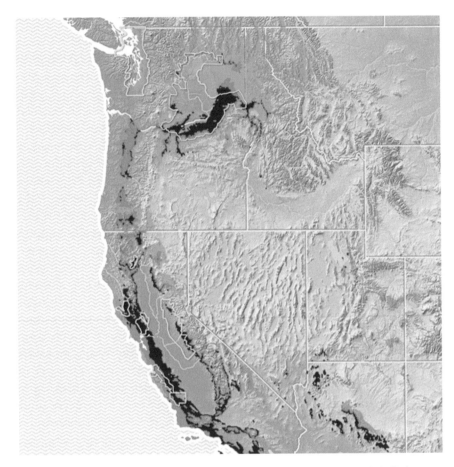

Fig. 8.11 Future outlook for wine producing regions in the Western US. *Red* = areas where wine grapes will become more difficult to grow. *Green* = areas where wine production will remain stable. *Blue* = areas that will become more suitable for wine production (From Hannah, L., Roehrdanz, P. R., Ikegami, M., Shepard, A. V., Shaw, M. R., Tabor, G., ... & Hijmans, R. J. (2013). Climate change, wine, and conservation. *Proceedings of the National Academy of Sciences*, **110**(17), 6907–6912

wine regions of the world.[107] Increasing temperatures will affect some of the most important wine producing regions (Fig. 8.11). Much of the wine producing area of California, including Napa and Sonoma, will reach what is called "loss of suitability."

Third, there are troubling questions concerning the continued feasibility of feeding large cities late into the twenty-first century. The food demand in megacities like New York and Los Angeles requires productive agricultural

---

[107] Hannah, L. et al. 2013. Climate change, wine, and conservation. *Proceedings of the National Academy of Sciences of the United States of America*, 110, 6907–6912.

Table 8.5 Ranking of lower
48 US states by farm outputs
(including crops and animal
products) and inputs in 2004
according to the USDA ERS

State	Rank in 2004 Farm output	Farm input
California	1	2
Iowa	2	3
Texas	3	1
Illinois	4	7
Nebraska	5	4
Minnesota	6	6
Kansas	7	5
North Carolina	8	10
Wisconsin	9	9
Indiana	10	13
Missouri	11	8
Arkansas	12	15
Florida	13	23
Ohio	14	14
Georgia	15	21
South Dakota	16	12
Washington	17	17
Michigan	18	18
Pennsylvania	19	16
North Dakota	20	19
Idaho	21	28
Oklahoma	22	11
Colorado	23	22
Mississippi	24	24
Kentucky	25	20
New York	26	26
Alabama	27	27
Oregon	28	31
Virginia	29	29
Tennessee	30	25
Arizona	31	34
Montana	32	30
Louisiana	33	33
South Carolina	34	35
New Mexico	35	32
Maryland	36	37
Utah	37	38
New Jersey	38	40
Delaware	39	41
Wyoming	40	36
Vermont	41	42
Maine	42	43
West Virginia	43	39
Connecticut	44	45
Massachusetts	45	46
Nevada	46	44
New Hampshire	47	47
Rhode Island	48	48

Table 8.6 Major producing regions and states for animal products in U.S. agriculture according to the 2007 USDA Census of Agriculture

Animal product	Major producing regions and states
Beef cows, cattle and calves	Spread throughout the lower 48, especially in the central portion of the country extending from North Dakota to Texas. Noteworthy high-density populations of cattle and calves also exist in Wisconsin, Nebraska, Idaho, Washington, California, Nevada, Arizona, Texas, Arkansas, Missouri, Louisiana, Florida, Kentucky, Tennessee, Virginia, Pennsylvania, New York, and Vermont.
Milk cows	California, Wisconsin, Minnesota, Pennsylvania, New York, Vermont.
Hogs and pigs	Concentrated primarily in North Carolina and the Midwest (especially Iowa and Minnesota).
Poultry (chickens and turkeys)	Spotty across the U.S. The vast majority of broiler chicken (sold for meat) are raised in the southeastern U.S. region stretching from Arkansas and eastern Texas east to Georgia and north to the Delmarva Peninsula. Egg-laying chicken populations are mostly in the eastern half of the country. States with particularly large populations of layers include Iowa, Indiana, Ohio, Pennsylvania, and Georgia. Turkey production is mostly confined to less than a dozen concentrated areas, the largest of which are in Minnesota, northwestern Arkansas/southwestern Missouri, the Carolinas, and land stretching between northern Virginia and southcentral Pennsylvania
Goats	Most US goats reside in Texas.

land areas far greater than what is available locally and even regionally,[108] along with the fuel, fertilizer, machinery, and other materials that support production on that land. Urban farming is a lovely idea that we are completely behind, but such gardens can provide only a small portion of the calories a megacity needs. Going back to the topic of percent employment in agriculture (Figs. 8.1a and 8.7), can we expect such a relatively small population of farm workers to continue to support immense populations of urban dwellers in a lower-energy U.S. future? One imaginable alternative to fossil fuel based machinery and human labor is of course draft animals. However, history provides a lesson on this point. Supporting horses and mules in 1915 required about 93 million acres of U.S. farmland for feed (primarily oats), or 27 % of the nation's total harvested acres. By 1960 this acreage had dropped to four million acres. This freed up about a quarter of the country's farmland for other purposes.[109] Going back to widespread use of draft animals would likely require a great expansion in the overall agri-

[108] Zumkehr, A., and J.E. Campbell. 2015. The potential for local croplands to meet US food demand. *Frontiers in Ecology and the Environment*, 13, 244–248.; Peters et al., 2007, ibid.

[109] Gardner, B.L. 2002. *American agriculture in the twentieth century: how it flourished and what it cost*. Harvard University Press. 400 p.

cultural land area (assuming we don't adopt less meat-intensive diets). We do not have the land to support this in the US. Abroad, intensive agriculture has already replaced substantial areas of the world's tropical forests, with further conversion expected in the future.[110] On a finite planet, achieving food security for over nine billion people without fossil fuel-dependent fertilization and mechanization is difficult to fathom. Our primary goals should be to increase food access to those currently food insecure, increase the efficiency of material and energy use in the food system, reduce over-consumption, and reduce food waste, all while limiting the further conversion of vital ecosystems to industrial agriculture. Taking the long view, an increase in the agricultural work force may eventually be our best strategy to cope with declining resource availability. However, this would significantly alter or even derail our economy. After all, food surpluses are a precondition for human specialization in non-sustenance activities, and thus the foundation upon which complex societies and their cities are built.[111] Emeril and Rachel Ray are right, it all begins with food.

---

[110] Gibbs et al. 2010 ibid.
[111] Diamond, 1997, ibid.

Chapter 9

Moving Away from a Ptolemic View of the Human Economy

> During the last two centuries we have known nothing but an exponential growth culture, a culture so dependent upon the continuance of exponential growth for its stability that is incapable of reckoning with problems of non-growth.—Marion King Hubbert

Introduction

Just before his death in 1543, Nicolaus Copernicus, the Polish mathematician and astronomer, published *De Revolutionibus Orbium Coelestium (On the Revolutions of the Celestial Spheres)*, one of the most important books in the history of science. By showing that the sun—rather than the earth—was at the center of the universe (actually the solar system, but this understanding came later), Copernicus overturned a thousand years of thought based on the writings of Claudius Ptolemy. This revelation helped usher in the modern era by restructuring human understanding of the cosmos and humanity's place in it. Ptolemy's earth-centric viewpoint required the development of very complicated models of celestial movements that had to be continually modified to fit new knowledge. Until Copernicus showed that the earth-centered view of the universe was wrong, Ptolemy was given almost legendary status. Is there a parallel with modern economics?

With the help of Alice Munro's short-story characters, we described the disconnect between modern city life and the average urban dweller's lack of recognition of the ecosystem services and primary economic production that takes place out of sight and out of mind. This disconnect and the many conveniences of oil-age technology has engendered a similar human-centric view of national and global economies as working according to our rules rather than as a part of the larger biosphere subject to physical laws and constraints. The latter view does not preclude an astonishing scope for human creativity, perseverance, and invention, but without fundamental change, it is unlikely that contemporary economics can contribute to a

© Springer Science+Business Media New York 2016
J.W. Day, C. Hall, *America's Most Sustainable Cities and Regions*,
DOI 10.1007/978-1-4939-3243-6_9

rational approach to the problems associated with the twenty-first century megatrends. Let's take a closer look at this far-reaching assertion, and the relationship between energy and other resources on the one hand, and the economy and the concept of economic growth on the other.[1]

In one form or another, we all earn and spend money. We are all toilet-trained on the idea that the earning and spending of this money is what the economy is all about. If you are fortunate enough to have a $100 bill in your wallet or purse, what does that piece of paper actually mean? Certainly, the paper is not literally worth $100. It cannot be eaten. If burned, it produces little heat. But that $100 can buy you a meal in a restaurant, fill your gas tank, get a ride in a taxi, purchase several shirts, pay someone to clean your house, pay your heating bill for a week or a month, or help to buy a myriad of other things. In early societies, when goods and services were exchanged without the benefit of money, each unit had direct real, usable value. A person might exchange a pig for materials to build a house, or for the labor for someone to help build that house. Another might exchange mushrooms gathered in the forest for a clay cooking pot. In this barter economy the surplus of a commodity generated by one person could be exchanged for that generated by another.

But exchanging things in this way was cumbersome, and those mushrooms had to be eaten rather quickly or they spoiled. So early on, different kinds of currencies were introduced to assist in the exchange of goods and services. These currencies were durable and transportable (they could be carried about, stored, and used later) and included things such as sea shells, beads, and wampum. Durable precious metals became literally the coin of the realm with the beginning of civilizations several thousand years ago. Even so, gold or silver alone, like the hundred dollar bill, cannot be eaten, burned for heat, or used to construct a house, at least not one that most people could afford.

All currencies are pegged to economic activity, and are a convenient way to facilitate economic exchange. In fact, given that all goods and services come into being only through the application of energy in some form

---

[1] There are numerous books about the shortcomings of contemporary economics. Herman Daly has written numerous books about steady state economics. The following two summarize many of his ideas. H.E. Daly. 2007. *Ecological Economics and Sustainable Development Selected Essays of Herman Daly*. Edward Elgar, Northampton MA. 270 p.; H.E. Daly. 2014. *From Uneconomic Growth to a Steady-State Economy*. Edward Elgar, Northampton MA. 253 p.; Kenneth Boulding. 1968. *Beyond Economics*. The University of Michigan Press, Ann Arbor. 302 p.; Tim Jackson. 2009. Prosperity Without Growth Economics for a Finite Planet. Earthscan, Washington DC. 276 p.; Michael Hudson. 2012. *The Bubble and Beyond*. ISLET, Dresden. 481 p.; Brain Czech. 2013. *Supply Shock*. New Society Publishers, Gabriola Island, Camada. 367 p.; Charles Hall and Kent Klitgaard. 2012. *Energy and the Wealth of Nations*. Springer, New York. 407 p.

(human or animal labor, water power, or combusted fuels), currencies are in fact a lien, meaning a claim, on energy use. If you take the currency out of a society, economies can still function, although awkwardly, through barter. Take the energy out and an economy slows proportionately or stops immediately, as Cuba found out in 1988 when Russia cut off its subsidized oil. All of our economy is based on a trust that these liens will be honored. This is not necessarily true, as is being sorted out in contemporary Greece as we write. Greece lost its soils and many of its fish thousands of years ago, has relatively few other resources beyond beautiful weather and coastlines, and interesting people. Yet, Greece has a large population who expect a European life style (and other problems, such as a lot of wealthy people who do not wish to pay their fair share of taxes to allow that lifestyle). Thus, money is a lien on energy based on our expectation of its application for the production or consumption of energy services, or their embodiment in the myriad of stuff that is manufactured from natural resources. Money can also buy other things, such as access to political power, prestige, subjective preferences and so on, but they are all dependent on our first definition. One of the fundamental problems of economics arises when money is viewed as having value in and of itself rather than as an exchange certificate for something that has real value, such as a gallon of gasoline or a truckload of lumber or the energy to generate those things. If the resources are not there in proportion to the money, we have increasingly worthless money, i.e. inflation, a problem as old as money itself.[2]

A few examples should help illustrate the relationship between "value" and the size of energy flows through the economy. The CEO of a large oil company doesn't do any more physical work in a day, and likely does much less, than a roustabout working on one of the company's oilrigs. But because the CEO has the responsibility for a multibillion-dollar corporation, and therefore manages more energy flow in the economy, he is entitled (in some people's minds) to a larger salary than the average worker in the company. Just how much more is currently a topic of hot debate.

So money is a way of valuing the various objects, activities, and jobs that make up the inter-workings of the economy. To at least some substantial degree this is in proportion to the energy controlled by, or used to train, the worker. These values can change according to the circumstances of the economy. The financial crash of 2008 was blamed, in part at least, on an inflated housing "bubble". The prices of some houses had increased by factors of two, three, four, or more in just a few years. How did a certain house

[2] Gibbon, E. 1776. *The decline and fall of the Roman Empire: VOLUME I.* Random House; Walker, D. R. 1974. *The metrology of the roman silver coinage. Part I: From Augustus to Domitian. Part II. From Verva to Commodus. Part III. From Pertinax to Uranius Antonius.* British Archaeology Report. Oxford. Supplementary Series 5, 22, 40.

worth, say $150,000, increase to $500,000 in a decade since it did not embody more energy (except possibly for certain upgrades)? Some blamed faulty lending practices that allowed people to buy houses that they could not really afford. And many believed that real estate was a safe investment, and that houses would only increase in value. But were these the only reasons? Bubbles have been recorded many times in history, like the infamous tulip mania in the Netherlands in the 1700s, where the exchange value of something like a rare tulip bulb greatly exceeded its value as a physical asset, at least for a short while. But over the long-term, value is related to the effort needed to produce an asset as well as demand for that asset, and prices that are far from this "equilibrium" tend to return to their real value. For example, Charles Hall and Kent Klitgaard found that the value of the Dow Jones Index was related to the actual energy used by U.S. industrial society, so that each followed a common increase for nearly 100 years. Although the stock market fluctuated much more it tended to come back to the path of actual physical work done.[3] Charles Hall and Carey King found that the cost of energy itself (i.e. oil) reflected the energy required to make it available to society.[4] Presumably real prices must reflect real worth over time, although they can stray from that for a few years or decades due to other factors, many psychological.

In his book, *The Wealth of Nature*,[5] John Greer calls the industrial revolution the greatest bubble in human history. This bubble has been fueled by hundreds of millions of years of solar energy stored as fossil fuels. In about three centuries, humans will burn through these fossil fuels. And in doing so, they created the enormous growth in population and economies that represents the economic work done by the huge energy resources of fossil fuels. But the fuels—and hence our modern supercharged economy—are finite, and presumably this bubble too will fail eventually. Whether some other source of energy is available on this scale is obviously of considerable discussion. The prime candidates are nuclear and renewables, but as we have discussed their potential role is not really clear. We have come to think of widespread affluence and its growth as natural and just the way the world is. In fact, they are rare and exceptional within the long history of human experience.

So how does all this relate to regional sustainability? The dominant, unlimited growth paradigm in the field of economics for the past hundred

---

[3] Hall and Klitgaard, 2012, ibid.

[4] Hall, C.A.S., King, C. 2011. Relating financial and energy return on investment. *Sustainability, Special Issue on EROI*, 1810–1832.

[5] John Greer. 2011. *The Wealth of Nature*. New Society Publishers, British Columbia, Canada.

years, called Neoclassical Economics, which flourished in an era of abundant, cheap energy, offers little hope for addressing or solving the problems arising from the emerging megatrends of the twenty-first century. The continuously increasing infusions of cheap energy needed to fuel growth and repay interest and debt—the influx at the heart of our economic system—are no longer available. As we noted earlier, even the recent infusions of shale oil and gas come at a very high price to the economy compared to the dwindling supplies of conventional oil (and a high price to the shale operators as well, many of whom, or their investors, were losing their shirts at the $40/barrel oil prices of early 2015, even when oil was $100 a barrel). With nature being in increasing control because of diminishing supplies of high-grade resources from oil to metal ores, to water, to cheap oil, we need an alternative economic model that is more compatible with our likely future.

What's Wrong with GDP?

Economist William Nordhaus stated that global warming would not have a significant impact on the U.S. economy because only agriculture would be impacted by climate change, and agriculture accounts for only about 3 % of GNP.[6] This conclusion is misleading on two counts. Climate change, as we have discussed earlier, can and most likely will have a far more pervasive impact on the economy. For example, climatologists predict the impacts of more strong storms like hurricanes Katrina and Sandy will grow as climate change continues. And growing water shortages, attributed by many scientists to climate change, will impact all aspects of the economy in the Southwest. We are most likely seeing these effects in California already. These points can be argued, but Nordhaus' argument is a poster child for how misleading GDP can be as a measure of sustainability. If agricultural activity on the farm accounts for only 3 % of GDP, it still supports the entire food system that includes transportation, processing, restaurants and stores, and the home dinner table, all of which constitutes somewhere around 15 % of energy use by the economy. Agriculture feeds us. If we cut agriculture in half, it would impact only 1.5 % of GDP, but it would also

---

[6] William Nordhaus, Science magazine's report on (Sept 14, 1991, p. 1206) of a National Academy of Sciences study on climate change. Later, Nordhaus' logic was repeated by two more distinguished economists: in 1995 by Oxford economist Wilfred Beckerman, and in 1997 by Nobel Laureate in economics Thomas Schelling, see Herman Daily's essay on page 188 in his book: Herman Daly. Ecological Economics and Sustainable Development – Selected Essays from Herman Daly. Edward Elgar, Northhampton MA. 2007.; GDP, Gross Domestic Product, and GNP, Gross National Product, are roughly equivalent measures of economic production.

halve the amount of food available. America needs to go on a diet, but this would be pretty dramatic. If we cut agriculture by 100 % then essentially everyone would die and 100 % of the economy would cease. So the value of various things in our economy is not simply what its money value is. We would guess that there are few or no other larger contributors to GDP that would be less dispensable than half of our food supply.

For example, although the secondary (manufacturing) sector of the economy is no longer as large as it was in the mid twentieth century, the products once produced in that sector are still widely available from man-ufacturing-intensive economies such as China. The longer supply chains now operating in the global economy are dependent on large amounts of cheap energy to maintain. As energy becomes scarcer, the maintenance of the global economy and the long supply chains will become even more expensive and difficult to maintain. As the supply and distribution chains of the global economy become more stressed due to energy cost, it will affect different regions of the US unequally. Increasing local dependence will mean that those strongly impacted by climate change and with low levels of natural resources will be at a disadvantage.

With this in mind, let's look at two fundamentally different models of the economy. The first is what can be called the conventional view of the economy. This economic paradigm is called the neoclassical synthesis or neoclassical economics. Neoclassical Economics (NCE) provides the pri-mary conceptual framework by which the human economy has been understood and managed for at least the past century. NCE is the dominant mainstream economic paradigm in the United States and the world today and has become the system whereby economic priorities of the state, and society in general, are determined and justified primarily by market forces. This kind of economics is usually considered a social science because it focuses on human psychological activity and their response to those who supply and sell goods and services. It essentially ignores resources except with respect to the prices of goods and services. NCE presumes that scar-city, when present, is reflected in high or increasing prices. Nevertheless, as we stated, it is but one way that economists have looked at economics. Others include classical, Marxist, Keynesian (a variant, perhaps, of NCE), behavioral, institutional, ecological, and biophysical.

The NCE diagram of the economy, called the circular flow model of eco-nomic production (Fig. 9.1), is taught in all introductory college courses in economics. In a sense all economists, indeed almost all citizens, are "toilet trained" on this model and on the concept that value is reflected in prices. In this model, households sell or rent land, natural resources, labor, and capital to firms in return for rent, wages, and profit. Firms produce goods and ser-vices for which they receive payment. Money flows in the opposite direction from goods, services, labor, land, and capital. This makes sense, but is very

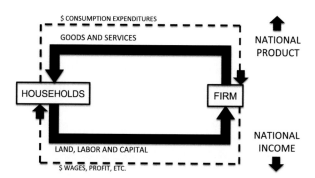

Fig. 9.1 Conventional neoclassical view of the economy showing the circular flow model of economic production (Redrawn from: Solow, RM. 1974. The economics of resources or the resources of economics. *American Economic Review* **66**: 1–14. Also Solow, R.M., 1994. Perspectives on growth theory. *Journal of Economic Perspectives* **8**, 45–54))

incomplete. People work at firms of one sort or another for which they receive money which they use to purchase goods and services produced by firms. Households include not only people in homes but also those working in firms that also purchase goods and services for themselves and their families.

What is missing in the diagram are the energy and materials that must flow into the economy from the natural world, as well as the waste material and heat that are produced by the economy that flow back out into the environment. For most of the past two centuries this did not matter much because these seemed unlimited. But as we discussed in earlier chapters on energy, climate, and ecosystem services, this is no longer the case, if indeed it ever was. Many non-renewable resources such as oil and many other important minerals will be depleted for all practical purposes in this century or even within decades. And as we discussed in the chapter on ecosystem services, natural systems are being degraded so that they are unable to provide the same level of renewable resources, including good soils and fresh water, as they did in the past; this is occurring at a time of growing population and increasing demand for these resources. The increasing level of carbon dioxide in the atmosphere and its impact on climate is perhaps the most notable example that the ability of the natural world to accept wastes from the economy is being overtaxed.

Biophysical economics, which provides a more accurate but still simplified view of the human economy, is represented below in Fig. 9.2. This diagram is much more accurate in representing how the real economy, with its four main sectors, operates. The term "biophysical" refers to the re-integration of biological and physical factors into a purely social neoclassical depiction of the economy.

Neoclassical economists argue that if one resource is in short supply, then another resource will substitute for it. They even argue that this ability to substitute one resource for another is essentially infinite, resulting in a lack of any absolute scarcity. But in the real world, this is not the case. For example, there is no real substitute for oil in most of its applications, such as heavy transport, tractors, the petrochemical industry, and air travel, because oil is so effective, transportable, widely available, and versatile. It is energy-dense and has been very cheap for more than a century. Electricity from solar and wind can power some things, but airplanes, big ships, 18 wheelers, much of agriculture, heavy industry, and much more require energy-dense fossil fuels, mainly oil. If oil becomes scarce, so will liquid fuels. And as we discussed in the energy chapter, biofuels have little net energy yield and compete with food supplies. We will come back to these concepts later on in this chapter.

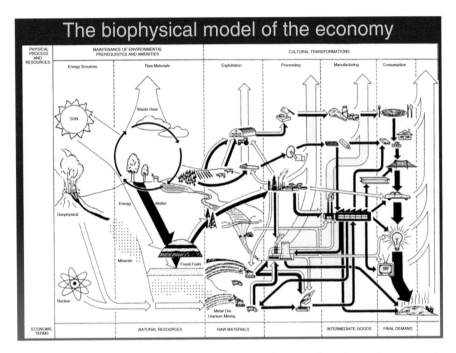

Fig. 9.2 A diagram of the biophysical economy showing that the real economy is ultimately dependent on flows of material and energy from the biosphere. Natural energies drive geological, biological, and chemical cycles that produce natural resources and public service functions. Extractive sectors use economic energies to exploit natural resources and convert them to raw materials, which are used by manufacturing and other intermediate sectors to produce final goods and services. These goods and services are distributed by the commercial sector to final demand. Waste materials (downward pointing *black arrows*) and waste heat (upward pointing *white arrows*) are returned to the environment as waste (From Hall, C.A.S., Cleveland, C., Kaufmann, R., 1986. Energy and Resource Quality: The Ecology of the Economic Process. Wiley, New York. Used by permission.)

The biophysical view encompasses the natural world in which the human economy exists. The diagram of the biophysical economy (Fig. 9.2) shows how the natural world is connected to the human economy. This model has the standard set of sectors that are considered in the conventional view of the economy but presented from a different perspective. The primary sector clearly shows that the overall economy is firmly based in the primary economic activities of farming, fishing, forestry, and mining. These activities are based, in turn, on the natural resources of the earth (soils, seafood, trees, ores, fossil fuels). The economy is totally and absolutely based on the existence and use of energy and other natural resources from the biosphere. These include a suitable climate and hydrological cycle, soil, fresh water, plant growth, aquatic organisms, metal ores, and fossil fuels. Included in the metal ores are nuclear isotopes that form the basis for nuclear energy.

Continuing energy inputs support the functioning of the biosphere, including the human economy that is nestled in it. These include solar, geophysical, and nuclear (the heat generated by the decay of radioactive isotopes below ground). Solar energy powers not only current climatic, hydrological, and ecological energy flows but was also responsible for generating fossil fuels, which are ancient, embodied solar energy that have been upgraded and concentrated by millions of years of geological processing. Another important aspect of the biophysical model is that energy is dissipated as it moves through the system. At each step, some energy is lost as heat (upward pointing white arrows) and can no longer perform useful work. This is the process of entropy production that we discussed in the chapter on energy. Finally, material waste is produced at every step in the economic process (black downward pointing arrows). Some of this waste can be, indeed must be, assimilated by the biosphere but as indicated above only within limits. The rest is pollution.

This view of the human economy dependent on, and contained within, the much more complex larger natural economy builds on a long intellectual history. Ernst Shumacher[7] considered the primary sector of the economy to be all the work nature does to provide energy and materials to humans. It is thus the biosphere that is the source of all ecosystem goods and services that support the human economy. In other words, the wealth of a nation starts with the wealth of nature.

Financialization of Wealth. When money dominates an economy, so does the world of finance, and money traded for money can be larger today than that being traded for real goods and services. Money can be made

---

[7] E. F. Schumacher. 2011. Small is Beautiful: Economics as if People Mattered. John Greer in The Wealth of Nature discusses Schumacher's book in great detail.

from nothing, such as the government printing more and more of it, or to a much larger extent, private banks lending it into existence at interest.[8] The money, if invested, can serve as a primer to draw more energy out of the ground and generate more wealth, as long as there are new energy and resources available. If no real wealth (enhanced energy flow) is generated then eventually the ratio of money to real resources is increased and inflation will occur, although other issues can influence this basic relation. Thus when a trader on Wall Street makes millions on a nanosecond trade, no real wealth is created, although it may change hands or, if invested, might generate real wealth eventually. In these cases, where money increases more rapidly than the energy that generates real wealth, the value of the currency degrades. This is inflation.

An Empty World to a Full World

In the empty world that prevailed until about 1800, the human economy was small relative to the global ecosystem that runs primarily on solar energy (Fig. 9.3).[9] Energy and matter flowed into the economy from the biosphere, and degraded energy and materials flowed back into the environment as waste where there was a high level of recycling. Human welfare was supported primarily by ecosystem services from the environment, and to a lesser extent, by economics services from the human-built economy.

In the full world that currently exists, the economy is large relative to the environment and support for human welfare is much more dependent on the economy, while ecosystem services are much reduced. But in the end the economy can continue to support human welfare only if the ecosystem can continue to supply essential goods and services from food to metals to hydrocarbon feedstocks for everything from plastics to pharmaceuticals.

Most of the concepts of modern neoclassical economics were developed in an empty world context with few resource constraints, when energy and material flows from the environment to the human economy, and the flow of wastes back to the environment, were small compared to the capacity of the environment to provide these flows and accept wastes. For example, in the late nineteenth century most fossil fuels were still in the ground and

---

[8] F. Soddy. 1922. *Cartesian Economics – The Bearing of Physical Science upon State Stewardship*. Henderson, London.; Daly, H. E. (Ed.). (1980). Economics, ecology, ethics: Essays toward a steady-state economy. San Francisco: WH Freeman.

[9] Note here that the circular flow model of the economy discussed earlier fits in the box called "Economy."

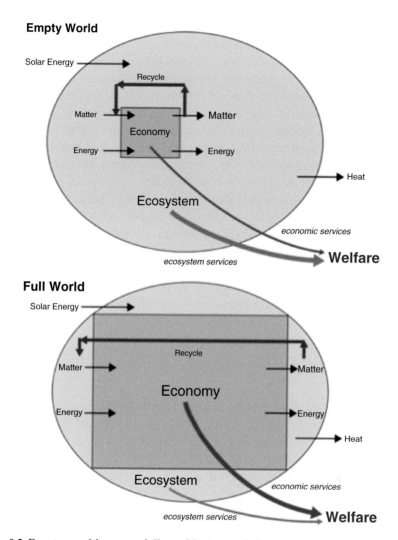

Fig. 9.3 Empty world versus full world views of the economy as visualized by Herman Daly. In the preindustrial world, the human economy was small compared to the biosphere and the ability of the biosphere to provide energy and materials and accept wastes seemed endless. The economy is now so large that it has compromised the ability of the earth to sustain it. In the diagram, human welfare should not be considered outside of the human economy, but that the major factors supporting welfare have shifted from ecosystem services to economic services. The energy output from the economy is mainly as waste heat as dictated by the second law of thermodynamics (H. Daly and J. Farley, *Ecological Economics – Principles and Applications*. Island Press, Washington DC, 2004, p.18. Figure courtesy of Herman Daly. Economics for a full world. Great Transitions Discussion, Tellus Insitute, May 2015. http://www.greattransition.org/publication/economics-for-a-full-world)

carbon dioxide levels in the atmosphere were 150 parts per million (ppm) lower than the 2015 level of about 400 ppm.

In a full world, global pollution problems have emerged from what were once local and regional problems. The burgeoning human population has led to a dramatic increase in demand for natural resources that has, in turn, resulted in widespread land use change and degradation, along with diminishing of ecosystem services.[10] The abundant concentrated high quality resources of the empty world, such as oil deposits and metal ores, high quality soils, large trees and rich coastal ecosystems, have been greatly depleted, and even supplies of lower quality fossil fuels will become increasingly limiting. Transitioning to a low energy society will require understanding the present limitations of the economic system and the dedication of a considerable amount of the remaining high quality fossil fuels to the transition that society has already begun. At the same time, this transition is not likely to become socially acceptable for some time because of the overwhelming dominance of neoclassical economics, and a continuation of the exponential growth paradigm, in our institutions and education.

A Two-Edged Sword

So we are stuck in a dilemma: the enormous growth in the use of fossil fuels and the other resources has underwritten the enormous and unprecedented increase in human numbers, and of the affluence of many of us in less than 200 years. The great wealth of past kings and other rich persons required the hard labors of tens of thousands of subjects. No longer. Today's wealth is built on what ecologist Howard Odum referred to as "energy slaves," most of which are non-renewable—unlike reproducing human populations. So there is a relation between energy use, mainly fossil fuels, and the astounding growth of population and the economy over the past couple of centuries.

Dr. James Brown and colleagues at the University of New Mexico published a paper more recently titled Energetic Limits to Economic Growth showing that there was a direct relationship between per capita energy consumption and per capita GDP.[11] This strong relationship is reflected in

---

[10] As discussed in Chap. 5, see: Vitousek, Peter M., Harold A. Mooney, Jane Lubchenco, and Jerry M. Melillo. 1997. "Human domination of Earth's ecosystems." *Science*, 277 (5325), 494–499.; Pimentel et al. 2005. Update on the environmental and economic costs associated with alien-invasive species in the United States. *Ecological Economics*, 52 (3), 273–288; Wackernagel, M., Schulz, N., Deumling, D., Callejas Linares, A., Jenkins, M., Kapos, V., Monfreda, C., Loh, J., Myers, N., Norgaard, R., & Randers, J., 2002. Tracking the ecological overshoot of the human economy. *Proceedings of the National Academy of Science*, 99(14), 9266–9271.; Elizabeth Kolbert. 2014. *The Sixth Extinction*. Henry Holt, New York. 319 p.

[11] Brown, J. et al. 2011. Energetic limits to economic growth. *BioScience*, 61, 19–26.

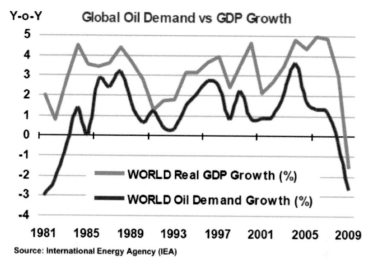

Fig. 9.4 The relation between global oil demand and GDP growth (IEA Data presented by Dave Cohen: Cohen, David. 2009. *The Aftermath of the Great Recession, Part I*. Peak Watch. Posted on September 17, 2009. http://peakwatch.typepad.com/peak_watch/2009/09/the-aftermath_of_the_great-recession-part-i.html; For a detailed analysis of the relation of oil to economic growth see: Aucott, M., & Hall, C. 2014. Does a Change in Price of Fuel Affect GDP Growth? An Examination of the US Data from 1950–2013. *Energies*, 7(10), 6558–6570)

the fact that the U.S. economy entered into a recession each time the cost of oil increased to over 5 % of GDP spending and total energy costs exceeded 10 % of GDP (Fig. 9.4).[12]

As of the end of 2015, oil prices had fallen to below $50 per barrel. Does this mean that expensive energy is a thing of the past? We think not. First of all, compared to historical values $50 a barrel is not cheap[13]. David Murphy and Charles Hall developed a diagram to show how energy prices and economic growth are inverse to one another (Fig. 9.5). In their diagram, economic growth increases oil demand. Higher oil demand increases the need for more oil, which will come from not only the cheap resources favored when less is used, but also from progressively more expensive

---

[12] Hamilton, J.D., 2008. Daily monetary policy shocks and new home sales. *J. Monetary Econ.* 55 (7), 1171–1190.; Hall and Klitgaard 2012. ibid.

[13] At $50 per barrel, the new "low" price for oil is more than double the average inflation adjusted price for oil from 1881 to 1971 of about $20 per barrel and significantly higher than the 1986 to 2003 average of $30 per barrel. https://en.wikipedia.org/wiki/Price_of_oil.

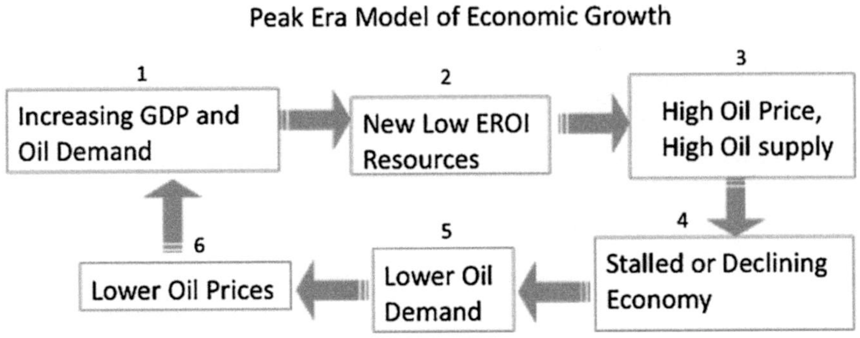

Peak Era Model of Economic Growth

Fig. 9.5 Peak era diagram of the economy showing the inverse relation between economic growth and the price of oil (David Murphy and Charles Hall. Adjusting the economy to the new energy realities of the second half of the age of oil. Ecological Modeling. 2011. doi:10.1016/j.ecolmodel.2011.06.022, used by permission)

resources needed to meet the higher demand, i.e., oil produced from lower EROI resources. These increasing extraction costs then lead to higher oil prices. Higher oil prices then stall economic growth or cause economic contraction, which leads to lower oil demand, and lower oil demand leads to less exploitation of the higher-priced resources and hence lower oil prices—which spurs another bout of economic growth until this cycle repeats itself. This system of insidious feedbacks is aptly described as a growth paradox: maintaining business as usual economic growth will require the production of new sources of oil, yet the only remaining sources of growth in oil supply require high oil prices, thus hampering economic growth. This growth paradox leads to a highly volatile economy that oscillates frequently between expansion and contraction periods, and as a result, there may be numerous peaks in oil production.

High Oil Prices and Regional Sustainability

Increases in the price of oil and total energy relative to GDP decreases the amount of discretionary income consumers have to spend.[14] Much of the economy, especially that connected to growth, is based on spending discretionary income, and cities and regions whose economies are based heavily on spending discretionary income will be more affected by decreases in discretionary income as energy prices rise. As we saw in Chap. 4, tourism-based economies are highly vulnerable in this respect as fewer travelers

---

[14] Hamilton, J.D., 2008. Ibid.

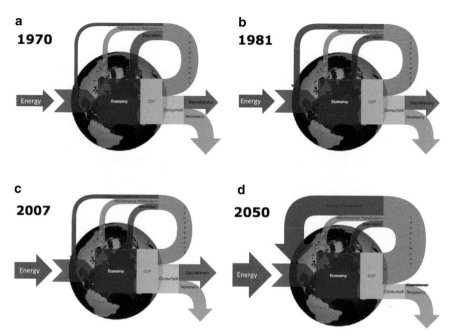

Fig. 9.6 A model of the economy showing use of energy to fuel the economy and the allocation of GDP to different activities in the economy from 1970 to 2050. GDP is allocated to investments, to obtain more energy (*dark blue line*) maintenance (*light blue line*) and discretionary societal spending (*red line*) and consumption by households, which is divide into discretionary (*red line*) and necessary (*pink line*) consumption. Discretionary investments and expenditures are possible only after allocation to required funding for investments and to feed, clothe and house the population. Values up to 2007 based on actual data, subsequently to extrapolation assuming EROI continues to decline (Hall and Klitgaard, ibid. Hall, C. A., Powers, R., & Schoenberg, W. (2008). Peak oil, EROI, investments and the economy in an uncertain future. In *Biofuels, solar and wind as renewable energy systems* (pp. 109–132). Springer Netherlands. We encourage readers to go online and run the model: https://forio.com/simulate/billy/cheese-slicer/simulation/#p=page0)

spend fewer dollars in cities such as Las Vegas and Orlando. Some of the most dramatic declines in housing price values since the 2007 mortgage crisis also occurred in these two markets. Estimates imply that a 1 % reduction in GDP (especially discretionary income) translates into a 2.6 % reduction in demand for new homes.[15] However, this decline in demand will not be the same across the country. For example, housing prices near the urban

---

[15] Cortright, J., 2008. *Driven to the Brink: How the Gas Price Spike Popped the Housing Bubble and Devalued the Suburbs*. In: Discussion Paper. CEOs for Cities, Chicago. Available online: www.ceosforcities.org

core of many cities rose during the worst part of the mortgage crisis, while homes in the suburbs with greater commuting distances declined in price and the number of foreclosures in these neighborhoods increased.[16] Thus, energy use is extremely important in maintaining the current growth-oriented economy and the increasing cost and scarcity of energy will have widespread impacts throughout the economy.

Charles Hall and colleagues assessed and then modeled the impact of increasing oil prices on discretionary income.[17] In Fig. 9.6, the economy is shown as a box inside of the earth to emphasize that the economy must operate within the biosphere. Energy and materials used by humans to maintain the economy that produces GDP must come from nature. Part of this GDP goes into investments to maintain the economy, and part of it is used for consumption. The investments are used to obtain more energy, for infrastructure maintenance (both of which are essential), and for societal level discretionary investments, i.e. those which are desired, but not essential, to keep the economy operating. The latter include such things as museums, universities, libraries, symphony orchestras, and parks as well as private investments for hotels, shopping centers and the like. The rest of the GDP is used for consumption by individuals and includes purchase of staples such as food, clothing, transportation, health care and the like and discretionary spending on things such as vacations, movies, and luxury goods.

Required investments (without which the economy would cease to function) include maintaining essential infrastructure (bridges, roads, etc.), basic requirements by individuals (food, housing, clothing, etc.), and investments required to get the energy that allows the economy to function. As it becomes more expensive and difficult to obtain energy, more of GDP must be diverted from other uses so that the energy input is maintained. The diagrams show that from 1970 to 2007, the economy grew moderately, but the proportion of GDP allocated to the different areas of the economy was relatively stable, although there was a transfer of wealth from investments in discretionary investments to energy extraction during the energy price increases from 1970 to 1980. In 2007, personal discretionary spending represented about a third of GDP. The model assumed that the energy returned from energy (and money) invested in energy production, or EROI, would continue to decline from 20:1 in 1970 to 5:1 in 2050. The single most important expenditure of GDP in 2050 was to obtain energy, which caused

[16] Hall, C. A., Powers, R., & Schoenberg, W. 2008. Peak oil, EROI, investments and the economy in an uncertain future. In Pimentel D. (ed) *Biofuels, solar and wind as renewable energy systems* (pp. 109–132). Springer Netherlands.

[17] Gowdy, J., 2000. Terms and concepts in ecological economics *Wildl. Soc. Bull.* 28 (1), 26–33.

a shrinkage of personal discretionary income by nearly 90 %. As more and more of the GDP is used to maintain energy input to the economy, the net (rather than gross) energy declines. Energy return on investment (EROI) falls from 20:1 in 1970 to 5:1 in 2050. Of course no one knows for certain if EROI will decline by so much, but it does seem very likely that it will decline by some significant amount, consequently restricting our economic options. Since spending discretionary income supports large parts of the economy (think tourism, airline travel, second homes, and fancy meals and cars), declining discretionary income bodes ill for the economy.

According to the assumptions of NCE, natural capital can be substituted with human or man-made capital, and the depletion of fossil fuels, fisheries, forests, and soils should be of little concern. This concept of changeability (i.e., infinite substitutability) creates a sort of perpetual impunity regarding the consequences and reality of economic activity.[18] One of the critiques of neoclassical economics is that it essentially reduces the complexity of real production processes to capital alone. According to this viewpoint, human capital, manmade capital, and natural capital are absolute substitutes for one another. This is akin to saying that if we are running out of cheap oil we should build more trucks or even drilling rigs, or train more engineers. No one would deny this might help a little, but it does little to change the amount of extractable oil in the earth. Substitute liquid fuels, such as ethanol, require large energy (fossil fuel) inputs and have very low EROI. In essence, the idea of absolute substitution is based on the false notion of presumed unlimited capital with infinite flexible use. There is little recognition that real capital is mostly the means of employing fossil fuels to generate structures and machines to use fossil fuels, and that the fuels are just as necessary as capital. Many people believe that technological progress is an all-powerful process that will replace resources with ideas indefinitely.[19] However, this is akin to saying that we can eat menus (which are cheap to produce and contain information on food but which have no nutritional value) instead of the food listed on the menu (which is very energy-intensive to produce and depends heavily on many natural resources).

The societal disconnect from the natural world was so large that by about 1960 the old production functions that were based on land, labor, and capital were replaced by new ones that did not even consider land— let alone energy, water, or other critical resources. The noted economist

---

[18] Gowdy, J., Klitgaard, K., Krall, L., 2010. Capital and sustainability. *Corporate Examiner* 37 (4–5), 16 March.; Granek, E.F., et al., 2010. Ecosystem services as a common language for coastal ecosystem-based management. *Conserv. Biol.* 24 (1), 207–216.

[19] Tainter, J. A., & Taylor, T. G. 2014. Complexity, problem-solving, sustainability and resilience. *Building Research & Information, 42*(2), 168–181.

Robert Solow posed the question that "If it is very easy to substitute other factors for natural resources, then there is in principle no 'problem' or in other words 'the world can, in effect, get along without natural resources.'"[20]

Although posed as a question, this reflects the view of many economists. The well-known steady-state economist, Dr. Herman Daly, who has frequently criticized conventional economics, commented that "As an 'if-then' statement this is no less true than saying. 'If wishes were horses, then beggars would ride.' But the facts are that wishes are not horses, and that natural resources and capital are not generally substitutes, but complements."[21]

Daly quoted his mentor, Nicholas Georgescu-Roegen, who also took issue with Solow's statement by noting that "One must have an erroneous view of the economic process as a whole not to see that there are no material factors other than natural resources. To maintain further that 'the world can, in effect, get along without natural resources' is to ignore the difference between the actual world and the Garden of Eden." As we stated earlier in this chapter, it is the role of the primary sector of the economy to provide natural resources that then provide the base for the whole rest of the economy.

Another problem with the neoclassical assumption of absolute substitutability is that the market creates "the illusion of decreasing scarcity".[22] In short, the market does not reflect or accurately price non-renewable resource depletion. This occurs for a variety of reasons including the lack of complete information, or disregard for credible information, regarding total recoverable resource reserves and the role of technology that may misinterpret resource abundance due to increased short-term extraction. Increased extraction can provide more immediate resources but also leads to faster depletion rates that eventually lead to lower extraction efficiency through time and increasing resource scarcity. This has been experienced repeatedly in oil and gas fields throughout the world as the high net energy yielding reserves are depleted and the lower net energy yielding reserves are all that remain. Therefore, the amount of energy and time required for the transition to the "new energy future" are unknowns. It is questionable that such a techno-utopian future can exist without relatively inexpensive

---

[20] H. Daly. 2007. Ecological Economics and Sustainable Development. Edward Elgar, Northampton, MA. 270 p.

[21] Gorgescu-Roegen. N. 1971. *The Entropy Law and the Economic Process.* Harvard University Press, Cambridge.; Reynolds, D.B., 2002. *Scarcity and Growth Considering Oil and Energy: An Alterna- tive Neo-Classical View.* Edwin Mellen Press, Lewiston, NY, pp. 232.; Cobb, K., 2008. *Will the Rate-of-Conversion Problem Derail Alternative Energy?* http://scitizen.com/future-energies/will-the-rate-of-conversion-problem-derail-alternative-energy-a-14-2020.html

[22] Reynolds, D.B., 2002. ibid; Cobb, K., 2008. ibid.

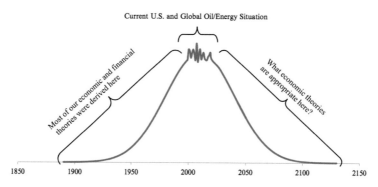

Fig. 9.7 Model of the relation of the development of economic and financial concepts shown in time compared to an approximation of the past and projected production of oil. In a sense, it easy to make all kinds of economic and financial models "work" when the real work done in an economy was expanding at 2 % or 3 % a year. Things may be difficult now that peak oil has been more or less reached. The fundamental question for economics is what economic concepts will work in an economy with declining energy? (Figure from: Hall and Klitgaard. ibid. Fig. 1.13, p. 37)

fossil fuels to construct, and probably to run, it.[23] Unless scarcity is priced into the market, price shocks will continue to occur that make the transition more expensive and difficult. Thus, the present price of gasoline does not reflect the fact that most oil will be gone within half a century or perhaps much less.[24] In addition, the price of fuel does not incorporate the costs to the environment and society that accompanies fuel combustion. It is therefore extremely important to consider the costs and benefits as well as the winners and losers when these policy considerations are being discussed and decided upon. Such decisions should be considered with regard to the short-term and long-term interests of society and sustainability. Sooner rather than later, nature will again assert control, and your city, town, or region must be in a position to cope with this ultimate megatrend.

Our overview of the current situation with respect to economics is summarized in Fig. 9.7. Most of our economic and financial theories, including that of growth, debt, financial return on investments and so on were developed during a period of expanding availability of energy to do work. Now

---

[23] Huesemann, M., Huesemann, J., 2011. *Techno-Fix: Why Technology Won't Save Us or the Environment*. New Society Publishers, Cabriola Island, BC, Canada, pp. 464.

[24] Hall C.A.S. and Day. J.W. 2009. Revisiting the limits to growth after peak oil. American Scientist.; Murray, J., & King, D. 2012. Climate policy: Oil's tipping point has passed. Nature, 481(7382), 433–435.

energy, at least of the highest quality (i.e. high utility and high EROI) is becoming much less available, and growth is declining. The economies of most OECD countries (e.g. Japan, Europe and to a lesser extent the U.S.) have essentially stopped growing despite large efforts to "prime the pump[25]". What lies ahead could be further restriction unless new energy technologies with high EROI can be developed.

---

[25] Galbarith, Jamie 2014. The end of normal: The great crisis and the future of growth. Simon and Shuster N.Y.

Chapter 10

Revisiting and Ranking the Cities and Regions

In Chaps. 5–9 we reviewed the major megatrends that will affect society in the twenty-first century. Now, let's take a look at how these trends may affect the cities and regions that we discussed in Chap. 4. The sustainability rankings are based on the following key factors and trends, and how they are expected to vary across the American landscape in coming decades.

Climate change means that temperatures will generally increase. In areas that are already hot, this will make life more difficult for both humans and ecosystems. In colder areas, this may be beneficial, but milder winters can also mean the spread of certain plant diseases and pests. Dry areas will likely become drier while wet areas will, on average, become wetter. Climate change will most dramatically impact the Southwest where extreme water shortages and wild fires will become more common.[1] Increased rain will fall in the upper Mississippi basin, Midwest and Northeast. More large floods are projected on the Mississippi River, a possible benefit for the Mississippi delta but a strain for flood control levees along the river. Sea-level rise is accelerating and is likely to reach a meter or more by 2100, which will have an enormous impact on property values and even space available in all coastal regions.[2] There is an increasing likelihood

---

[1] See *The bug that's eating the woods* by Hillary Rosner in the April 2015 National Geographic for a description of how high temperatures are leading to widespread tree death due to beetle infestations and resulting forest fires.

[2] E. Rignot, J. Mouginot, M. Morlighem, H. Seroussi, B. Scheuch. Widespread, rapid grounding line retreat of Pine Island, Thwaites, Smith, and Kohler glaciers, West Antarctica, from 1992 to 2011. *Geophysical Research Letters*, 41, 3502–3509. Doi:10.1002/2014GL060140.; IPCC 2013, ibid.

© Springer Science+Business Media New York 2016
J.W. Day, C. Hall, *America's Most Sustainable Cities and Regions*,
DOI 10.1007/978-1-4939-3243-6_10

of more frequent and stronger hurricanes and other storms. Climate in general is likely become much more erratic.

Our primary energy supply, fossil fuels, especially conventional oil, will become scarce and much more expensive. By mid-century, affordable conventional oil and natural gas will be in steep decline and coal will be the dominant fossil fuel for the second half of the century. By the end of the century, the sun will be setting on the fossil fuel age. Non-conventional oil (shale oil and gas, tar sands, ultra deep water oil) may provide a short-term respite but will not make up for decreases in conventional oil because of their relatively low energy return on energy and money invested in recovery and processing. As stated recently by the IEA,[3] "the short-term picture of a well-supplied oil market [buoyed in part by the U.S. shale boom] should not disguise the challenges that lie ahead as reliance grows on a relatively small number of producers." Any fossil fuel extraction operations that are still around at the end of the century will be very capital intensive. Big oil as we know it will cease to exist long before then, and renewables probably can fill only a fraction of the impending supply gap. All renewables have a much lower net energy yield than fossil fuels did, especially for much of the twentieth century in the glory days of unprecedented economic growth and wealth accumulation. Solar and wind energy and other renewables can offer some additional energy to society if developed and used wisely. But we cannot conceive that they can support the patterns of energy consumption that currently exist among the affluent in today's globalized industrial economy. Diminishing energy returns to society will result in fewer resources available to devote to other economic activities. As Gail Tverberg recently pointed out,[4] the result of changes in the energy economy has tended to be fewer well-paying jobs, leading to a disparity between what consumers can afford to pay for oil and the amount it costs to extract the oil. Therefore, even if supplies of liquid fuels in all forms, from processed tar sands to coal liquefaction, can be maintained for years, people and societies are likely to find them much more expensive, and the environmental costs in air and water pollution are certain to be high as well. As we have already witnessed, oil prices will likely exhibit wild swings in coming years as declining conventional supplies and fluctuations in demand interact, but two facts are undeniable: (1) society's energy return on investment for energy is declining and (2) we are continuing to deplete finite resources at a rapid pace, leaving future generations with bleak

---

[3] International Energy Agency (IEA). 2014. *World Energy Outlook 2014*. London.
[4] Tverberg, G. 2014. Oil price slide – no good way out. *Our Finite World*. http://ourfinite-world.com/2014/11/05/oil-price-slide-no-good-way-out/

prospects. Keep these points in mind when media outlets such as NPR[5] and the Wall Street Journal[6] suggest that "peak oil" has been debunked.

Rich natural ecosystems offer a fall back for the support of society as fossil fuels are depleted. Although society lived on the goods and services provided by natural systems for thousands of years, nature alone cannot sustain the energy intensive economic system that exists now. But at least nature is a cushion. If we learn to husband and protect natural systems, they offer a sustained flow of goods and services that society can use, and indeed has always used. Ecosystem services are high in the eastern US and in a narrow band of the Northwest. They are especially high in coastal zones and river valleys. Ecosystem services decrease from the southern Great Plains to the Southwest where they are lowest, and climate change will further reduce ecosystem services in this region. The high levels of ecosystem services of coastal regions will be impacted by climate change due to sea-level rise, stronger storms, and in some cases decreased freshwater discharge. Natural ecosystems have been degraded over wide areas of the earth. An important goal for society is restoring these systems so that they can continue to supply ecosystem goods and services. The poet Ezra Pound got it right when he wrote[7]:

> Pull down thy vanity, it is no man
> Made courage, or made order, or made grace,
> Pull down thy vanity, I say pull down.
> Learn of the green-world what can be thy place.

Today's industrialized food systems run on immense inputs of fossil fuel energy in the form of machinery, fertilizers, pesticides, and irrigated water. In the future, price and supply shocks for key inputs (as occurred in 2008) will require higher food prices for production to remain profitable. Potential for food system failure will likely be greatest in regions where intensive resource use is absolutely necessary to maintain productivity (e.g., arid regions reliant on irrigated water, or regions with poor soils reliant on relatively large fertilizer inputs). The tropics will remain ground zero in the debate between supporters of conventional intensification and ecological intensification to meet projected global food demand. Conventional intensification modeled on the current U.S. industrial food system (masquerading as "sustainable intensification") still holds appeal to many due to high yields per acre that can produce immediate food security and economic benefits.

---

[5] Geewax, M. 2014. *Predictions of 'peak oil' prove slippery. NPR.* http://www.npr.org/2014/10/17/356713298/predictions-of-peak-oil-production-prove-slippery

[6] Gold, R. 2014. Why peak-oil predictions haven't come true. *The Wall Street Journal.* http://online.wsj.com/articles/why-peak-oil-predictions-haven-t-come-true-1411937788

[7] Ezra Pound, *Canto LXXXI*

However, this is a short-term solution to a long-term problem. Conventional intensification has contributed to the unsustainable drawdown of fossil fuels, high-quality phosphate rock, non-replenished ground water, and other finite resources. A transition to agricultural systems inspired by ecosystems that recycle materials and require relatively lower energy and material inputs will eventually be imperative. Such systems are already sprouting. Local food production, in some cases, may serve an important role in decreasing communities' reliance on uncertain industrial food systems. Smaller cities located outside of the most densely populated regions in suitable environments could feed themselves from more local foodsheds. However, there will not be enough land to support most megacities with local production. Urban agriculture is a wonderful pursuit and may play a role in food security, but its capacity for supporting cities, especially very large ones, should not be overestimated. Eating less meat and reducing systemic food waste are two ways that we can begin to transition to a more sustainable food system. Currently, 41 % of the plant food calories produced globally do not end up on our plates due to the production of meat, waste, and biofuels.[8] The following question should motivate our future efforts in the U.S. and abroad: How can we minimize per capita land requirements, produce sufficient affordable food, eliminate malnourishment (both hunger and obesity), and greatly reduce waste and environmental destruction using techniques that are not dependent on high inputs of fossil fuels, inorganic fertilizers, other chemicals, and irrigated water? Achieving this at the scale needed to support billions around the globe is a daunting task.

The U.S. population stood at just over 320 million in Spring 2015,[9] most of which is concentrated in 10 or 11 mega regions. The large U.S. and world population demands tremendous amounts of resources. The U.S. population is predicted to increase to over 350 million by 2050, resulting in a greater demand for energy, goods, and services even as extraction of these resources from the biosphere on such a massive scale becomes more difficult. It will become increasingly difficult to maintain densely populated areas, especially in the southwest with growing water problems and low natural resources. Climate change also threatens densely populated areas near the coast. Thus we cannot over emphasize our conclusion that decreasing population levels is a critical component of any solution.

The size of the economy that can be supported relates to energy consumption. As society moves through the twenty-first century, it almost certainly will come to rely more and more on harder to obtain energy sources that have a lower net yield and cost more. Certainly our most

---

[8] Cassidy et al. 2014. Redefining agricultural yields: from tonnes to people nourished per hectare. *Environmental Research Letters*, 8, 034015.

[9] http://www.census.gov/popclock/

valuable sources of oil and gas are likely to be strongly depleted over the rest of this century, and it is not clear to what degree PV and wind energy can fill in the lacunae. It is possible that technology will make renewables more efficient and cost effective so that these sources can provide a basic amount of energy for society. Maybe, but it will be a different society than what exists now.

As energy likely gets more expensive, it will depress the economy, making it more difficult for many people to pay high prices for energy for heating and cooling, transportation, etc. (especially for oil-derived products and services—both directly and indirectly). Thus, as we showed in the energy and economics chapters, the economy will be caught in an increasingly vicious cycle. High energy prices will be a drag on the economy, leading to less oil used and then likely falling energy prices, which then leads to increased economic growth and increasing energy prices (Fig. 9.6). Each turn through the cycle uses vast quantities of energy. For example, between the high oil price of nearly $150 per barrel in 2008 and prices less than $50 per barrel in 2015, about 200 billions barrels of oil were burned, or between 10 and 15 % of proved remaining conventional oil. Higher energy costs also means that it will become harder to recover from natural disasters, climate change-induced or not, such as Hurricanes Katrina and Sandy. In this likely scenario, discretionary income will shrink, perhaps dramatically. Consequently, much of the urban economy will shrink as well, as it is particularly dependent on spending discretionary income. Areas that produce the basic commodities for the economy (food, forest products, minerals— especially energy), and that are less dependent on discretionary income spending, existing in rich natural resource areas, will likely be better off economically.

With these comments in mind, let's revisit the cities and regions introduced earlier in the book and rank them according to sustainability.

Sustainability of the Cities and Regions

In this book, we are using sustainability as the ability of some process (food production) or some system (e.g., a city) to maintain itself through time. Webster's New Universal Unabridged Dictionary defines the term sustain as "to maintain, keep in existence, keep going, to keep supplied with necessities, to provide for," and sustainable as "capable of being sustained or maintained." All of these definitions imply persistence through time, and although it doesn't specifically say it, the implication is persistence of a certain state or situation. In the sense of this book, "to keep supplied with necessities" can be taken to mean maintaining the flows of resources from nature.

However, these definitions do not imply a static, unchanging condition. There will always be adaptation, migration, and transient stability. This is clear from the descriptions of how the cities and regions discussed in this book developed and changed over time. Thus, sustainability is not a final state, but instead a process. So sustainability is the dynamic durability of systems and processes. In terms of human development, the Brundtland Commission of the United Nations defined sustainable development as "development that meets the needs of the present without compromising the ability of future generations to meet their own needs."[10] But current human development has already seriously compromised the ability of future generations to meet their needs. Within the lifetime of people living now, humans will have used most of the affordable, readily available fossil fuels, degraded soils in many agricultural regions, caused widespread environmental deterioration, appropriated much of the plant growth on land for direct or indirect human use, dramatically changed the climate, and mined most of the high quality metal ores. The technological advancements allowing access to previously untapped nonrenewable energy sources that have been invoked by some as evidence that human ingenuity can defy limitations may serve us now, but will leave future generations with even fewer options. So it is within the context of such dynamic change through time that we are considering the future sustainability of cities and regions.

Ecologists speak of a climax community for an ecosystem. This is a long-lived, relatively stable and healthy ecosystem that will persist in a particular location given soil, climate and other environmental conditions that exist there. Many people tend to think of cities being like ecosystems; as relatively stable, and most people think of the normal or virtuous ecosystem or city as one they knew during their childhood. But ecosystems, like cities, are materially open systems where disturbance is common, equilibrium states are transient, and there are multiple pathways of system dynamics. For example, in much of the eastern United States, a deciduous hardwood forest is the climax community for the region unless something intervenes to change the climax, such as a forest fire or a pest such as the fungus that killed Elm trees throughout the eastern U.S. One could call this system, which had persisted for centuries, sustainable. But it is clear that it is not static but rather dynamic. Neither ecosystems nor cities are

---

[10] United Nations General Assembly (March 20, 1987). *"Report of the World Commission on Environment and Development: Our Common Future*; Transmitted to the General Assembly as an Annex to document A/42/427 - Development and International Co-operation: Environment; Our Common Future, Chapter 2: Towards Sustainable Development; Paragraph 1"*. United Nations General Assembly.

unchanging. Flint is an example of a city that is dynamic and changing. Its early development was based on rich natural resources, but then it became an industrial powerhouse of the automobile age. With changing conditions, it lost its industry and half its population. Flint is now actually greening as plants replace structures, perhaps in a way that puts it on a trajectory for a more sustainable future. New Orleans and environs grew and prospered for two centuries while living with a dynamic Mississippi River. In the twentieth century, the river was tamed (albeit for a relatively short period), ultimately with disastrous consequences, as the delta that protected the city from hurricanes disappeared. The future sustainability of the city and region, both socially and ecologically, will require a new accommodation with the river.

Make no mistake about it. The megatrends of the twenty-first century will impact all of society in fundamental and sometimes dramatic ways. But some cities and regions will do better than others. Figuring out which ones will do better requires educated guesswork. We next present our thoughts as to which of these cities and regions will be most sustainable Although the impacts of the trends are based on considerable quantitative information, these rankings are based on our judgment of how each area will be impacted by the different trends. We asked the following questions about the individual cities and regions:

- How strongly will climate change impact the city or region?
- Does the region have a rich natural environment, including fertile soils and adequate rainfall that supports high levels of ecosystem services such as food production?
- What is the total population and the population density of the city or region?
- How much is the economy of the city and region dependent on people spending discretionary income?
- Does the city or region produce economic goods and services that support the base of the economy?
- Is there the potential for catastrophic economic and/or population decline, and why?

We then classified the cities and regions into three broad categories:

- Likely Sustainable
- Moderately Sustainable. By working hard and adapting, life can go on.
- Severely compromised sustainability. Difficult or impossible to maintain in anything like the current situation.

In Table 10.1, we rank the different cities and regions based on how the mega trends will impact them.

Let's explore the prospects for each city and region in more detail.

Table 10.1 Ranking of sustainability of the cities and regions discussed in this book

Sustainability of cities and regions

City	Climate impact	Regional level ecosystem services	Potential for urban ag	Population		Dependence disc income	Importance of producer based economy	Compromised sustainability/Reason
Cedar Rapids	Low-moderate	High	High	Low	250,000	Low	Very high	No
Flint	Low-moderate	Mod-high	High	Low	102,000	Low	Mod	No
Portland	Moderate	Mod-high	Low	Mod-high	2,265 000	Mod-high	Low-mod	Low-mod/decline in discretionary income/ inadequate food
Baton Rouge	Moderate	Mod-high	High	Mod	603,000	Low-mod	Mod	Low-mod/climate, industrial decline
Asheville	Low-moderate	Moderate	Mod-high	Low-mod	226,000	Very high	Low-mod	Yes low /decrease in discretionary income, decline in tourism
Houston	Mod-high	Mod-high	Low	High	4,669,300	Mod-high	Mod	Yes mod/climate, industrial decline, high pop, inadequate food
Orlando	High	Moderate	Low	Mod-high	1,644,000	Very high	Very low	Yes/failure of tourism economy, climate
New Orleans	Very high	High	Mod-low	Mod-high	1,340,000	Mod-high	Mod	Yes/climate (hurricane, slr), economic decline, high pop

New York	Mod-high	Mod-high	Very low	Very high	21,200,000	Very high	Very low	Yes/climate, decrease in discretionary income, inadequate food, high pop
Amarillo	Very high	Low	Low	Low-mod	218,000	Low	Mod-high	Yes/climate-drought
Los Angeles	Very high	Very low	Very low	Very high	13,700,000	Very high	Very low	Yes/climate-lack of water, decline in discretionary income, tourism decline, high pop, in adequate food
Las Vegas	Very high	Very low	Very low	High	1,563,000	Very high	Very low	Yes/climate, decrease in discretionary income, decline in tourism, inadequate food

Note: Population figures are for 2000 from the U.S. Census Bureau for metropolitan areas. The population for southern California (combined Los Angeles and San Diego meto areas) is 19,186,000

Population numbers are to the nearest thousand

The Census Bureau lists Flint as part of the Detroit metropolitan area with a population of 5,456,000 in 2000

New York City

New York City is arguably the most important city in the world. It is a global financial and cultural center and one of the ten largest metropolitan areas in the world. The 21 million strong New York metro area is part of the northeast U.S. megalopolis of nearly 60 million people, stretching from Washington DC to Boston.

The economy of New York, like most large cities, is sustained by enormous inputs of materials and energy (and money) that have been subsidized for over a century by cheap fossil fuels. Much of what supports the economy of New York and other large cities is based on people spending discretionary income. As cheap fossil fuels became more and more abundant in the nineteenth and especially twentieth centuries, discretionary income grew very large. Using this income, people come to New York as tourists and business people. Over ten million tourists visit NYC yearly.[11] These tourists, and million of business visitors, pour hundreds of millions of dollars into the economy. People also send large amounts of discretionary wealth to New York because it is a financial center. Indirectly, profits from enterprises all over the country and the world flow to wealthy individuals who own much of the financial and industrial means of production. As the net energy available to society diminishes and the economy shrinks, so will discretionary income and the flows to cities like New York, probably dramatically. Many of the millions of tourists and business people who come to New York each year do so on airplanes. The airline industry as it currently exists will most likely shrink considerably and possibly even disappear due to a combination of increasing energy prices and reduced ability of consumers to pay for air travel. An analysis by Christian Kerschner of the University of Barcelona and colleagues found that the sectors of the economy that will be most impacted by increasing energy prices are fertilizer production, transport by air, and iron mills.[12]

The many restaurants, hotels, museums, sports centers, and other business and tourist attractions will have to shrink in number and many people who work in these businesses will lose their jobs. Large urban areas like New York will have trouble feeding themselves. New Yorkers, like all citizens in the U.S., have benefited from an industrial agricultural system supercharged on fossil fuels. The study by Kerschner and colleagues mentioned above showed that fertilizer production will be one of the sectors of the economy most impacted by decreased energy availability and/or higher

---

[11] Elizabeth Becker. 2013. *Overbooked*. Simon & Schuster, NY. 449 p.

[12] C. Kerschner, C. Prell, K. Feng, and K. Hubacek. 2013. Economic vulnerability to peak oil. *Global Environmental Change*, 23, 1424–1433.

cost. This will impact the cost of food since fertilizers are important for agricultural production. Similar trends will happen in the other large cities in the Northeast such as Boston, Philadelphia, Baltimore, and Washington, as well as in the rest of the country.

The 21 million people who live in the New York metropolitan area, and the whole population of the nearly 60 million-strong northeast megalopolis, consume an enormous amount of food. For the current food system and typical Western diet, it takes about 1.25 acres (0.5 hectares) of productive agricultural land (pasture and cultivated land) to feed each person in the U.S. Thus, about 75 million acres of land are needed to feed the population of the Northeast megalopolis. However, there are only about 34 million acres of agricultural land in the states from Virginia to Maine that encompass the northeastern megalopolis. New York and the Northeast currently consume much more food than the local agriculture of the region can produce.

New York and the other cities of the Northeast megalopolis will also be strongly impacted by climate change. All of these large cities are on the coast and have significant areas within a meter or two of sea level. Sea level is projected to rise by about a meter or more by 2100 and continue rising thereafter. Coastal areas will also be impacted by hurricanes and other coastal storms such as Irene in 2011 and Sandy in 2012 that generated storm surge of almost 14 feet at Battery Park, the highest ever measured. People will either have to move to higher ground or construct coastal defense systems. The costs to protect coastal urban areas in the nation would likely run into the hundreds of billions, if not trillions, of dollars. And as energy prices increase, costs will rise and such coastal protection may become prohibitively expensive.

Clearly, maintaining the densely populated northeastern corridor through the twenty-first century in its current state will be difficult if not impossible. Decreasing discretionary income, economic contraction, inability of local food production to feed the population, the inability of the economic system to maintain the enormous inputs to large cities, and climate change will combine to bring dramatic change to this region.

William Rees contends that if cities are to be sustainable in the future, they must rebalance production and consumption, abandon growth, and re-localize.[13] The trajectory of megatrends of the twenty-first century will make this difficult for large urban regions like the northeast megalopolis. People face choices and can move and/or reorganize their activities, but it is difficult to see from the present with any clarity what the overall results

---

[13] William Rees. 2012. Cities as dissipative structures: Global change and the vulnerability of urban civilization. In: M. Weinstein and R. Turner (editors), *Sustainability Science*. Springer, New York.

will be. The world has become increasingly urbanized over the last century, with more than half of the population now living in urban areas. It will be difficult to maintain this trajectory given twenty-first century mega trends. It will be interesting to see how large urban areas respond to these coming changes.

Flint

Flint is widely viewed as a failed or more challenged legacy city in comparison with other cities of the east coast and elsewhere. We don't ascribe to this view. In fact, we believe that Flint may be a harbinger of how cities can survive through the twenty-first century. The information developed in this book indicates that densely populated urban areas like New York will have a number of difficulties maintaining their size and level of inputs, given the megatrends of this century. But, one might argue—to the contrary—that large cities are less vulnerable because they are connected via diverse networks. If, for example, a port gets knocked out due to a hurricane, other infrastructure such as roads, rail and airports can still maintain flows into and out of the city. These same networks also buffer cities from local conditions such as seasonality or drought. Smaller cities with fewer connections may therefore be less likely to "weather storms." This may be correct in the medium term, but as the impacts of twenty-first century megatrends become more and more severe, the diverse networks, which are often energy intensive to maintain, may begin to become less functional. Despite this, small-to-medium sized cities are much larger in number than major cities, and addressing their plight in terms of sustainability may have an equal or greater impact on the sustainability of the nation.

Flint is already decreasing in size, having lost about half of its population. This is characteristic of a number of the rust belt cities, including Flint's nearest major city neighbor Detroit. While this is usually considered very negative, in fact it is one way that a city can adjust to reduced resources, and hence might help prepare a city for the future we envision. A considerable amount of green space exists within the city, and there is ample farmland in the region, so Flint might be able to feed itself largely from its local and regional foodshed. However, it is unclear if this available green space and land, much of which may still require remediation before planting, has been fully realized in terms of its full capabilities. Although Flint is part of the Great Lakes megaregion, it is near the northern boundary in Michigan. There is abundant farmland with fertile soils in the lower populated portion of the state north of Flint. Thus there is considerable opportunity for food production. The Great Lakes area is the largest megaregion in the U.S. but it is also located in the Midwest bread basket, one of the most important agricultural regions of the world. Good farmland is dispersed throughout the area. In general, this region has high ecosystem services with good soils, many rivers, and ample precipitation.

Climate change in this region will be somewhat benign compared to other areas. All regions of the U.S. will be impacted by climate change, but Flint is in an area that will not suffer the extreme drought as in much of the western U.S. or the effects of sea-level rise and hurricanes on coastal areas. How Flint and other similar "failing" cities manage the coming transition may provide examples for other cities that have not yet suffered as Flint has.

Some readers may find it puzzling that we even included Flint in this book. After all, it is an example of what many see as a failing city and hardly an example of a sustainable urban area. But others have looked at small to medium sized cities as examples of a promising urban future. Catherine Tumber in her book *Small, Gritty, and Green* sees promise in smaller industrial cities in a resource-scarce future.[14] She writes about rust-belt cities in the northeast and Midwest; cities like Syracuse, Akron, Worcester, Buffalo, Peoria, and Youngstown, their deterioration and grow-ing invisibility as cities, but also their potential for urban renewal. But her book could just as well be about similar cities in other parts of the country. These cities, like many others, once had vibrant downtowns. But the forces of deindustrialization, outsourcing, and globalization led to loss of jobs, population loss, and poor school systems. Many of these trends are rein-forcing and could be reversed via the joint benefits of city and regional greening and land reuse policies that, for instance, produce food and other necessities but also revitalize abandoned urban spaces and draw popula-tions back into once forgotten spaces as residents and a local workforce. In terms of population, Tumber is writing about small cities in the range of 50,000–500,000 persons. For the cities we discuss in this book, Amarillo, Asheville, Baton Rouge, Cedar Rapids, Flint, and Orlando fall within or near this population range.

In a climate challenged world dealing with energy scarcity, Tumber argues that small to mid-sized cities have certain advantages. They are large enough to maintain significant urban amenities, they have abundant open space that could produce food and other natural amenities (especially in the Midwest, which has some of the best soils in the world), and they have the potential and often the infrastructure for some level of manufac-turing. She admits that her viewpoint is at odds with much conventional

---

[14] Catherine Tumber. 2012. *Small, Gritty, and Green*. The MIT Press, Cambridge. 211 p.

wisdom that the future lies in large, super dense cities, such as Edward Glaeser argues in *Triumph of the City*. The decline of small to medium sized cities had much to do with abundant cheap energy. Distance didn't matter. This allowed globalization, the outsourcing of whole industries, and cheap travel for vacations and business. But the age of cheap energy is ending. Distance is beginning to matter again. It is for these reasons that Flint has potential to become a hopeful vision of the future rather than a bad memory of the past.

Asheville

Asheville, North Carolina, is located on the north-central edge of the Piedmont Atlantic megaregion in the foothills of the Smoky Mountains at an elevation of about 2000 feet. Because of the benign climate and friendly, progressive culture, it has become a popular area to settle. The area has a number of attributes that will make it less susceptible to the megatrends of the twenty-first century. Because of its elevation, it does not suffer the extreme heat of much of the southeast, and it has mild winters. Even though it is part of the Piedmont Atlantic megaregion, it is not located in the most densely populated parts of this region such as Atlanta, Birmingham and Charlotte. The local population density is relatively low and the region to the north and west has a low population density. The natural areas surrounding Asheville are generally rich and productive, so there are high ecosystem services. However soils in the region are relatively poor and require soil amendments for productive agriculture. Sloping land may further limit the sustainability of food production in and around this city among mountains.

Economic well-being is perhaps the greatest challenge for Asheville's future. The economy is dependent to a large degree on discretionary wealth and income. There is an important tourist industry associated with its bohemian culture (and the legacy of Thomas Wolff), the attractions of the surrounding mountains, and the Vanderbilt's Biltmore house. Asheville has long been a destination for southerners escaping the summer heat. Thomas Wolff's mother ran a boarding house that catered to the lowbrow end of this clientele. The house is now a state historical site that draws aficionados of Wolff's fiction. Asheville is also an important educational center with several colleges (e.g., Warren Wilson) and universities (a branch campus of the University of North Carolina). Many people have retired there and brought with them their accumulated wealth. Presumably as society moves through this century, societal and individual discretionary income and wealth will decrease. Therefore, the great challenge for Asheville is to develop an economy that depends increasingly on the natural resources and people of the region while maintaining the amenities that will attract such retirement wealth as remains. People will continue to be drawn to Asheville because of its climate and culture, but already well-paying employment can be hard to come by for some of the recent arrivals. It seems possible that a hybrid economy based on local natural resources and moderate discretionary spending may be a sustainable model for the future of this region and many other smaller cities located in relatively rich natural resource regions.

Orlando

Orlando, like many other tourist cities, has boomed over the past several decades. But can such an economy continue in a time of emerging scarcity and decreasing discretionary income? The worldwide tourism industry, estimated at nearly $7 trillion in 2012, is huge and supports 1 out of 12 global jobs. However, it is very energy intensive.[15] The dramatic growth of the industry is indicated by the fact that in 1950, there were 25 million international tourist trips compared to one billion in 2012. People who go to the attractions in Orlando typically do so by air. Increased potential for energy price shocks threaten to end airline travel as we know it. Many also travel to Orlando by personal auto, but this too will become less common in coming decades as gasoline prices continue to increase and/or people's ability to pay decreases. Both cheap transportation and abundant discretionary income are absolutely necessary for Orlando to continue to function as it does now. The large tourism complex centered around Orlando will likely shrink significantly.

Orlando is in the center of the Florida mega region, the fifth largest in the U.S. This includes the Atlantic coast from Daytona to Miami, the Keys, Everglades National Park, and the southwest coast. In general, the region is dependent on discretionary income and discretionary wealth. A large population of snow birds descend on the state each winter to escape northern cold. These people spend large sums of money buoying up the economy. Also, a large resident population of retirees spend their pensions, social security income, and accumulated wealth. On average, this retired population requires greater medical care. This brings in additional money from Medicare and secondary medical insurance, much of which comes from out of state. South Florida is an international destination for a large

---

[15] Becker, 2013, ibid.

number of well-off individuals from Latin America, many of whom have second homes in the Miami area and elsewhere. The net result of this is an enormous subsidy from the rest of the nation and beyond that is critical for Florida's economy. As energy and resource scarcity emerges, as we think quite likely, this subsidy will decline and with it the economy.

Climate change will also impact the Florida peninsula. Water for drinking and irrigation comes almost entirely from ground water, which is replenished by frequent rainstorms. The extremely high demand for both agriculture and direct human use and periodic drought already leads to periodic water shortages as well as salt water intrusion into drawn-down drinking water aquifers. Some climate estimates project more droughts leading to even more water stress in the future for South Florida. Seventy-five percent of Florida's 18 million people live in coastal counties, and these areas will be impacted by both rising sea level and increased intensity of stronger hurricanes. Because south Florida is a peninsula, essentially the whole area is coastal and regularly affected by hurricanes. For example, in August, 1995, hurricanes Humberto and Iris and tropical storm Jerry were all visible on the same image as they tracked towards Florida (Fig. 10.1). It is also likely that the region will become hotter. The February 2015 issue of National Geographic graphically illustrates the impact climate change

Fig. 10.1 Satellite image from August 24, 1995, showing hurricanes Humberto and Isis and tropical storm Jerry approaching Florida. (*Source*: National Oceanographic and Atmospheric Administration)

will have on Florida including rising seas, salt water intrusion into ground-water, eroding beaches, and huge amounts of threatened infrastructure.[16]

Soils in the Orlando region are relatively poor and the limestone bed-rock limits the movement of surface water. Given the high population liv-ing on the Florida peninsula, it is doubtful that all required food could be produced locally in the region. Sea-level rise and hurricanes will threaten the rich coastal ecosystems of the state, as well as the interior Everglades.

Cedar Rapids

We believe that Cedar Rapids likely has the greatest potential for long-term sustainability out of all of the cities considered in this book. Cedar Rapids is a relatively small city compared to others that we have discussed, and agriculture has been an important economic activity from the time of its settlement. It is situated in one of the richest farming regions in the world. In the past, the Midwest breadbasket had a more diverse group of crops, but now grows primarily corn and soybeans. But the region could become more agriculturally diverse again. The rich agricultural productivity reflects high levels of ecosystem services in this area with rich soils and abundant rainfall. The area enjoys a low dependence on the spending of discretionary income because it is a producer region. Most products of the Cedar Rapids regions go into staple goods. The international aspect of the food system is reflected in the fact that about ten large ships carrying grain from the Midwest leave the mouth of the Mississippi each day.

Climate change impacts to the Cedar Rapids region will be moderate compared to other regions of the country such as the Southwest and coastal areas. Cedar Rapids will see some temperature increase, but rainfall should remain adequate for agriculture.

For the foreseeable future, Cedar Rapids could easily feed itself from its local foodshed, and could provide a surplus for other regions, as it has for over a century. Cedar Rapids is an example of the many small to moderate sized cities spread across the landscape in resource rich areas with suffi-cient precipitation that are, or could be, producers of the stuff (food, timber, minerals, energy) that supports the rest of society. Much of the eastern U.S. away from densely populated areas and not too near the coast could sup-port such small to moderate sized cities that could produce a surplus for the rest of society. But the Eastern U.S. generally does not have soils as fertile as the Midwest (see Fig. 6.3 in the Food chapter).

---

[16] Laura Parker. Treading Water. *National Geographic*. February 2015. pp. 107–125.

The Lower Mississippi River and the North Central Gulf–New Orleans–Baton Rouge–Houston

Although New Orleans and Baton Rouge are two distinct cities and differ culturally, the area of south Louisiana encompassing the two cities is part of an extremely important regional ecological and economic system. The Gulf Coast megaregion stretches for nearly a thousand miles from the Florida Panhandle to the Rio Grande Valley in South Texas. This is an area of low regional population density, generally abundant rainfall, and high natural resource productivity. New Orleans, Houston, and Baton Rouge are all major ports. Four of the largest ten port authorities in the nation are located between Baton Rouge and the mouth of the Mississippi River. The lower Mississippi ships more tonnage than any other port system in the world, and is connected by barge, truck, and rail to all parts of the U.S. The map below shows that truck traffic carrying goods from the ports of the north central Gulf of Mexico region reaches every state in the nation (Fig. 10.2). Similar maps could be displayed showing the flow of oil and natural gas from the Gulf region, and coal and grain from the northern plains and Midwest to the lower Mississippi. Clearly this region is an integral component of the U.S. economy.

The Mississippi delta stretching from Chandeleur Sound south of Biloxi to Galveston Bay, Texas is the largest coastal system in North America, and supports one of the largest fisheries in the world. This is a major producer region with high production in all four primary sectors—fishing, farming, forestry, and mining of oil and natural gas. The north central Gulf region is one of the most important energy producers in the nation. In addition, much imported oil is also refined in—and transshipped through—the region. New Orleans is a major national and international tourist center. What goes on in south Louisiana and the north central Gulf supports hundreds of billions of dollars of economic activity in the U.S. economy. An important question is can the region sustain this economic production?

South Louisiana is an extremely rich natural resource area with the highest regional value of ecosystem goods and services in the nation. The Mississippi River subsidizes, and has subsidized, this region for thousands of years, and this subsidy has underwritten the Louisiana economy for centuries. Both New Orleans and Baton Rouge got their starts as ports. But human activity has dramatically altered and degraded this region. In an attempt to tame the wild Mississippi, humans constructed levees and cut off many channels that have isolated the vast Mississippi alluvial floodplain and delta from the life-giving waters of the river. Deprived of river input, the Mississippi delta is falling apart, and both its non-human and human inhabitants are threatened. If nothing is done, the delta will be largely gone

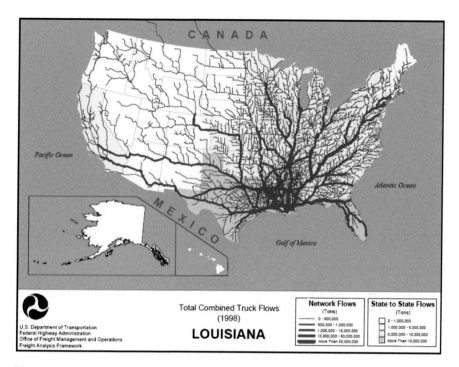

Fig. 10.2 Total combined truck flows from the north-central Gulf of Mexico centered around the New Orleans-Baton Rouge-Houston Region. This region is effectively the heart of commodity transport in the U.S. The the *red* pathways resemble arteries pumping blood out to the extremities. The width of the line is proportional to the amount of truck traffic. The *thickest* lines are more than 10 million tons per year (Source: U.S. Dept. of Transportation, Federal Highway Administration 2006.)

in this century (Fig. 10.3). People have been leaving the coastal area for decades, and the New Orleans metropolitan area still has not regained its pre-Katrina population.

In response to the deterioration of the delta, the State of Louisiana has embarked on an ambitious 50-year, $50 billion dollar plan to restore the delta. But the emerging megatrends of the twenty-first century will severely challenge the success of this plan, and will force substantial changes. Climate change is already strongly impacting the area. The Mississippi delta has millions of acres that are at or only slightly above sea level. New Orleans and other areas of the delta are below sea level and protected by levees. Anything that affects sea level affects this area strongly; and sea level is predicted to increase by a meter or more by 2100. Actual water level rise will be greater because much of the land is sinking. On top of this long-term gradual rise, hurricanes bring short-term water level surges that can reach as high as 10 m. And the frequency of the strongest hurricanes is projected to increase.

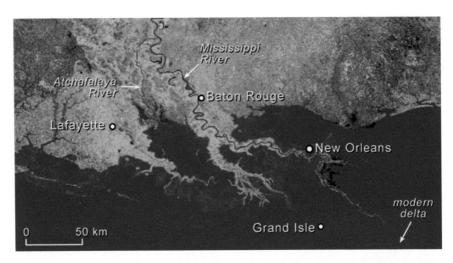

Fig. 10.3 A Geospatial map rendering of the Mississippi Delta in south Louisiana showing potential land submergence given 1 m of sea level rise with no input of sediment and no catastrophic events. (From: Blum, M. D., & Roberts, H. H. 2012. The Mississippi delta region: past, present, and future. *Annual Review of Earth and Planetary Sciences*, **40**, 655-683. http://wvue.images.worldnow.com/images/18672826_BG4.gif)

Before the levees were built, the Mississippi River overflowed its banks almost every year. But the nearly continuous levees from Cairo, Illinois to the mouth of the river have stopped this. In south Louisiana during high flood stages, the river can be tens of feet higher than houses on the other side of the levee. Some climate projections predict that there will be more large floods on the river threating the integrity of the flood control system. There are hundreds of miles of levees along the river and around populated areas in the coast to protect against flooding. Building and maintaining such a system is extremely expensive and energy intensive. It is not difficult to imagine a time in this century when the combination of climate forces, decreasing societal discretionary income, and scarce and expensive energy will lead to an inability to maintain this flood control system as it currently exists. Hurricanes, such as Katrina (Fig. 10.4) and river floods such as occurred in 2011 may overwhelm the system. This would be a disaster; or this threat could be an opportunity. If the river breaks the hold the levees have, then it could go about rebuilding the coast and reinvigorating the flood plain on its own. So one approach to living with the river and coast in an energy scarce future is for people to pull back to safe, mainly higher areas and to "let her rip" as some along the coast say. This could be managed in a way that protects important areas like New Orleans.

Such a prospect might not be all that bad, especially if it is the only choice. People would be protected. The region generally has a relatively low

Fig. 10.4 Hurricane Katrina making landfall near New Orleans, Louisiana (http://commons.wikimedia.org/wiki/File:Katrina_2nd_landfall.jpg Image from NASA, [Public domain], via Wikimedia Commons)

population density surrounded by a rich productive ecosystem that supports farming, forestry and fisheries. Since there is not enough current agriculture in the immediate region to feed the local population, food production would need to expand considerably. The river would begin to rebuild the faltering ecosystem services of the delta and floodplain as well as continue to be an avenue for river-borne transport of goods. Oil and natural gas probably will last several decades more before they play out and could help to build a more sustainable future during the coming transition, if used wisely. One can imagine people again living more sustainably with the system that has supported the economy of the area, and of the nation, for the past three centuries and indeed for thousands of years. But society would look much different than it does now. Any "solution" will need to be dynamic, involve continual adaptation, and accept nature's transience. This is true for any region where a more sustainable status is achieved. But unless a more sustainable approach is adapted, the future of New Orleans is in doubt.

The future of Baton Rouge is tied to most of the same factors as New Orleans. The greater part of economic activity of Louisiana is located in the southern part of the state and especially in the southeast. The rich resources of the Mississippi delta and activities in and along the river are critical to the economy of the region. Fisheries, forestry, agriculture, shipping, petrochemicals, the oil and gas industry, and tourism buoy the economy of both cities. But the two cities differ in one critical aspect—elevation. While New Orleans is susceptible to, and has experienced, both riverine and hurricane flooding, most of Baton Rouge sits on the bluffs overlooking the river. Even if the levees failed, no river flood or hurricane surge can reach most of Baton Rogue. Continued sea-level rise may make the extreme southeastern part of the city coastal by 2100, but even the highest projections for sea-level rise in the next two to three centuries will not flood much of the city. An important source of population increase has been people leaving the coastal zone. As sea-level rise accelerates and stronger hurricanes become more common, Baton Rouge along with Lafayette, which sits on the bluff on the western side of the flood plain, and areas north of Lake Pontchartrain are increasingly becoming areas of refuge.

One of the largest oil refineries in the nation is in Baton Rouge, and there are numerous petrochemical industries along the river. These will continue to hum along for at least the next several decades during the waning fossil fuel era. The city is the seat of state government and home to the largest university in the state, which will continue to support the economy. Baton Rouge is also an important market city for the region. All of these factors may make Baton Rouge and other "higher elevation" areas more sustainable than downriver New Orleans.

Houston, several hundred miles west of the Mississippi River, is also a major center for trade and part of the nexus of high natural resource productivity, energy production, and international trade of the north central Gulf. Like the lower Mississippi ports, Houston is also in the top ten U.S. ports. The major difference between Houston and the lower Mississippi ports is that Houston does not have the major water connection to the center of the country so that trade may be more based on the local region. Houston is part of both the Gulf Coast and Texas Triangle megaregions. Both of these megaregions span the boundary of the transition from a moist east to a dry west. The eastern edge of the map of the arid region of the U.S. produced by John Wesley Powell (Fig. 3.1) is on the boundary of these megaregions. The 100th meridian running north–south through central Texas is the dividing line between the wet east and the dry west. Drought is a factor that must be considered in the western part of this region.

Economically, Houston is in the middle of an important producer region. Like Louisiana, this region of Texas is an important producer of oil and natural gas, both onshore and offshore. It is also a center for the new boom of oil and gas produced from hydraulic fracking. Although much of the hype about fracking appears quite overblown, Texas—like Louisiana—likely has several decades more of enormous available mineral wealth, and a long tail of importance as production falls. Also like Louisiana, Texas is a center for refining and petrochemicals. If the controversial Keystone pipeline is ever approved, this will keep the refineries humming even though there is little net energy in Canadian tar sands. However, past mid century, the fossil fuels of the northern Gulf coast will have played out so the city (and the nation in general) should use these remaining resources to move towards greater sustainability.

Compared to most other large cities in the nation, Houston is moderately dependent on discretionary income spending. The area serves an important regional market center and attracts vacationers and conventions.

Texas is also a very important agricultural and fisheries producer. The northern Gulf fisheries region that supports the Louisiana fishery also stretches along the Texas coast, and the two states vie for the largest shrimp fishery in the nation. Texas agriculture supports cattle, dairy farming, and cotton. The lower Rio Grande Valley is an important producer of citrus, sorghum, and other crops. All of this rich production is important regionally and helps support Houston.

Houston's population is nearly five million people, but it is spread out, and the city incorporates much green space. The region surrounding Houston has a moderate population density, generally productive soils where excessive soil moisture is not present, and it may be possible for the city to largely feed itself locally if low-meat diets are adopted. But this is a tall, Texas-sized order. Currently, eastern Texas generally has sufficient rainfall to support agriculture and the region has high ecosystem services.

As with the rest of the north central Gulf, Houston will be strongly impacted by climate change. Hurricanes regularly hit this coastal region. Ike was the last major storm to affect the Houston area. It flooded parts of south Houston, Galveston Island, and wiped out the Bolivar Peninsula to the east of Galveston. The infamous hurricane of 1900 was the deadliest hurricane in U.S. history, causing the death of more than 6000 people and destroying much of Galveston.[17] As a result, the construction of what is called the "Ike" dike is being proposed to protect Galveston and Houston. Like such systems in Louisiana, the Ike Dike would be expensive and probably could not be maintained in a time of decreasing energy availability. The low-lying areas of south Houston (where much of the industrial activity is concentrated) are particularly susceptible to Hurricane surge. Accelerated sea-level rise and more strong hurricanes will strongly impact such low-lying areas in the region. The Houston region and east Texas are

---

[17] John Edward Weems. 1957. *A Weekend in September*. Texas A&M University Press, College Station. 180 p.

the wettest parts of the state, but climate change may make this area more prone to drought. Drought impacts will be much more extreme further to the west, as will be discussed in the next section on Amarillo.

Amarillo

Amarillo, in the northwest corner of Texas, lies in a region very different from that of Houston, over 700 miles away to the southeast. While Houston is in the wettest part of the state with an average rainfall of about 60 in., Amarillo is in one of the driest areas, averaging less than 20 in. a year.

Amarillo is in the middle of an important agricultural region with quite fertile soil that supports production of cotton, wheat, corn, and cattle. But unlike Cedar Rapids to the northeast, this area is dependent on irrigation from impounded surface runoff and groundwater pumped from the Ogallala Aquifer. Both of these are likely to decrease. The drought of the 1930s that led to the infamous dust bowl strongly impacted the area, and climate projections are for such droughts to become more common in coming decades. The Ogallala is being overdrawn over much of its extent, but the Texas Panhandle is one of the worst areas, with ground water falling by as much as 150 feet (see Fig. 6.9). The need for irrigation is a reflection of low values of ecosystem services due to the low natural productivity of this semi-arid region. On the other hand, it is a producer region with little dependence on discretionary income. But growing climate impacts and increasing energy costs will threaten agriculture and the economy of this region.

Las Vegas

Las Vegas is arguably the least sustainable large city in the nation. The city is located in a desert with very low rainfall and low soil fertility. Las Vegas has among the lowest ecosystem services in the nation. The whole economy of the city is underwritten by tourism. Increasing energy prices will likely make it impossible to maintain the high levels of tourism, without which the city cannot exist in anything like its present form. In other words, the city is entirely dependent on people spending discretionary income, and this will decrease significantly in this century. The city will also be strongly impacted by climate change. The Colorado River is already over-allocated, and the area faces a strong possibility that Lake Mead may dry up in the next couple of decades. Thus Las Vegas is a dense urban area in a region that has very low natural resources that give rise to ecosystem goods and services. Everything that is used in the city, including water, is imported. There is nothing locally to fall back on. This is the most threatened of the cities and regions that we discuss in this book.

Orlando and Las Vegas are examples of two cities that will be hard pressed to maintain their economic status because of their great dependence on tourism, which—as we have emphasized—is dependent on high levels of discretionary income and cheap travel. Elizabeth Becker discusses the tourism travel industry in her book *Overbooked*. Tourism is a big business. According to Becker, it produces $6.6 trillion for the global economy and employs one of every 12 working people on earth. In terms of overall global economic impact, tourism is in the same class as oil, energy, finance, defense, and agriculture. But this is likely to shrink significantly as twenty-first century mega trends impact society. Other examples of activities and locations where tourism-dependence may challenge sustainability include ski resorts such as Sun Valley, Aspen, Vail, Crested Butte, Telluride, and Steamboat Springs, stock car racing meccas such as Talladega, Georgia, variety shows and musical attractions in Nashville, Tennessee and Branson, Missouri, ecotourism such as rafting in the Asheville area, and the Shakespeare Festival in Medford Oregon. Many large cities also strongly depend on tourism and business conventions such as New York, New Orleans, Los Angeles, Portland and Houston.

Los Angeles

Los Angeles is the second largest metropolitan area in the United States after New York. But unlike NYC, which is in a rich and productive natural region, L.A. is in a desert. The area has fertile soils, but low rainfall limits agriculture unless there is irrigation. New York blossomed about two centuries before L.A., and only with the coming of cheap fossil fuels was L.A. able to grow into a major city. L.A. had to overcome the obstacles of very little water, low ecosystem services due to the arid conditions, and being far from other population centers, especially the populated east where the major markets exist for all of the products produced by Los Angeles, and indeed the rest of California. Cheap energy was essential to overcoming these obstacles.

The Los Angeles region is densely populated. About 12 million people inhabited the area in 2000, and the Southern California megaregion is projected to grow to nearly 29 million by 2025. There is almost no production of basic commodities locally, and much of the region is based on government spending and people spending discretionary income.

Water is critical to survival of L.A. and southern California in general. And L.A. has had to reach far afield to the Colorado River, Owens Valley, and northern California to obtain sufficient water. All of these water sources are threatened by climate change. Even as population grows in the Southwest, water availability is decreasing. The Southwest suffers chronic water shortages and this is projected to grow worse. In 2013–2015, drought raged over the region and the snow pack in the Sierras was very low. A USA Today headline proclaimed "California's 100-year drought. Fierce fires, agricultural losses – severe water shortage a threat to civilization."[18] In the summer of 2014, California imposed statewide mandatory water restrictions.[19] Drought conditions persisted through the winter of 2014–2015 and the state imposed additional restrictions. In a very readable and insightful book on climate in the west (*The West Without Water*), Lynn Ingram and Frances Malamud-Roam discussed the last 20,000 years of climate history, and projections for the future.[20] They reviewed a large body of scientific evidence about past climate based on such measures as tree rings, presence of charcoal as evidence of past fires, sedimentation rates of small organisms and pollen in lake and coastal ocean sediments and chemical composition of these sediments, and the location of old shorelines of lakes. The book is an excellent demonstration of how scientists synthesize a wide variety of information to paint a picture of climate for a region, in this case the Southwest. This information provides a record of periods of both dry and wet climates, both droughts and floods, and shows that in the past there were decades-long, even centuries-long droughts interspersed with very wet periods and gigantic floods that dwarf anything in the historical record.

What is extremely interesting is how this climate variability affected Native Americans living in the Southwest. A period of extended and severe drought lasted from about 900 AD to about 1400 AD. This period included two mega-droughts separated by a wet interval between 1100 and 1200 AD. Many paleoclimatologists studying the climate of this period consider that these droughts may be a harbinger of the future of the Southwest with climate change. Prior to these droughts in a wetter period, Native Americans developed a sophisticated agricultural society in the Southwest called the Ancestral Pueblo or Anasazi. Well-known pueblo and cliff dwelling sites dot the Four Corners region at Chaco Canyon in New Mexico, Mesa Verde in Colorado, and Canyon de Chelly in Arizona.

---

[18] USA Today. Sept. 3, 2014. P. 1A.

[19] http://online.wsj.com/articles/california-expected-to-set-mandatory-water-curbs-1405367707.

[20] Lynn Ingram and Frances Malamud-Roam. 2013. *The West Without Water*. The University of California Press, Berkeley. 256 p.

As the climate became increasing dry and unpredictable, the Pueblo farmers developed increasingly sophisticated ways of managing and storing water. At Chaco Canyon, inhabitants constructed an extensive system of canals and dams for crop irrigation. Some of the canals were masonry-lined. In southern Arizona, members of the Hohokam tribe developed a canal irrigation system. The largest of these canals were 24 feet wide and 19 feet deep, and were capable of moving large volumes of water over long distances.

After existing for centuries, all of these societies apparently collapsed rather quickly during the very dry periods from thirteenth to fifteenth century. Archaeologists have documented many signs of a population in decline, including malnutrition, starvation, disease, decreased life spans, high infant mortality rates, and signs of violence.

In a number of ways, the predicament of the modern Southwest and California is history writ large in the story of the Ancestral Pueblo. They were a people who flourished when water was abundant, who constructed a sophisticated system to manage, store, and move water. But in the end, when conditions became too dry and further adaptation was limited or impossible, their population crashed and people abandoned the pueblos. Some recent research predicts that droughts in this century will be more severe than those experienced by the Ancestral Pueblo people.[21]

Ingram and Malamud-Roam concluded that for a brief time, humans have had relative climate stability, but this "is slipping away, and we are entering a period of drier and more erratic conditions." Without sufficient water, L.A. cannot survive, and this is true for the whole of the Southwest. And this seems to be the direction the state and the region are heading. In the past, Californians used cheap energy to overcome water scarcity, develop some of the most productive agriculture in the country, and become the most populous state. But just as water is becoming scarcer, energy prices will rise as fossil fuel supplies are depleted. Ingram and Malamud-Roam summarized the future of the west in terms of climate: "less snow in the mountains, earlier snowpack melt in the spring, drier summers, increased forest fires, and shifted storm tracks." All of this will bring less water to the west. This does not bode well for the region. Our technology makes us different to a degree, but remember that this technology to be useful is strongly dependent on the availability of a diminishing commodity, large amounts of affordable energy.

Ingram and Malamud-Roam suggest a number of ways that water can be conserved in a time of growing water scarcity. For individuals, these

---

[21] Cook, B. I., Ault, T. R., & Smerdon, J. E. 2015. Unprecedented 21st century drought risk in the American Southwest and Central Plains. *Science Advances*, 1(1), e1400082.

include more water efficient appliances, drought tolerant landscaping, water recycling, and rainwater harvesting. But agriculture accounts for nearly 80 % of water use in the state. Here such things as using a tiered pricing structure, adjusting the amount and timing of irrigation, using more efficient irrigation methods (e.g., drip irrigation), and recycling irrigation water can be used to achieve higher efficiency in agriculture. During a severe drought in Australia, the use of such techniques reduced water use by 37 %.[22]

But in his book "A Great Aridness" (discussed in Chap. 2), William deBuys argues that water conservation often doesn't do what people think it does. He says that "as currently practiced, it doesn't relieve long-term shortage; it can actually make it worse." If one person or activity saves water, it is then available for some other use. When all is said and done and each person or activity, public or private, conserves as much as possible, demand for water "hardens." This means that the demands that remain are essential and can't be turned off. If the drought continues, perhaps for decades as happened many in the past in the Southwest, then there is nowhere to go and further adaptation is difficult, if not impossible. When there are no more options, no more water to be imported from afar, then something essential has to give. The current human population of the Southwest is probably just too large. California and the Southwest are projected to continue to grow in population even as the climate, energy, and natural resources that support the state shrink. Something has to give, and systemic failure in the region seems likely.

Los Angeles is situated in the larger context of California, and its fate is intricately tied to the future of the state. The state is a draw to millions of tourists who come and spend their discretionary income on the attractions of Los Angeles, San Francisco, Napa and other parts of the wine country, Santa Barbara, the Monterrey Peninsula, Yosemite, and similar areas. But, as we have mentioned already, tourism will decline as discretionary income shrinks. And climate change will affect all some of these areas. Wildfires will burn ever-greater areas. Recent studies indicate that the iconic wine industry will falter as areas such as Napa and Sonoma become suboptimal for grapes and the fine vineyards may produce grapes of wrath.[23]

California leads the nation in agriculture with $44.7 billion in sales in 2012.[24] The other five leading states in terms of sales are Iowa ($31.9 billion), Nebraska ($24.4 billion), Texas ($22.7 billion) and Minnesota ($20.9 billion)

---

[22] Ingram and Malamud-Roam, 2013, ibid.

[23] Hannah, L. et al. 2013. Climate change, wine, and conservation. *Proceedings of the National Academy of Sciences of the United States of America*, 110, 6907–6912.

[24] National Geographic magazine, November 2014 and references cites therein.

The Central Valley alone produces about half the vegetables, fruits, and nuts of the U.S. On average, there are about 2.6 million registered truck-tractors on U.S. roads on a typical day. Trucks traveling east tend to carry food while those traveling west generally carry non-perishables. So it is safe to assume that hundreds of thousands of semis carrying food leave California each year. But California agriculture will suffer two major body blows. First, as noted, climate change is leading to drying in the region. It is unlikely that that there will be enough water to sustain the current level of agricultural output; farmers are already being forced to cut water use and thus curtail production. Modern industrial agriculture is high tech and energy intensive, and the Central Valley is the highest of the high tech. The emerging water-energy crunch will threaten California's agricultural sector to a greater degree than most other areas. And without water, there is not much of a natural system to fall back on (see falling groundwater levels in Fig 6.9).

So the fates of Los Angeles and California, and indeed the rest of the Southwest, are intertwined. The emerging megatrends of the twenty-first century will threaten this region extensively.

Portland

Portland is promoted as one of the most "green" cities in the U.S. Because of this, we have used Portland to look into the meaning of "green" and its relationship to sustainability. The region in which it is situated is actually green (e.g., the color of the verdant vegetation), a reflection of a rich natural environment with high ecosystem services. Do the characteristics that make cities green contribute to sustainability in the face of the mega trends we are discussing in this book? Portland provides a context to view the issues of green and sustainable. In Chap. 3, we discussed the work of Dr. Robert Burger and colleagues showing that Portland as a whole was not that different from other cities of similar size in terms of energy and materials used and wastes generated.

So let's take a look at what "green" means. In a general sense, green refers to a set of characteristics that makes a place more pleasant, enjoyable, and healthy to live in; and perhaps more energy efficient and respectful of the natural world. But perhaps not. There are really no highly quantitative definitions of green, but carbon emissions tend to be evoked. Consider the following statement: "Although the EPA has not established official criteria for ranking the greenness of a city, there are several key areas to measure for effectiveness in carbon footprint reduction. These include air and water quality, efficient recycling and management of waste, percentage of LEED-certified buildings, acres of land devoted to greenspace, use of renewable energy sources, and easy access to products and services that make green lifestyle choices (organic products, buying local, clean transportation methods) easy."[25]

According to the Wikipedia entry on green cities:

> A **sustainable city**, or **eco-city** is a city designed with consideration of environmental impact, inhabited by people dedicated to minimization of required inputs of energy, water and food, and waste output of heat, air pollution - CO_2, methane, and water pollution. ...There remains no completely agreed upon definition for what a sustainable city should be or completely agreed upon paradigm for what components should be included. Generally, developmental experts agree that a sustainable city should meet the needs of the present without sacrificing the ability of future generations to meet their own needs. The ambiguity within this idea leads to a great deal of variation in terms of how cities carry out their attempts to become sustainable. However, a sustainable city should be able to feed itself with minimal reliance on the surrounding countryside, and power itself with renewable sources of energy. The crux of this is to create the smallest possible ecological footprint, and to produce the lowest quantity of pollution possible, to efficiently use land; compost used materials, recycle it or convert waste-to-energy, and thus the city's overall contribution to climate change will be minimal, if such practices are adhered to.[26]

In his book entitled _Urban Regions Ecology and Panning Beyond the City,_ Richard Forman includes Portland in an analysis of a several dozen cities throughout the world.[27] Some of the features that Forman says make an urban area green include location adjacent to a water body such as a lake, river, or the ocean, abundant green space, an urban growth boundary, market gardening and cropland concentrated near the urban area, and a protected water supply system. Dr. Forman has produced a valuable book with many interesting insights about the factors that make the urban environment more livable and enjoyable. His characteristics of a green city parallel many of those in the definitions given above. It is clear that green urban characteristics do contribute to sustainability in a number of ways, such as supplying clean drinking water and providing areas and opportunities for people to meet and think about sustainability issues facing urban areas. But few of the characteristics of a green city—as defined by Foreman and others—address the enormous energy and material inputs required to maintain modern, large, and especially first-world cities.

[25] http://www.mnn.com/health/allergies/photos/top-10-green-us-cities/what-makes-a-city-green#ixzz383BBmkyN.

[26] http://en.wikipedia.org/wiki/Green_cities.

[27] Richard Forman. 2008. _Urban Regions Ecology and Planning Beyond the City._ Cambridge University Press, Cambridge UK.

The characteristics of green as outlined by Foreman and most others leave out many key things related to sustainability, and the unsustainable nature of modern, especially first-world cities. We focus here on Portland because it is an icon of green, but these comments could be directed at all urban areas in the U.S., including all those we treat in this book. All people in Portland eat, but most food is not produced in Oregon. Most people in Portland have a car, but cars are not produced in Portland or Oregon. Vehicles may come from the eastern U.S., Japan, Korea, or Mexico. Almost none of the gasoline used to power these vehicles is refined in Portland, and the oil from which gasoline is made may come from the Gulf Coast (remember the rich fields off Texas and Louisiana), some other part of the U.S., Canada, Mexico, Venezuela, or the Middle East. Likewise, many other things used and consumed by Portlanders likely come from far away, such as clothes (Bangladesh, Vietnam, Thailand), computers and smart phones (China), and tomatoes (California or Mexico). The same can be said for mattresses, plastic plates, windows, kitchen tables, toilets, books, refrigerators, washing machines, carpeting, computers, glass jars, and a myriad of other things. And as we say, Portland is certainly not unique in all this.

So how will Portland and other cities in the Cascadia region of the Northwest fare as the twenty-first century megatrends play out? From a climate point of view, the impacts of climate change will be relatively benign compared to the Southwest, southern Great Plains, and many coastal regions. There may be problems meeting summer water needs as winter snow pack declines and more winter precipitation evaporates or sinks into the ground. The Northwest has one of the highest levels of natural productivity and ecosystems services in the nation. Thus, there is an ecological cushion to fall back on compared to cities like Las Vegas and Los Angeles. The region has significant potential for food production, but agriculture in the Cascadia region cannot meet current food needs. Agriculture in eastern Oregon and Washington is irrigated and may be impacted by climate change. The economies of Portland, Seattle, and Vancouver will likely face looming threats as energy and resource constraints impact the national and global economies. All three of these cities, and the region in general, are highly dependent on people spending discretionary income. There is a huge tourism industry based on people visiting the three cities and the rich natural environment of the Northwest. Seattle and Vancouver are departure points for the international cruise industry. All three cities are meccas for business conventions. One of the major industries in the region is the video gaming industry. All of these economic activities are likely to decline as energy and resources become scarcer. When the next great earthquake occurs, will there be energy to rebuild?

So Portland and the Northwest will have to change in fundamental ways as the mega trends play out. One positive benefit is that much of the electricity used in the Northwest comes from hydropower. The region possesses a rich natural system to fall back on.

Global Versus Local Constraints on Local Sustainability

Many cities have published reports outlining goals for enhancing sustainability. Typical actions proposed include producing most to all electricity from renewable sources by a certain date, say 2050; improving options for public transit, walking, and bicycles; improving efficiency in commercial, public, and private buildings; reducing runoff; local food production; recycling, and the like. All of these are designed to improve urban living and reduce greenhouse gas emissions, and all of them are laudable goals. But will such goals help individual cities achieve sustainability and prosper over this century? Cities are not closed systems. Global constraints on sustainability must also be considered.

Plans for achieving sustainability at the local level often fail to consider broader global constraints as a result of the megatrends that we have discussed throughout this book. An implicit assumption of many local sustainability initiatives seems to be that for the most part the global economic system will continue to function more or less as it has in the past. We must consider a number of global constraints in discussions of sustainability.

In Chap. 7, we have seen that replacing fossil fuels with renewables, especially wind and solar, includes a large dash of wishful thinking. While wind and solar produce electricity, much of global energy use presently requires direct fossil fuel use for transportation, food production, industrial activity, as well as feedstocks for producing petrochemicals, plastics, and other materials. Liquid fossil fuels are especially important in transportation, agriculture, and industry and power activities that underwrite important sectors of the economy (tourism, trade, and food production). As we discussed in the energy chapter, there are also questions about how fast renewables can be brought on line, and whether resource constraints such as metals will limit growth in renewables.

Most of the cities we discuss in this book are dependent on large numbers of business travelers or tourists to sustain their economies (Syracuse, Flint, Cedar Rapids, Baton Rouge, and Amarillo are exceptions). These visitors bring billions of dollars into the economies of these cities and regions. Tourism is the dominant economic activity in Orlando, Asheville, and Las Vegas while Portland, Los Angeles, Houston, New Orleans, and New York have large numbers of both business travelers and tourists. They arrive primarily by air and to a lesser extent by private auto. The international cruise industry depends strongly on both large ships and airplanes that bring people to their ports of embarkation. All of this travel is energy intensive and sensitive to fuel prices, and will likely decline as energy and resource prices rise and discretionary income declines.

Many cities are critically dependent on high levels of global trade that in turn depends on relatively cheap energy and abundant natural resources. Portland, Los Angeles, Houston, New Orleans, and New York are all major ports shipping hundreds of millions of dollars worth of commodities and manufactured goods. All of this is part of the huge globalized trade system supercharged on abundant and relatively cheap energy and natural resources. As energy becomes more expensive and scarce, trade will likely slow. Distance will matter again, and what goods we consume will more and more be produced closer to home. The millions of shipping containers or TEU's (twenty foot equivalent units) that flow into these ports, especially Los Angeles/Long Beach, will diminish. The large shipments of oil and coal flowing through the northern Gulf Coast will also decline as these energy sources are exhausted. Discretionary goods in TEU's will decline first, but ultimately so will coal and oil shipments.

The current global industrial agricultural system is highly energy intensive and embedded in the global trade system. Agricultural products make up almost half of the commodities shipped from Gulf Coast ports, primarily grain from the midwestern breadbasket. As energy prices rise, so will the cost of food and long distance shipments of food will likely decrease. Large cities that can't feed themselves from their local region and are dependent on massive inputs of food from the global system will be relatively more impacted.

The globalized economic system is dependent on abundant and cheap renewable and non-renewable natural resources to sustain the entire economy. Renewable resources include agricultural products, fisheries, and forest products that feed us and provide a wide variety of building materials. Natural ecosystems are a backbone of the global tourism industry as millions of people flock to see nature in its many and varied forms; think of the Great Barrier Reef in Australia, tropical rain forests in Costa Rica, temperate rain forests, whales, and eagles in British Columbia, immense herds of wildebeest and zebra in East Africa. It is unlikely that visitors who go to see the rainforest in the Amazon would pay the same amount to see a 10,000 acre soybean field that replaced the forest. Nature also provides non-renewable resources such as ores that are processed into a wide variety of essential metals. As the richest ore bodies are exhausted, it takes more energy, and produces more pollution, to obtain these metals.

Finally, the climate system is approaching tipping points that will impact much of the global economic system. Drought in the Southwest is affecting water availability and agriculture, and will likely significantly impact the economy of the region. Accelerated sea-level rise and more frequent, stronger storms will impact coastal regions, especially along the Gulf and Atlantic coasts.

All of these considerations indicate that individual cities will have to look beyond their local regions if they are to fully analyze the potential for achieving sustainability.

Sustainability at the National Scale

The megatrends described in this book will impact both human and natural systems differentially across the land (Fig. 10.5). Diminishing returns from energy production will pervasively impact the economy, leading to unstable prices for many essential goods and reducing discretionary income. This will interact with climate change, ecosystem services, food production, and population density to differentially compromise the sustainability of the various regions. Figure 10.5 is a generalized, broad-scale depiction of how sustainability may vary across the landscape as impacted by the twenty-first century megatrends.

Large urban areas are at risk because they have such high demands for continuous inputs of energy and materials. By 2025, it is estimated that 165 million people, or about half the U.S. population, will live in four megaregions; the Northeast, Great Lakes, Southern California, and San Francisco Bay regions. An additional 45 million will live in south Florida and the Houston-Dallas region. Large cities are often touted as more energy efficient than suburban or rural areas, but all megaregions have extensive suburban and exurban areas and city size is only poorly related to per capita greenhouse emissions.[28] A far-flung network of supply lines supports these megaregions with food, energy, and other materials that stretch for long distances across the global landscape. Areas dependent on longer, energy intensive supply lines are vulnerable to the rising costs of energy for transportation.

---

[28] Fragkias, M. Lobe, J., Strumsky,D., and Seto, K. 2013. Does size matter? Scaling of CO_2 emissions and U.S. urban areas. *PLOS one*, 8(6): e64727.; Jones, C. and Kammen, D. 2011. Quantifying carbon footprint reduction opportunities for U.S. households and communities. *Environmental Science and Technology*. x.doi.org/10.1021/es102221h; Jones, C. and Kammen, D. Spatial distribution of U.S. household carbon footprints reveals suburbanization undermines greenhouse gas benefits of urban population density. *Environmental Science and Technology*. dx.doi.org/10.1021/es4034364.

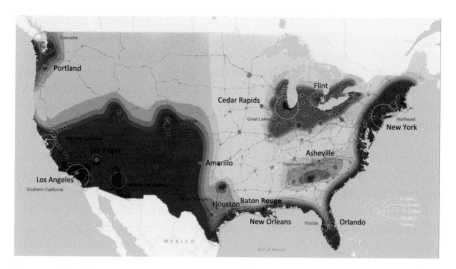

Fig. 10.5 Based on our analysis, several areas of the U.S. will have compromised sustainability in the twenty-first century. These include the southern Great Plains, the Southwest, the southern half of California, the Gulf and Atlantic coasts (especially southern Louisiana and Southern Florida), and areas of dense population such as south Florida and the Northeast. The cities analyzed and the 11 megaregions from Chap. 2 are also shown on the map (John Day et al. Sustainability and Place. How emerging megatrends of the twenty-first century will affect humans and nature at the landscape level. Ecological Engineering. 65: 33–48. 2014.)

The economies of urban areas, especially the currently most economically successful ones based on the human, financial, and information service sectors, are strongly dependent on discretionary income and wealth, which is predicted to decrease substantially over the twenty-first century.[29] But many cities have lost population, especially those that were based in the manufacturing sector of the economy during the twentieth century. Detroit and Flint, Michigan, are often cited as examples but there are many others.[30] Between 1950 and 2000, St. Louis lost 59 % of its population and Pittsburgh, Buffalo, Detroit, and Cleveland each lost more than 45 %. It is possible that many of the rust belt cities that have experienced population decreases will be more sustainable than more "successful" cities in the northeast and elsewhere. They now have a lower population density and tend to exist in rich agricultural regions. Indeed, abandoned land is being used for food production in a number of depopulating cities. By contrast, the northeast is the most densely populated region of the country. The population is expected to reach almost

[29] Hall and Klitgaard, 2012, ibid.

[30] Jordan Rappaport. 2003. U.S. *Urban Decline and Growth, 1950 to 2000. Federal Reserve Bank of Kansas City.* 44 p. www.kc.frb.org

60 million by 2025. The states that make up the region have about 34 million acres of farmland, or about a half acre per person. By contrast, it currently takes about 1.25 acres per person to provide the food consumed in the U.S.[31] If agriculture becomes more local and less productive—as some predict— due to increasing energy costs,[32] then it will be a challenge to maintain the current food supply to the northeast, especially at the price society is used to paying.

The least sustainable region will likely be the southwestern part of the country from the southern plains to California that includes Amarillo, Las Vegas, and Los Angeles. Climate change is already impacting this region, and it is projected to get hotter and drier. Models predict that precipitation will decrease, falling more as rain and less as snow. These trends will lead to less water for direct human consumption and for agriculture as well as for natural ecosystems. This is critical since practically all agriculture in the region is irrigated. The Southwest has the lowest level of ecosystem services of any region in the U.S. because it is so dry. California is the most populous state in the nation, with most people living in the southern half of the state (San Francisco and south). This area experiences high water stress. The Los Angeles metro area is the second largest in the nation. But population density is low over much of the rest of the region and is concentrated in large urban areas such as Las Vegas, Phoenix, Tucson, San Diego, and Albuquerque. California is one of the most important food producing states in the nation, but this productivity will be threatened by water scarcity and increasing energy costs. Much of the region is strongly dependent on tourism and spending discretionary income, especially Las Vegas, so future economic health will likely be threatened in coming decades. Many cities and regions whose economy is dependent on tourism will have compromised sustainability.

The Gulf and Atlantic coasts have high ecosystem services but are also highly vulnerable to climate change in terms of accelerated sea-level rise and more intense hurricanes. The southern parts of Florida and Louisiana have large areas near or below sea level that are threatened by climate change. Miami and New Orleans are highly vulnerable, but all coastal cities have significant areas near sea level. Hurricanes Katrina, Irene, and Sandy are likely harbingers of the future for these coasts.

In summary, the emerging megatrends of the twenty-first century will result in large challenges for sustainability in the U.S. We believe that the most difficult areas to maintain are likely to be the southwest—especially

[31] Pimentel, D., Williamson, S., Alexander, C.E., Gonzalez-Pagan, O., Kontak, C., Mulkey, S.E., 2008. Reducing energy inputs in the U.S. food system. *Hum. Ecol.*, 36, 459–471.
[32] William Rees, 2012, Ibid.

southern California—coastal regions of the Gulf and Atlantic, and large urban regions especially those in the northeast, southern Florida, and the southern half of California.

How does the long history of humankind fit with current ideas of green and sustainability? Preindustrial cities were not like modern first world cities with soaring skyscrapers of steel and glass and high levels of services. Prior to the industrial revolution, cities were dirty, with little attention to sanitation and clean water. There was little understanding of the causes of disease, and people often fled cities during times of epidemics. Paraphrasing Thomas Hobbes, life in cities was often poor, nasty, brutish and short. Without radical transformation—which will very likely require lower human populations in many cases—many of today's cities in the U.S. and elsewhere may return toward such an undesirable state later this century. That is, unless humans recognize coming limitations and actively adapt.

Chapter 11

Summing It Up. Alternative Routes for the Way Forward

What is our vision for the future? The analysis in this book strongly indicates that systemic change is inevitable and that the rising affluence (of some, but certainly not all) during the past century cannot be sustained. This understanding is not new. For most of the history of mankind, humans lived mainly on the annual solar energy input that powered the climate, hydrologic cycle, and ecosystems that determined their wealth. This allowed most humans to capture enough food and fiber to feed and clothe themselves, and to harvest a very small amount of the mineral wealth stored in the surface of the earth (metal ores, clay for pottery, and even a bit of fossil fuels) to construct various tools, weapons, and other kinds of artifacts and icons of civilization such as buildings, boats, public works projects, and works of art. But it did so for only a small number of humans compared to now. The global population grew from about four to six million people 10,000 years ago[1] to about one billion in 1800. Practically everything done during that period was done with solar energy manifest in one form or another, including moving water, rain, wood, crops, fish and other seafood, as well as metal ores reflecting the work of people who mined them and combustible material used to process them. The unexploited state of these resources allowed earlier humans to exploit them with far less energy than currently needed, when stocks of soils, fish, metals and so on were much richer than is the case today.

Given the dilute nature of solar energy, humans were able to achieve a remarkable amount. Just think about it: practically all of the paintings, sculptures, and other works of art in the Louvre, the Vatican, and most other great museums of the world recording human achievement prior to 1800;

---

[1] http://www.scottmanning.com/content/year-by-year-world-population-estimates/

© Springer Science+Business Media New York 2016
J.W. Day, C. Hall, *America's Most Sustainable Cities and Regions*,
DOI 10.1007/978-1-4939-3243-6_11

the music of Bach and Mozart; and great monuments such as the pyramids at Gaza and Mexico, Angkor Watt, the Taj Mahal, the Great Wall of China, the Erie Canal, St. Basil's Cathedral in Moscow, and the Potala Palace in Lhasa, Tibet, all were built on solar energy in one form or another. The works of Plato, Sophocles, Aristotle, Beethoven, Michelangelo, Botticelli, Newton, Darwin, Shakespeare, Austin, Tolstoy, and many others that we rightfully consider the highest human achievements were developed by people and societies operating on solar energy. Jesus, Mohammad, the Buddha, Krishna, and other religious leaders trod their earthly path and brought their message to mankind using solar energy. Even the most fundamental aspects of science, such as the laws of motion, the movement of the spheres, the nature of elements and the periodic table, the nature of evolution and even the laws of thermodynamics were derived from mostly solar-driven cultures. Of course, available power and wealth were concentrated in few hands.

Empires were built throughout history on stored solar energy stolen from conquered regions. This stored energy came in various forms that included food (such as wheat from Egypt for the Romans), spices, wood, slaves, information, precious metals, and other goods. Plutarch, the great writer of ancient heroes, celebrated men who stole embodied solar energy from others. The great age of European colonization was in essence a grand, and in many ways immoral, transfer of embodied solar wealth from Asia, Africa, and the Americas to continental powers. The incredible growth of population, wealth, information, and technology of the nineteenth and twentieth centuries was underwritten primarily by ancient stores of solar energy in the form of fossil fuels.

But, almost certainly, the fossil fuel age will largely run its course during what is left of the twenty-first century. Solar and nuclear energy sources have their advocates, but the full replacement of fossil fuels—even while increasing numbers of people will be wanting their benefits—does not seem likely to us. Wrenching change is coming, and humans need to plan for this transition. We need to be realistic about what is and is not possible, and where. A main message of this book is that some areas will be much worse off than others and, indeed, some places will hardly be able to continue. For us, a mindset conditioned by the oil age must change. Cheap energy will no longer be available to offset the impact of degraded ecosystems or to power purely technological solutions to our sustainability challenges.

This mindset of the need for change is not embodied in the dominant neoclassical economic system that pervades society today. Neoclassical economics has become almost completely disconnected from its biophysical base, and as presently practiced cannot offer comprehensive solutions dealing with future sustainability in an age of scarcity. Analyses that grow out of the assumptions of neoclassical economics, such as Glaeser's "Triumph of the City," are also not the way forward. As we write, the US, Europe, and Japan (not to mention many smaller countries) have essentially ceased economic growth despite many efforts to "prime their pumps", while Greece, Puerto Rico and many other areas have publically acknowledged that they cannot possibly pay back the debts by which they had continued a life of affluence beyond their real resources. What we see is a world in which the "limits to growth" are playing out not all at once but region by region.

If indeed this is the future, as we believe, the continued growth of cities, and of the economy in general, is not practically possible or even conceptually feasible as society moves toward the end of the age of cheap energy and resources. Cities will probably have to become smaller and reintegrate with their local regions.

What should or can be done?

Time is of the essence. Society must quickly recognize that we are not in complete control, and it must also recognize what is and is not possible in the context of our dependent relationship with the physical world. Above all, we need to learn to avoid folly. In her 1984 book, *The March of Folly*,[2] Barbara Tuchman defined folly as the pursuit of policies contrary to long-term public interest by a large group—governments or industry, or as far as the Mississippi delta is concerned, the population of a state. In the Southwest, society has pursued water policies that are clearly unsustainable. According to Tuchman, for an event or series of events to be considered as folly three criteria must be met: The policy must be perceived as counter-productive in its "own time." That is, a relatively large number of thoughtful people know that the policies are counter-productive and damaging. Feasible alternative courses of action must have been proposed. These suggested alternatives should come from a group, not an individual leader, and should persist beyond one political lifetime. And finally, the alternative courses are ignored and the pursuit of the counter-productive policies persists. Modern society satisfies all three of these conditions for folly.

The failure to recognize the transitory nature of the fossil fuel age is one of the grandest follies of humans. In terms of the issues we are addressing in this book, folly is easy to discern. It is the nexus of unlimited aspirations within a world of finite resources, all driven by an economic system that depends on the economic folly of endless growth in a kind of societal perpetual motion machine.

The folly of our time has been perceived as counter-productive in its "own time" for a half century or more, and we have documented the extensive

---

[2] Barbara Tuchman. 1984. *The March of Folly: From Troy to Vietnam*. Random House.

literature on this topic throughout this book. These include many publications on peak oil and energy scarcity going back to the work of M. King Hubbert in the 1950s, an extensive literature on climate change, many thoughtful critiques of current neoclassical economics, data showing that the human footprint is now greater than the ability of the earth to sustainably support the current economic system, and the *Limits to Growth* study that predicted (now accurately it seems) that the trajectory of society was unsustainable and likely leading to catastrophe.[3] Our analysis fits well within the larger context of scarcity clearly outlined in the *Limits to Growth*. A number of other authors have addressed the issue of sustainability and collapse in past societies, including Joseph Tainter and Jared Diamond.[4] In almost all cases of collapse of earlier societies, resource scarcity played a central role. We have also documented abundant information on alternative ways forward that can make the coming transition more humane, an issue that we perceive as being especially important. The problem of dealing with folly is succinctly summed up in the words of the great eighth century Chinese poet Tu Fu; "Easy to discern the trend in the drift of life, hard to compel one creature from its course."

Contemporary Economics Cannot Solve These Problems

In the twentieth century, the "science" of economics was able to solve many problems, because the immense source and sink functions of the earth were essentially limitless. The earth could provide all the natural resources society needed and accept all the wastes that society generated, and at a relatively low price. The issue was allocating abundant resources or wealth, and economics' role was to do this, even when it increasingly ignored natural resources. Whether it was done well depends on whom you talk to. But by the end of the twentieth century, the great source and sink functions

---

[3] Donella Meadows, Dennis Meadows, Jorgen Randers, and William Behrens. 1972. The Limits to Growth. Universe Books, New York.; Donella Meadows, Dennis Meadows, and Jorgen Randers. 1992. *Beyond the Limits*. Chelsea Green Publishing, White River VT. 300 p.; Donella Meadows, Jorgen Randers, Dennis Meadows. 2004. *Limits to Growth – The 30-Year Update*. Chelsea Green Publishing, White River VT. 338 p.; Ugo Bardi. 2011. *The Limits to Growth Revisited*. Springer, New York. 119 p.; Graham Turner. 2008. *A Comparison of the Limits to Growth with Thirty Years of Reality*. Socio-Economics and the Environment in Discussion. CSIRO Working Paper Series. Canberra Australia.; Graham Turner. 2012. *On the cusp of global collapse?* Updated comparison of The Limits to Growth with historical data. Gaia. 21/2:116–124. www.oekom.de/gaia.; Charles Hall and John Day. 2009. Revisiting the Limits to Growth after Peak Oil. *American Scientist*. 97: 230–237.

[4] Joseph Tainter. 2005. *The Collapse of Complex Societies*, Cambridge University Press. 1998; Jered Diamond, *Collapse*. Penguin Books, New York. 589 p.

were no longer limitless. Humans had drawn down many non-renewable natural resources, especially fossil fuels but also soils, water, forests, mineral ores and fish, and the ability of the earth to accept the wastes generated by society without repercussions was severely compromised. In the words of Wendell Berry, "it is not the area of a country that makes its value or its most meaningful strength, but its life, the depth and richness of its topsoil." This is another way of stating that the value of ecosystem goods and services is extremely important for sustaining society. This is what supported society for millennia.

The grand challenges of the twenty-first century are primarily ecological and environmental problems and will be solved, for better or worse, using ecological, as well as social principles. Ecology is the science of how ecosystems function and how organisms deal with scarce resources and their allocation. The economic ideas of continued growth, and the concept of lack of absolute scarcity, and infinite substitutability will no longer work. Humans will have to learn how to deal with allocation of ever-scarcer resources. We humans are not set apart from populations of other organisms. We live by the same basic rules as all our fellow species on earth— those of thermodynamic and biophysical constraints. The abundance of the twentieth century has lulled us into thinking that the future will always be more and more. But this time is over. Humans will have to live with less.

The Limits to Technology

Technology has grown exponentially during the past century. This growth of technology is directly tied to the rapid growth of cheap and abundant energy and to the exponential growth of population and the economy, indeed the whole globalized industrial lifestyle. It is assumed by many that technology will continue to grow and will solve all future problems related to climate and scarcity of resources, especially energy. Techno-optimists have faith in the great breakthrough. But is this likely to happen?

Deborah Strumsky and colleagues question this assertion.[5] They report that, "scientific fields undergo a common evolutionary pattern. Early work establishes the boundaries of the discipline, sets out broad lines of research, establishes basic theories and solves questions that are inexpensive but

---

[5] Strumsky, D., Lobo, J., & Tainter, J. A. (2010). Complexity and the productivity of innovation. *Systems Research and Behavioral Science, 27*(5), 496–509. Tainter has written extensively on this topic and why societies collapse.See. Tainter, 2005, Ibid. J. Tainter and T. Patzek. 2012. *Drilling Down – The Gulf Oil Debacle and Our Energy Dilemma*. Springer, New York. 242 p.

broadly applicable. Yet this early research carries the seeds of its own demise. As pioneering research depletes the stock of questions that are inexpensive to solve and broadly applicable, research must move to questions that are increasingly narrow and intractable. Research grows increasingly complex and costly as the enterprise expands from individuals to teams, as more specialties are needed, as more expensive laboratories and equipment are required, and as administrative overhead grows." Strumsky and her colleagues showed that patents per inventor decreased over time. They found that the production of knowledge has continued to increase. However, this was not because science was as productive as ever, but because more and more resources have been devoted to it. They argue that it will take ever-increasing resources to keep technology as productive as it has been. The evolution of technology is characterized by what is called logistic growth: slow innovation at first, then acceleration, and finally a slow down in the growth of innovation. Early innovations of technology give the largest increments of improvement. Joseph Tainter and Tad Patzek call this the energy-complexity spiral.[6] If energy is added to human societies, as has been the case for fossil fuels over the past couple of centuries, then society will become more complex and will develop new technologies, new institutions, new kinds of information. But it takes increasing energy to keep this process on the same trajectory. In other words, society found, very quickly, an extremely abundant and cheap energy source and used it to become more complex. As energy becomes more expensive and scarce, it will be more difficult, if not impossible, to maintain the same level of complexity. Tainter and Patzek put it succinctly: "Having surplus energy is a rare experience in human history." This leads "many people to think that today's conditions are normal. In fact, they are not. Today's conditions of inexpensive energy are highly unusual, an aberration of history…This, then, is the energy-complexity spiral: complexity grows because we have extra energy, complexity grows because we must solve problems, and complexity requires that energy production increase still more." But this is becoming impossible. The only way out of the energy-complexity is to simplify.

Adaptation: Learning from History

Like a wayward child that inherits his parents' enormous wealth and spends it recklessly, society is exhausting the highest-grade resources of our Mother Earth. More and more, we are going to have to live on our annual income. But how can we do this? There are now more than seven times the number of people who lived on the earth in 1800, just as the industrial revo-

---

[6] Tainter and Patzek, 2012, ibid.

lution was beginning and humans were living off the continuous flows of the biosphere (which itself had been compromised). In the context of the broad sweep of history, the ways of the past two centuries are not sustainable and will end whether we plan for it or not.

In *Powerdown*, Richard Heinberg offers four views on how society may react to the megatrends that portend resource scarcity in this century.[7] He calls these four options *last one standing, waiting for a magic elixir, powerdown,* and *building lifeboats*.

Last one standing is essentially the continuation of the trajectory that humanity has been on for the last two centuries—that of exponential population growth, massive resource consumption fueled mainly by fossil fuels, and continued degradation of the earth's natural systems. Michael Klare called this *The Race for What's Left* in his book of the same name. But this option appears unsustainable, whether because of diminishing net energy reserves or climate dysfunction, even on a relatively short time frame of no more than a few decades. Humans have mined the easiest and most inexpensive mineral ores and fossil fuels, degrading natural systems on a global scale. Even now the cost of fossil fuels has become a drag on global economic growth. Heinberg asserts that following this course will lead to wars, economic crises, and environmental catastrophe. Unfortunately, this describes, for the most part, our current societal trajectory.

Waiting for a magic elixir is the techno-utopian vision that technology and dependence on market forces will lead humanity into a new age of abundance. Technology can undoubtedly help in the coming transition, but there is very little hope that technology can maintain the current highly consumptive system in a time of natural resource depletion. There has always been a human economy, even if it wasn't called economics, and there will be one in the future. But present day neoclassical economics is simply not up to the task of dealing with real resource scarcity, environmental degradation, and massive overpopulation. Heinberg maintains that this approach serves primarily as a distraction from what really needs to be done.

Powerdown is a vision of cooperation, conservation and sharing. This path will require a reduction of per-capita resource consumption, especially in rich countries, development of alternative energy sources, a more equitable distribution of wealth, and a reduction in the world's population. All of these things will happen whether we want them to or not. But Heinberg argues that humans must consciously embrace this alternative if society is to avoid the negative impacts of the race for what's left.

---

[7] Richard Heinberg. 2004. *Powerdown – Options and Actions for a Post-Carbon World.* New Society Publishers, Gabriola Island, BC Canada. 208 p.

Building lifeboats assumes that industrial civilization cannot survive in anything like its present condition. Heinberg argues that humans must go about preserving the most worthwhile information developed over the past few centuries.

Howard and Elizabeth Odum plot a potentially more optimistic vision of the coming transition in *A Prosperous Way Down*.[8] Howard Odum was one of the leading ecologists of the twentieth century and a pioneer in the field of holistic systems ecology. His 1971 book *Environment, Power, and Society* is still a fresh, understandable read on the current problems facing society. In *A Prosperous Way Down*, the Odums show that the scientific principles of energetics and ecology that apply to all systems provide guidance for society for the coming transition to a lower energy society. They liken human systems to natural systems and argue that the transition can be smooth if humans learn from nature. They note that because people don't generally understand the noisy complexity of the world, they pursue "beliefs in randomness, indeterminacy, miracles, money, magic, predestination, divination, anarchy, and selfhood."

Another thing that society can do to lessen the impact of the coming transition is to address the problem of growing wealth inequality. In their extremely illuminating book entitled *The Spirit Level*,[9] Richard Wilkinson and Kate Pickett show that the degree of wealth and income inequality affects both the depth of a wide range of social problems (including various indices of health, child wellbeing, level of trust, the status of women, mental illness, drug use, infant death rates, life expectancy, education attainment, high school graduation, teen pregnancy, homicide rate, and incarceration rates among others) and the difficulty of solving them. The findings of Wilkinson and Pickett have important implications for society at a time of decreasing resources, as does the analysis of Lambert and colleagues, who show that many indices of human well being are highly correlated with the availability, EROI, and equity of energy distribution.[10] These trends will give rise to stress and discord with a strong potential to lead to social unrest. Clearly, reducing inequity will not solve the problem of scarce

---

[8] Howard and Elizabeth Odum. 2001. *A Prosperous Way Down*. University of Colorado Press, Boulder. 326 p.; Howard Odum. 1971. *Environment, Power, and Society*. John Wiley, New York. 331 p.

[9] Richard Wilkinson and Kate Pickett. 2009. *The Spirit Level*. Bloomsbury Press, New York. 375 p. Others have recently addressed problems associated with inequality. See Thomas Piketty. 2004. *Capital in the Twenty-First Century*. The Belknap Press of Harvard University Press, Cambridge MA. 685 p.; Joseph Stiglitz. 2013. *The Price of Inequality: How Today's Divided Society Endangers our Future*. W.W. Norton, New York.

[10] Lambert et al. 2014, ibid.

resources. But if most people perceive that this burden is shared more fairly by all members of society, this will reduce the potential for unrest. For the states where the ten cities and regions discussed in this book are located, the ranking of most to least inequality are as follows: New York, Louisiana, California, Texas, Florida, North Carolina, Michigan, Oregon, Nevada, and Iowa. So other things being equal, the solution of problems associated with the mega-trends of the twenty-first century will be more difficult in cities and regions with greater inequality.

Regionalism in a Globalized World

We end this complex book with some personal observations about how the megatrends we have discussed already seem to be impacting the world, and even our personal lives. As we mentioned earlier, many countries are having financial problems, even China that has been an economic power house for years. The same is true for many States. We believe that these economic problems are related in a fundamental way to growing resource and energy scarcity.[11]

The limits of resources are appearing first in individual areas like Greece and Puerto Rico, where populations are attempting to live the affluent life style characteristic of other "first world" countries without the rich resource or industrial base that makes such lives possible. The U.S. is not immune to these problems, but for now the world is willing to support our debt. But debt is not a solution; sooner or later, the piper must be paid.

One of the authors (Charles Hall) worked in Puerto Rico for many years and is often asked what he would suggest to revitalize the economy. Unfortunately there is no real solution that will support 3.5 million people on a small island with few resources, other than marvelous weather, beaches, and mountains that can be turned into a relatively few tourist dollars. Puerto Rico had its moment in the economic sun due to section 936 of the US tax code of 1976 which subsidized industrial development and increased the expectations of Puerto Ricans to an unsustainable level. But that opportunity is long gone as pharmaceutical, electronic and other industries moved to the cheaper labor of Asia.[12]

---

[11] See for example: McRae, Hamish. "How can we explain why productivity is lagging behind overall economic growth." *The Independent*, June 25, 2015.

[12] Surowiecki, James. "The Puerto Rican problem." New Yorker April 6. Surowiecki, who asked "What can Puerto Rico offer that other locations can't?" did not have a solution either.

The larger question that we must ask, and that is asked in this book, is whether Greece and Puerto Rico are special cases or rather harbingers of conditions that are likely to become more common in future decades? Is our temporary surplus of premium fossil fuels unrealistically pumping up the expectations of the whole world? And if so, what can or should be done? How useful are things labeled "green" to a future with a different climate and a lack of premium fuels? Unfortunately, much of the "green" economics literature is not very helpful, because it does not analyze how many people can be employed doing green things, or whether these green things can feed or otherwise support entire cities and regions. And it is rarely understood that many green activities are highly subsidized by fossil fuels.

As in Greece and Puerto Rico, cheap energy and various policies that encourage and subsidize the use of large quantities of fossil fuels can pump up an economy, and hence affluence, to levels unheard of prior to the industrial revolution. Affluence takes many forms, such as a university education, automobiles, and leisure, that were formerly available only to the wealthy few. Can this be continued in an increasingly resource-constrained world? One answer is a tentative yes, if remaining resources are more equitably shared and if people's expectations for hyper affluence are lowered.

The time is now to discuss issues of resource scarcity and equitability, before constraints leave us with even fewer choices. It is also time to consider to what degree most "green" approaches can provide the essentials that humans need (jobs, food, transport, temperature regulation) on the scale that is required. Discussion of population growth and desired population levels have to come out from their hiding places and be made central to discussions about the future. And finally, the global megatrends will be expressed differently for different regions. They will affect all of the U.S., but some cities and regions will fare better. Not all is predictable, but we hope that this book provides an understanding of how the megatrends and primary drivers of change will affect your city and region. Based on a better understanding of how the world works, humans can see what is successful, right, and worthwhile. We hope that this book has offered a meaningful step in that direction.

Biographical Information

John W. Day, Jr. is distinguished professor emeritus in the Department of Oceanography and Coastal Sciences, School of the Coast & Environment at Louisiana State University, where he has taught since 1971. He has published extensively on the ecology and management of coastal and wetland ecosystems and has over 200 peer-reviewed publications. He received his Ph.D. in marine sciences and environmental sciences from the University of North Carolina in 1971 working with Dr. Howard T. Odum. Since then, he has conducted extensive research on the ecology and management of the Mississippi Delta and deltas and coastal systems in the Mediterranean and Latin America.

Dr. Charles Hall is professor emeritus in the School of Environment Sciences and Forestry, State University of New York, Syracuse, New York. He is a systems ecologist who received his Ph.D. under Howard T. Odum at the University of North Carolina at Chapel Hill. Dr. Hall is the author or editor of seven books and more than 250 scholarly articles. He is best known for his development of the concept of EROI, or energy return on investment, which is an examination of how organisms, including humans, invest energy into obtaining additional energy to improve biotic or social fitness. He has applied these approaches to fish migrations, carbon balance, tropical land use change, and the extraction of petroleum and other fuels in both natural and human-dominated ecosystems. He is active in developing a new field, biophysical economics, as a supplement or alternative to conventional neoclassical economics, while applying systems and EROI thinking to a broad series of resource and economic issues.

© Springer Science+Business Media New York 2016
J.W. Day, C. Hall, *America's Most Sustainable Cities and Regions*,
DOI 10.1007/978-1-4939-3243-6

Dr. Eric Roy is assistant professor of Environmental Science focusing on Ecological Design in the Rubenstein School of Environment and Natural Resources, University of Vermont, Burlington, Vermont.

Dr. Mathew Moerschbaecher is an environmental and coastal scientist who received his doctorate from the School of Renewable Natural Resources at Louisiana State University, Baton Rouge, Louisiana.

Dr. Christopher D'Elia is professor in the Department of Oceanography and Coastal Sciences and Dean of the School of the Coast and Environment at Louisiana State University, Baton Rouge, Louisiana.

David Pimentel is professor emeritus in the Department of Ecology and Evolutionary Biology and the Department of Entomology at Cornell University, Ithaca, New York.

Dr. Alejandro Yáñez-Arancibia is distinguished professor in the Department of Environment and Sustainability in the Institute of Ecology A.C., Xalapa, Mexico.

Hall's work on EROI has been featured in such media outlets as Scientific American (most recently in April, 2013), Forbes Magazine, and the Discovery Channel.

Bibliography

Ackerman-Leist, P. 2013. *Rebuilding the foodshed: How to create local, sustainable, and secure food systems.* Post Carbon Institute, Chelsea Green Publishing, VT, USA.

Aerts, J.C.J.H., Botzen, W.J. and De Moel, H. 2013. Cost Estimates for Flood Resilience and Protection Strategies in New York City. Annals of the New York Academy of Sciences.

Aleklett, K., 2012. Peeking at Peak Oil. Springer, New York, 336 p.

Alexandratos, N., and J. Bruinsma. 2012. World agriculture towards 2030/2050. The 2012 revision. ESA Working Paper No. 12–03. Agricultural Economics Division, Food and Agriculture Organization of the United Nations. http://www.fao.org/docrep/016/ap106e/ap106e.pdf.

Ambrose, Stephen. 1997. *Undaunted Courage: Meriwether Lewis, Thomas Jefferson, and the Opening of the American West.* Simon & Schuster, 521 p.

Anderegg, W. R., Prall, J. W., Harold, J., & Schneider, S. H. 2010. Expert credibility in climate change. Proceedings of the National Academy of Sciences, 107(27), 12107–12109.

Ashworth, W. 2007. *Ogallala blue: Water and life on the Great Plains.* Countryman Press, Woodstock, VT, 330 p.

Aucott, M., & Hall, C. 2014. Does a Change in Price of Fuel Affect GDP Growth? An Examination of the US Data from 1950–2013. Energies, 7(10), 6558–6570.

Aulbach, L. 2011. *An Echo of Houston's Wilderness Beginnings.* CreateSpace Independent Publishing Platform, 752 p.

Balogh, S.B., C.A.S. Hall, A.M. Guzman, D.E. Balcarce, and A. Hamilton. 2012. The potential of Onondada County to feed its own population and that of Syracuse, New York: Past, present, and future. In: Pimentel, D. (ed). Global economic and environmental aspects of biofuels. CRC Press.

Barbara Tuchman. 1984. *The March of Folly: From Troy to Vietnam.* Random House

Barber, N.L. 2009. Summary of estimated water use in the United States in 2005. U.S. Geological Survey Fact Sheet 2009–3098, 2 p.

Bardi, Ugo. 2011. The Limits to Growth Revisited. Springer, New York, 119 p.

Barkley, A. P. 1990. The determinants of the migration of labor out of agriculture in the United States, 1940–85. American Journal of Agricultural Economics, 72(3), 567–573.

Barnett, Tim P., Pierce, David. 2008. When will Lake Mead go dry? Water Resources Research. Vol. 44 Issue 3. March 2008.

Barry, John M. 1997. Rising Tide: The Great Mississippi Flood of 1927 and How It Changed America. New York, Simon & Schuster.

Barthel, S., C. Folke, and J. Colding. 2010. Social-ecological memory in urban gardens – retaining capacity for management of ecosystem services. Global Environmental Change 20, 255–265.

Batker, D., et al. 2014. The importance of Mississippi Delta Restoration on the Local and National Economies. In: Day J.W., Kemp G.P., Freeman A.M., Muth D.P. (eds). Perspectives on the Restoration of the Mississippi Delta: The Once and Future Delta. Springer, Netherlands, 155–173.

Beck, T.B., M.F. Quigley, and J.F. Martin. 2001. Emergy evaluation of food production in urban residential landscapes. Urban Ecosystems 5, 187–207

Becker, Elizabeth. 2013. Overbooked. Simon & Schuster, NY, 449 p.

Benjamin Cook, Toby Ault, and Jason Smerdon. Unprecedented 21st century drought risk in the American Southwest and central plains. Science Adv doi 10.1126/sciadv.1400082

Bennington J. B. and Farmer C. 2014. Learning from the Impacts of Superstorm Sandy, 1st Edition. Academic Press, 140 p.

Benson L. et al. 2007. Anasazi (Pre-Columbian Native-American) Migrations During The Middle12[Th] and Late13[th] Centuries – Were they Drought Induced? Climatic Change Vol. 83. Issue 1, 187–213.

Benson, L. V., Kashgarian, M., Rye, R. O., Lund, S. P., Paillet, F. L., Smoot, J., Kester, C., Mensing, S., Meko, D., and Lindstrom, S. 2002. 'Holocene multidecadal and multicentennial droughts affecting Northern California and Nevada', Quaternary Science Reviews 21, 659–682.

Berry, Wendell. 1972. A Continuous Harmony – Essays Cultural and Agricultural. Shoemaker and Hoard, Washington, DC.

Boulding, Kenneth. 1968. Beyond Economics. The University of Michigan Press, Ann Arbor. 302 p.

Brady, N.C., and R.R. Weil. 2007. The nature and properties of soils, 14th edition. Prentice Hall.

Brandt, A.R., Englander, J. and Bharadwaj, S. 2013. "The energy efficiency of oil sands extraction: Energy return ratios from 1970 to 2010." Energy 55, 693–702.

Brown, J. et al. 2011. Energetic limits to economic growth. BioScience 61, 19–26.

Brown, L.R. 2012. Full planet, empty plates: the new geopolitics of food scarcity. W.W. Norton and Company, New York, USA.

Bucheli, T.D., F. Blum, A. Desaules, and O. Gustafsson. 2004. Polycyclic aromatic hydrocarbons, black carbon, and molecular markers in soils of Switzerland. Chemosphere 56, 1061–1076.

Burger J. R., et al. 2012. The Macroecology of Sustainability. PLOS Biology. June 2012 Vol. 10 Issue 6 Pgs. 1–7.

Campbell, C. J., & Wöstmann, A. 2013. *Campbell's Atlas of Oil and Gas Depletion.* Springer.

Campbell, C.J., and Laherrère J.H., 1998. The end of cheap oil. Scientific American (March), 78–83.

Canning, P., Charles, A., Huang, S., Polenske, K.R., and Waters, A. 2010. Energy use in the U.S. food system. Economic Research Report no. 94. United States Department of Agriculture Economic Research Service

Cao, S., Xie, G., and Zhen, L. 2010. Total embodied energy requirements and its decomposition in China's agricultural sector. Ecological Economics 69, 1396–1404.

Carbajales-Dale M., et al. 2013. Household Solar Photovoltaics: Supplier of Marginal Abatement, or Primary Source of Low-Emission Power? Energy Environmental Sciences, DOI:10.1039/c3ee42125b.

Carbajales-Dale, M., S. Krumdieck, and P. Broder. 2012. Global energy modeling – A biophysical approach (GEMBA) Part 2: Methodology. Ecological Economics 73, 158–167.

Carlson, P.H. 2006. *Amarillo: The Story of a western Town.* Texas Tech University Press. 283 p.

Cassidy et al. 2014. Redefining agricultural yields: from tonnes to people nourished per hectare. Environmental Research Letters 8, 034015.

Chandler, T. 1987. *Four thousand years of urban growth: an historical census.* St. David's University Press.

Chapman, L. 2007. Transport and climate change: a review. Journal of Transport Geography, 15(5), 354–367.

Clark, H.F., D.J. Brabander, and R.M. Erdil. 2006. Sources, sinks, and exposure pathways of lead in urban garden soil. Journal of Environmental Quality 35, 2066–2074.

Cleveland, C.J. 1995. The direct and indirect use of fossil fuels and electricity in USA agriculture, 1910–1990. Agriculture, Ecosystems and Environment 55, 111–121.

Coastal Protection and Restoration Authority of Louisiana. 2012. Louisiana's Comprehensive Master Plan for a Sustainable Coast. Coastal Protection and Restoration Authority of Louisiana. Baton Rouge, LA.

Cole, C.J., Friesen, B.A., and Wilson, E.M. 2014. Use of satellite imagery to identify vegetation cover changes following the Waldo Canyon Fire event, Colorado, 2012–2013: U.S. Geological Survey Open-File Report 2014–1078, 1 sheet.

Colten, C.E. 2000. "Too Much of a Good Thing: Industrial Pollution in the Lower Mississippi River," in *Transforming New Orleans and Its Environs* (Pittsburgh,), 148–49.

Colten, C.E. 2006. The Rusting of the Chemical Corridor. Technology and Culture, 47(1).

Cook, B. I., Ault, T. R., & Smerdon, J. E. 2015. Unprecedented 21st century drought risk in the American Southwest and Central Plains. Science Advances, 1(1), e1400082.

Cook, E. L. 1971. The flow of energy in an industrial society. Scientific American, (225), 135–42.

Cook, J., Nuccitelli, D., Green, S. A., Richardson, M., Winkler, B., Painting, R., ... & Skuce, A. 2013. Quantifying the consensus on anthropogenic global warming in the scientific literature. Environmental Research Letters, 8(2), 024024.

Cordell, D., and S. White. 2013. Sustainable phosphorus measures: strategies and technologies for achieving phosphorus security. Agronomy 3, 86–116.

Cortright, J., 2008. Driven to the Brink: How the Gas Price Spike Popped the Housing Bubble and Devalued the Suburbs. In: Discussion Paper. CEOs for Cities, Chicago. www.ceosforcities.org.

Costanza, R. and Daly, H. E. 1992. Natural Capital and Sustainable Development. Conservation Biology, 6, 37–46.

Costanza, R. et al. 1997. The value of the World's Ecosystem Services and Natural Capital. Nature Vol. 387. 15 May 1997.

Costanza, R. et al. 1998. Special Section: Forum on Valuation of Ecosystem Services. The value of ecosystem services: putting the issues in perspective Ecological Economics 25 (1998) 67–72.

Costanza, R., Cumberland, J. H., Daly, H., Goodland, R., Norgaard, R. B., Kubiszewski, I., & Franco, C. 2014. *An introduction to ecological economics*. CRC Press. the updated estimates in table 3.

Czech, Brain. 2013. *Supply Shock.* New Society Publishers, Gabriola Island, Camada. 367 p.

Daly, H.E. and Farley, J. 2004. *Ecological Economics* (Principles and Applications) Island Press, Washington DC, p.18.

Daly, H.E. (Ed.). 1980. *Economics, ecology, ethics: Essays toward a steady-state economy.* WH Freeman, San Francisco.

Daly, H.E. 2007. *Ecological Economics and Sustainable Development Selected Essays of Herman Daly.* Edward Elgar, Northampton MA, 270 p.

Daly, H.E. 2014. *From Uneconomic Growth to a Steady-State Economy.* Edward Elgar, Northampton MA. 253 p.

Daly, H.E. *Ecological Economics and Sustainable Development, selected essays from Herman Daly.* Edward Elgar, Northhampton MA. 2007.

Daly, H.E. Economics for a full world. Great Transitions Discussion, Tellus Insitute, May 2015.

Davidsson, S., Grandell, L., Wachtmeister, H., & Höök, M. 2014. Growth curves and sustained commissioning modelling of renewable energy: Investigating resource constraints for wind energy. *Energy Policy.*

Day J.W. et al. Sustainability and Place. 2014. How emerging megatrends of the 21st century will affect humans and nature at the landscape level. Ecological Engineering, 65, 33–48.

Day, J., D. Boesch, E. Clairain, P. Kemp, S. Laska, W. Mitsch, K. Orth, H. Mashriqui, D. Reed, L. Shabman, C. Simenstad, B. Streever, R. Twilley, C. Watson, J. Wells, D. Whigham. 2007. Restoration of the Mississippi Delta: Lessons from Hurricanes Katrina and Rita. Science, 315, 1679–1684.

Day, J., J. Gunn, W. Folan, A. Yanez, and B. Horton. 2012. The influence of enhanced post-glacial coastal margin productivity on the emergence of complex societies. The Journal of Island and Coastal Archaeology, 7, 23–52.

Day, J., P. Kemp, A. Freeman, and D. Muth. Editors. 2014. Perspectives on the Restoration of the Mississippi Delta: The Once and Future Delta. Springer, New York. 194 p.

De Bon, H., L. Parrot, and P. Moustier. 2010. Sustainable urban agriculture in developing countries. A review. Agronomy for Sustainable Development, 30, 21–32.

deBuys, William. 2011. *A Great Aridness*. Oxford University Press, Oxford. 369 p.

Decker, Ethan, Scott Elliott, Felisa Smity, Donald Blake and Sherwood Rowland. 2000. Energy and material flow through the urban ecosystem. Annual Review of Energy and the Environment, 25, 685–740

Deffeyes, KS. 2001. Hubbert's Peak: The Impending World Oil Shortage. Princeton University Press, Princeton, NJ.

Desjardins, E., R. MacRae, and T. Schumilas. 2010. Linking future population food requirements for health with local production in Waterloo Region, Canada. Agriculture and Human Values 27, 129–140.

Despommier, D. 2011. *The vertical farm: feeding the world in the 21st century*. Picador. New York, NY.

Diamond, J.M. 1997. *Guns, germs, and steel: the fates of human societies*. W.W. Norton and Company, New York, USA

Diamond, Jared. *Collapse*. Penguin Books, New York, 589 p.

Dobrovolski et al. 2011. Agricultural expansion and the fate of global conservation priorities. Biodiversity Conservation, 20, 2445–2459.

Doran, P., and Zimmerman M.K. 2009. Examining the scientific consensus on climate change. EOS, 90, 22–23.

Dukes, J.S. 2003. Burning buried sunshine: human consumption of ancient solar energy. Climatic Change, 61(1–2), 31–44.

Edwards-Jones, G., L.M. Canals, N. Hounsome, M. Truninger, G. Koerber, B., … D.L. Jones. 2008. Testing the assertion that 'local food is best': the challenges of an evidence-based approach. Trends in Food Science & Technology, 19, 265–274.

EIA. 2014. Annual Energy Outlook. Energy Information Agency, U.S. Department of Energy.

Erb, K.H., F. Krausmann, W. Lucht, and H. Haberl. 2009. Embodied HANPP: Mapping the spatial disconnect between global biomass production and consumption. Ecological Economics, 69, 328–334.

Farrell, William R. 2002. Classical Place Names in New York State: Origins, Histories and Meanings. Pine Grove Press.

Farrelly, Elizabeth 2008 *Blubberland: The Dangers of Happyness*. The MIT Press, Cambridge MA. 218 p.

Fischetti, M. 2011. "*The efficient city.*" Scientific American, 305, 74–75.

Fitzgerald, D. M., et al. 2009. Coastal Impacts Due to Sea-Level Rise. Annu. Rev Earth Planet Sciences 36, 601–47.

Forman, Richard. Urban Regions Ecology and Planning Beyond the City. Cambridge University Press, Cambridge UK. 2008

Fragkias, M., Lobe, J., Strumsky, D., and Seto, K. 2013. Does size matter? Scaling of CO_2 emissions and U.S. urban areas. PLOS one, 8(6), e64727.

Francou B. et al. 2000. Glacier Evolution in the Tropical Andes during the Last Decades of the 20th Century: Chacaltaya, Bolivia, and Antizana, Ecuador. Ambio. Vol. 29, No. 7, Research for Mountain Area Development: The Americas. 416–422.

Francou, B. et al. 2003. Tropical climate change recorded by a glacier in the central Andes during the last decades of the twentieth century: Chacaltaya, Bolivia, 16° S. Journal of Geophysical Research, 108.

Galloway, J.N., Aber, J.D., Erisman, J.W., Seitzinger, S.P., Howarth, R.W., Cowling, E.B., Cosby, B.J. 2003, The nitrogen cascade, BioScience, 54, 341–356.

Galzki, J.C., D.J. Mulla, and C.J. Peters. 2014. Mapping the potential of local food capacity in Southeastern Minnesota. Renewable Agriculture and Food Systems. DOI:10.1017/S1742170514000039

Gardner, B.L. 2002. *American agriculture in the twentieth century: how it flourished and what it cost.* Harvard University Press. 400 p.

Gibbon, E. 1776. *The decline and fall of the Roman Empire*: Volume I. Random House.

Gibbs et al. 2010. Tropical forests were the primary sources of new agricultural lands in the 1980s and 1990s. PNAS, 107, 16732–16737.

Giombolini, K.J., K.J. Chambers, S.A. Schlegel, and J.B. Dunne. 2011. Testing the local reality: does the Willamette Valley region produce enough to meet the needs of the local population? A comparison of agriculture production and recommended dietary requirements. Agriculture and Human Values, 28, 247–262.

Glaeser, E. 2011. *Triumph of the city: how our greatest invention makes us richer, smarter, greener, healthier, and happier.* Penguin, New York NY, 338 p.

Gold, R. 2014. Why peak-oil predictions haven't come true. The Wall Street Journal. http://online.wsj.com/articles/why-peak-oil-predictions-haven-t-come-true-1411937788

Gomez G.M. Transforming New Orleans and Its Environs. C.E. Colton Editor. University of Pittsburgh Press. 120 p.

Gorgescu-Roegen. N. 1971. The Entropy Law and the Economic Process. Harvard University Press, Cambridge.

Gowdy, J., 2000. Terms and concepts in ecological economics. Wildlife Society Bulleton 28(1), 26–33.

Gowdy, J., Klitgaard, K., Krall, L., 2010. Capital and sustainability. Corporate Examiner 37(4–5), 16 March.

Grafton, Q. 2010. Adaptation to climate change in marine capture fisheries. Marine Policy, 34, 606–615.

Granek, E.F., et al., 2010. Ecosystem services as a common language for coastal ecosystem-based management. Conserv. Biol. 24 (1), 207–216.

Greer, John. 2011. *The Wealth of Nature.* New Society Publishers, British Columbia, Canada.

Griffin, T., Z. Conrad, C. Peters, R. Ridberg, and E.P. Tyler. 2014. Regional self-reliance of the Northeast food system. Renewable Agriculture and Food Systems. DOI:10.1017/S1742170514000027

Guilford, Megan C., Charles AS Hall, Peter O'Connor, and Cutler J. Cleveland. 2011. "A new long term assessment of energy return on investment (EROI) for US oil and gas discovery and production." Sustainability 3(10), 1866–1887.

Hagler Y., Yaro R.D., and Ronderos L.N. 2009 . New Strategies for Regional Economic Development. America 2050 Research Seminar Discussion Papers and Summary. Healdsburg, California – March 29–31.

Hall C.A.S, Tharakan PJ, Hallock J, Cleveland C, Jefferson M. 2003. Hydro-carbons and the evolution of human culture. Nature, 425, 18–322.

Hall, C.A.S., & Day, J. W. 2009. Revisiting the Limits to Growth After Peak Oil In the 1970s a rising world population and the finite resources available to support it were hot topics. Interest faded—but it's time to take another look. American Scientist, 97(3), 230–237.

Hall, C.A.S., Cleveland, C., Kaufmann, R., 1986. Energy and Resource Quality: The Ecology of the Economic Process. Wiley, New York.

Hall, C.A.S., King, C. 2011 Relating financial and energy return on investment. Sustainability, Special Issue on EROI, 1810–1832.

Hall, C.A.S., Klitgaard, K.A., 2012. Energy and the Wealth of Nations: Understanding the Biophysical Economy. Springer Publishing, New York, 407 p.

Hall, C.A.S., Powers, R., & Schoenberg, W. 2008. Peak oil, EROI, investments and the economy in an uncertain future. In *Biofuels, solar and wind as renewable energy systems*. Springer, Netherlands, 109–132.

Hall, Charles AS, Stephen Balogh, and David JR Murphy. 2009. What is the minimum EROI that a sustainable society must have? Energies, 2(1), 25–47.

Hall, Charles AS. 1972. Migration and metabolism in a temperate stream ecosystem. Ecology, 585–604.

Hallock, J. L., Tharakan, P. J., Hall, C. A., Jefferson, M., & Wu, W. 2004. Forecasting the limits to the availability and diversity of global conventional oil supply. Energy, 29(11), 1673–1696.

Hamilton, A., S. Balogh, A. Maxwell, and C. Hall. 2013. Efficiency of edible agriculture in Canada and the U.S. over the past three to four decades. Energies, 6, 1764–1793.

Hamilton, A.J., K. Burry, H.F. Mok, F. Barker, J.R. Grove, and V.G. Williamson. 2014. Give peas a chance? Urban agriculture in developing countries. A review. Agronomy for Sustainable Development, 34, 45–73.

Hamilton, J.D., 2008. Daily monetary policy shocks and new home sales. J. Monetary Econ. 55 (7), 1171–1190.

Hannah, L. et al. 2013. Climate change, wine, and conservation. Proceedings of the National Academy of Sciences of the United States of America, 110, 6907–6912.

Haughwout A, Orr J, and Bedoll D. 2008. The price of land in the New York Metropolitan Area. Federal Reserve Bank of New York. Current Issues in Economics and Finance, 14(3).

Hayhoe, K., et al. 2004. Emissions pathways, climate change, and impacts on California. Proceedings of the National Academy of Sciences of the United States of America, 101, 12422–12427.

Heady, D., and S. Fan. 2008. Anatomy of a crisis: the causes and consequences of surging food prices. Agricultural Economics, 39, 375–391

Heinberg, Richard. 2004. *Powerdown – Options and Actions for a Post-Carbon World*. New Society Publishers, Gabriola Island, BC Canada. 208 p.

Heinberg, Richard. 2012. Snake Oil – How Fracking's False Promise of Plenty Imperils Our Future. Post Carbon Insitute, Santa Rosa , CA. 251 p. www.postcarbon.org.

Heller, M., and G. Keoleian. 2000. Life cycle-based sustainability indicators for assessment of the U.S. food system. The University of Michigan Center for Sustainable Systems, pub. No. CSS00-04.

Heller, M.C., G.A. Keoleian, and W.C. Willett. 2013. Toward a life cycle-based, diet-level framework for food environmental impact and nutritional quality assessment: a critical review. Environmental Science and Technology, 47, 12632–12647.

HLPE. 2014. Sustainable fisheries and aquaculture for food security and nutrition. A report by the High Level Panel of Experts on Food Security and Nutrition of the Committee on World Food Security, Rome.

Hu, G., L. Wang, S. Arendt, and R. Boeckenstedt. 2011. An optimization approach to assessing the self-sustainability potential of food demand in the Midwestern United States. Journal of Agriculture, Food Systems, and Community Development, 2, 195–207

Huang, W.Y. 2009. Factors contributing to the recent increases in U.S. fertilizer prices, 2002–08. Economic Research Service Report AR-33, United States Department of Agriculture.

Huesemann, M., Huesemann, J., 2011. *Techno-Fix: Why Technology Won't Save Us or the Environment*. New Society Publishers, Cabriola Island, BC, Canada, 464 p.

Hughes, D. 2013. *Drilling California: a reality check on the Monterey Shale*. Post-Carbon Institute.

Hughes, D. 2014. *Drilling Deeper: A Reality Check on the U.S. Government Forecasts for a Lasting Tight Oil and Gas Boom*. Post Carbon Institute. October 2014. www.postcarbon.org

Imhoff, M.L., and Bounoua L. 2006. Exploring global patterns of net primary production carbon supply and demand using satellite observations and statistical data. Journal of Geophysical Research, 111, D22S12.

Imhoff, M.L., Bounoua L, Ricketts T, et al. 2004. Global patterns in net primary productivity (NPP). Data distributed by the Socioeconomic Data and Applications Center (SEDAC). http://sedac.ciesin.columbia.edu/es/hanpp.html.

Ingram, Lynn and Frances Malamud-Roam. 2013. The West Without Water. University of California Press, Berkeley. 256 p.

International Bank for Reconstruction and Development / The World Bank and International Cryosphere Climate Initiative. 2013. On Thin Ice: How Cutting Warming Can Slow Pollution and Save Lives. A Joint Report of the World Bank and The International Cryosphere Climate Initiative.

IPCC. 2013. Climate Change 2013: The Physical Science Basis. Contribution of Working Group I to the Fifth Assessment Report of the Intergovernmental Panel on Climate Change. Stocker, T.F., D. Qin, G.-K. Plattner, M. Tignor, S.K. Allen, J. Boschung, A. Nauels, Y. Xia, V. Bex and P.M. Midgley (eds). Cambridge University Press, Cambridge, UK and New York, NY, 1535 p.

Jackson, T. 2009. Prosperity Without Growth Economics for a Finite Planet. Earthscan, Washington DC. 276 p.

Jones, Christopher and Daniel Kammen. 2011. Quantifying carbon footprint reduction opportunities for U.S. households and communities. Environmental Science and Technology, 45, 4088–4095.

Jones, Christopher and Daniel Kammen. 2013. Spatial distribution of U.S. household carbon footprints reveals suburbanization undermines greenhouse gas benefits of urban population density. Environmental Science & Technology. DOI: 10.1021/es4034364.

Joyce, B.A., V.K. Mehta, D.R. Purkey, L.L. Dale, and M. Hanemann. 2011. Modifying agricultural water management to adapt to climate change in California's central valley. Climatic Change, 109(1), S299–S316.

Junk, W. J. et al. 1989. The flood pulse concept in river-floodplain systems. Pages 110–127 in D. P. Dodge, editor. Proceedings of the International Large River Symposium. Canadian Special Publications Fisheries Aquatic Sciences, 106.

Kay, R.T., T.L. Arnold, W.F. Cannon, and D. Graham. 2008. Concentrations of polycyclic aromatic hydrocarbons and inorganic constituents in ambient surface soils, Chicago, Illinois: 2001–2002. Soil and Sediment Contamination, 17, 221–236. Johnson, D.L., and J.K. Bretsch. 2002. Soil lead and children's blood levels in Syracuse, NY, USA. Environmental Geochemistry and Health, 24, 375–385.

Kerschner, C., C. Prell, K. Feng, and K. Hubacek. 2013. Economic vulnerability to peak oil. Global Environmental Change, 23, 1424–1433.

Khoury, C.K. et al. 2014. Increasing food homogeneity in global food supplies and the implications for food security. Proceedings of the National Academy of Sciences of the United States of America. In Press. DOI:10.1073/pnas.1313490111

Klare, Michael. 2012. *The Race for What's Left*. Holt and Company, New York. 306 p.

Klein, Marvin. 1967. Thermodynamics in Einstein's thought. Science, 157, 509–516.

Ko Jae-Young, et al. 2012. Policy adoption of ecosystem services for a sustainable community: A case study of wetland assimilation using natural wetlands in Breaux Bridge, Louisiana Ecological Engineering, 38, 114–118.

Ko Jae-Young. 2007. The Economic Value of Ecosystem Services Provided by the Galveston Bay/Estuary System. Final Report. Texas Commission on Environmental Quality Galveston Bay Estuary. files.harc.edu/Projects/Nature/GalvestonBayEconomicValue.pdf

Kolbert, Elizabeth. 2014. *The Sixth Extinction*. Henry Holt, New York. 319 p.

Kremer, P., and Y. Schreuder. 2012. The feasibility of regional food systems in metropolitan areas: an investigation of Philadelphia's foodshed. Journal of Agriculture, Food Systems, and Community Development. DOI: 10.5304/jafscd.2012.022.005

Kunzig, Robert. 2011. The City Solution: Why cities are the best cure for our planet's growing pains. *National Geographic*, December.

Lal, R. 2013. Food security in a changing climate. Ecohydrology & Hydrobiology, 13, 8–21.

Lambert, J. G., Hall, C. A., Balogh, S., Gupta, A., & Arnold, M. 2014. Energy, EROI and quality of life. Energy Policy, 64, 153–167.

Lambert, J., Hall, C. A. S. and Balogh. S. 2013. EROI of Global Energy Resources: Status, Trends and Social Implications. Report to Division of Foreign Investment, United Kingdom.

Larson, Erik, and Isaac Monroe Cline. 1999. *Isaac's storm: A Man, a Time, and the Deadliest Hurricane in History.* Vintage.

Laura Parker. Treading Water. National Geographic, February, 107–125.

Lewis, Pierce. 2003. *New Orleans, the Making of an Urban Landscape.*

Lobell, D.B., C.B. Field, K.N. Cahill, C. Bonfils. 2006. Impacts of future climate change on California perennial crop yields: model projections with climate and crop uncertainties. Agricultural and Forest Meteorology, 141, 208–218.

Maggio, G., and Cacciola, G. 2012. When will oil, natural gas, and coal peak? Fuel, 98, 111–123. DOI:10.1016/j.fuel.2012.03.021

Mariola, M.J. 2008. The local industrial complex? Questioning the link between local foods and energy use. Agriculture and Human Values, 25, 193–196.

McRae, Hamish. 2015. "How can we explain why productivity is lagging behind overall economic growth". THE INDEPENDENT June 25, 20015

Meadows, Donella, Dennis Meadows, and Jorgen Randers. 1992. Beyond the Limits. Chelsea Green Publishing, White River VT. 300 p.

Meadows, Donella, Dennis Meadows, Jorgen Randers, and William Behrens. 1972. The Limits to Growth. Universe Books, New York.

Meadows, Donella, Jorgen Randers, Dennis Meadows. 2004. Limits to Growth – The 30 Year Update. Chelsea Green Publishing, White River VT. 338 p

Melillo, J.M., Richmond, T.C., and Yohe G.W., (Eds.) 2014: Climate Change Impacts in the United States: The Third National Climate Assessment. U.S. Global Change Research Program, 841 p.

Merk, Frederick and Bannister, Lois. 1963. *Manifest destiny and Mission in American History.* Harvard University Press.

Michael Brune, Executive director of the Sierra Club. January 2015 National Geographic Magazine.

Michael Hudson. 2012. *The Bubble and Beyond.* ISLET, Dresden. 481 p.

Millenium Ecosystem Assessment, 2005. *Ecosystems and Human Well-being: Synthesis.* Island Press, Washington D.C.

Millennium Ecosystem Assessment. 2003. *Ecosystems and Human Well-being: A Framework For Assessment.* Island Press, Washington. 266 p.

Milly, P. C., Dunne, K. A., & Vecchia, A. V. 2005. Global pattern of trends in stream-flow and water availability in a changing climate. Nature, 438(7066), 347–350.

Moersbaecher, M., and J. Day. 2011. Ultra-deepwater Gulf of Mexico oil and gas: Energy return on financial investment and a preliminary assessment of energy return on energy investment. Sustainability, 3, 2009–2026. DOI:10.3390/su3102009..

Mok, H.F., V.G. Williamson, J.R. Grove, K. Burry, S.F. Barker, A.J. Hamilton. 2014. Strawberry fields forever? Urban agriculture in developed countries: a review. Agronomy for Sustainable Development, 34, 21–43.

Monro, Alice. 2004. *Runaway – Stories.* Vintage Books, Random House, New York. 335 p.

Murphy, D.J., and Hall C.A.S. 2011. Adjusting the economy to the new energy realities of the second half of the age of oil. Ecological Modeling. DOI:10.1016/j.ecolmodel.2011.06.022

Murray, J., & King, D. 2012. Climate policy: Oil's tipping point has passed. Nature, 481(7382), 433–435.

Nixon SW, et al. 1996. The fate of nitrogen and phosphorous at the land sea margin of the North Atlantic Ocean. Biogeochemistry, 35, 141–180.

Odum H.T., 1971. Environment, Power, and Society. Wiley-Interscience, John Wiley and Sons, New York, 331 p.

Odum, H.T. 2007. *Environment, power, and society for the twenty-first century: the hierarchy of energy*. Columbia University Press.

Odum, H.T. and Elizabeth Odum. 2001. *A Prosperous Way Down*. University of Colorado Press, Boulder. 326 p.

Oliveira, V. 2013. The food assistance landscape: FY 2012 annual report. United States Department of Agriculture, Economic Research Service, Economic Information Bulletin No. 109.

Owen, David. Green Manhattan. *The New Yorker*, October 18 2004.

Palma, I.P., J.N. Toral, M.R.P. Vasquez, N.F. Fuentes, F.G. Hernandez. 2014. Historical changes in the process of agricultural development in Cuba. Journal of Cleaner Production. DOI:10.1016/j.jclepro.2013.11.078

Palmer, Graham. 2014. *Energy in Australia: Peak Oil, Solar Power, and Asia's Economic Growth*. Springer, New York, NY.

Parry, M.L., C. Rosenzweig, A. Iglesias, M. Livermore, and G. Fischer. 2004. Effects of climate change on global food production under SRES emissions and socio-economic scenarios. Global Environmental Change, 14, 53–67.

Pauly, D. et al. 1998. Fishing down marine food webs. Science, 279, 860–863.

Pelletier, N., et al. 2014. Energy prices and seafood security. Global Environmental Change, 24, 30–41.

Perlin, J. 2005. *A forest journey: The story of wood and civilization*. The Countryman Press.

Peters, C.J., J.L. Wilkens, and G.W. Fick. 2007. Testing a complete-diet model for estimating the land resource requirements of food consumption and agricultural carrying capacity: The New York State example. Renewable Agriculture and Food Systems, 22, 145–153.

Peters, C.J., N.L. Bills, A.J. Lembo, J.L. Wilkins, and G.W. Fick. 2009. Mapping potential foodsheds in New York State: a spatial model for evaluating the capacity to localize food production. Renewable Agriculture and Food Systems, 24, 72–84.

Petit, J.R. et al. 1999. Climate and atmospheric history of the past 420,000 years from the Vostok Ice Core, Antarctica. Nature, 399, 429–436.

Piketty, Thomas. 2014 *Capital in the Twenty-First Century*. The Belknap Press of Harvard University Press, Cambridge MA, 685 p.

Pimentel D and Pimentel M. 2003. Sustainability of meat-based and plant-based diets and the environment. American Journal of Clinical Nutrition, 78, 660S-663S.

Pimentel D. et al. 1997.Water Resources: agriculture, the environment, and society. BioScience, 47, 97–106.

Pimentel et al. 2005. Update on the environmental and economic costs associated with alien-invasive species in the United States. Ecological Economics, 52(3), 273–288.

Pimentel, D. 2006. Soil erosion: a food and environmental threat. Environment, Development and Sustainability, 8, 119–137.

Pimentel, D., and M. Pimentel. 2008. Corn and cellulosic ethanol cause major problems. Energies, 1, 35–37.

Pimentel, D., B. Berger, D. Filiberto, M. Newton, B. Wolfe, et al. 2004. Water resources: agricultural and environmental issues. BioScience, 54, 909–918.

Pimentel, D., et al. 1997. Water resources: agriculture, the environment, and society. BioScience, 47, 97–106.

Pimentel, D., P. Hepperly, J. Hanson, D. Douds, and R. Seidel. 2005. Environmental, energetic, and economic comparisons of organic and conventional farming systems. BioScience, 55, 573–582.

Pimentel, D., Williamson, S., Alexander, C.E., Gonzalez-Pagan, O., Kontak, C., Mulkey, S.E., 2008. Reducing energy inputs in the U.S. food system. Hum. Ecol. 36, 459–471.

Pollan, M. 2009. *In defense of food: an eater's manifesto*. Penguin Books, London, UK.

Pound, Ezra. *Canto LXXXI*

Powell, Lawrence. 2012. *The Accidental City: Improvising New Orleans*. Harvard University Press.

Powers, Bill. 2013. Cold, Hungry and in the Dark Exploding the Natural Gas Supply Myth. New Society Publishers. Gabriola, Canada, 312 p.

Powlson, D.S., P.J. Gregory, W.R. Whalley, J.N. Quinton, D.W. Hopkins, A.P. Whitmore, P.R. Hirsch, and K.W.T. Goulding. 2011. Soil management in relation to sustainable agriculture and ecosystem services. Food Policy, 36 (Suppl. 1), S72-S87.

Prieto, P. A., & Hall, C. 2013. *Spain's photovoltaic revolution: the energy return on investment*. Springer Science & Business Media.

Ramírez, C.A., and E. Worrell. 2006. Feeding fossil fuels to the soil: an analysis of energy embedded and technological learning in the fertilizer industry. Resources, Conservation and Recycling, 46, 75–93.

Rappaport, Jordan. 2003. U.S. Urban Decline and Growth, 1950 to 2000. Federal Reserve Bank of Kansas City, 44 p. www.kc.frb.org

Rees W. 2012. Cities as dissipative structures: Global change and the vulnerability of urban civilization. In: M. Weinstein and R. Turner (eds), Sustainability Science. Springer, New York, pp 243–274.

Rees, W.E. 2006. "Ecological Footprints and Bio-Capacity: Essential Elements in Sustainability Assessment."Chapter 9 in Jo Dewulf and Herman Van Langenhove (eds) Renewables-Based Technology: Sustainability Assessment, John Wiley and Sons, Chichester, UK, 143–158.

Reisner, Marc. 1993. *Cadillac desert: The American West and its disappearing water*. Penguin.

Reynolds, D.B., 2002. Scarcity and Growth Considering Oil and Energy: An Alternative Neo-Classical View. Edwin Mellen Press, Lewiston, NY, 232 p.

Rignot E., Mouginot J., Morlighem M., Seroussi H., Scheuch B. 2014. Widespread, rapid grounding line retreat of Pine Island, Thwaites, Smith, and Kohler glaciers, West Antarctica, from 1992 to 2011. Geophysical Research Letters, 41, 3502–3509. DOI:10.1002/2014GL060140.

Rignot, E., Mouginot, J., Morlighem, M., Seroussi, H., & Scheuchl, B. 2014. Widespread, rapid grounding line retreat of Pine Island, Thwaites, Smith, and Kohler glaciers, West Antarctica, from 1992 to 2011. Geophysical Research Letters, 41(10), 3502–3509.

Riskin, S.H., S. Porder, M.E. Schipanski, E.M. Bennett, and C. Neill. 2013. Regional differences in phosphorus budgets in intensive soybean agriculture. BioScience, 63, 49–54.

Robert N. McMichael, "Plant Location Factors in the Petrochemical Industry in Louisiana" (Ph.D. diss., Louisiana State University, 1961).

Roberts, Sam. "New York doesn't care to remember the Civil War". New York Times, December 26, 2010.

Rodrigue, Jean-Paul, and Michael Browne. 2002. International Maritime Freight Transport and Logistics. In Transport Geographies: An Introduction: pp 156–178. Whiley-Blackwell Publishing.

Rose, S.L., H. Akbari, and H. Taha. 2003. Characterizing the fabric of the urban environment: a case study of Metropolitan Houston, Texas. Lawrence Berkeley National Laboratory. Report no. LBNL-51448.

Rosner, Hillary The bug that's eating the woods National Geographic April 2015

Roy, E.D., J.R. White, and M. Seibert. 2014. Societal phosphorus metabolism in future coastal environments: insights from recent trends in Louisiana, USA. Global Environmental Change, 28, 1–13.

Sallenger Jr. et al. 2012. Hotspot of accelerated sea-level rise on the Atlantic coast of North America. Nat. Clim.Change, 2 884–888.

Satterthwaite, David, Gordon McGranahan, and Cecilia Tacoli. 2010. Urbanization and its implications for food and farming. Philosophical Transactions of the Royal Society, 365, 2809–2820.

Scanlon, B. R., Faunt, C. C., Longuevergne, L., Reedy, R. C., Alley, W. M., McGuire, V. L., & McMahon, P. B. 2012. Groundwater depletion and sustainability of irrigation in the US High Plains and Central Valley. Proceedings of the national academy of sciences, 109(24), 9320–9325.

Schaetzl, R.J., F.J. Krist, and B.A. Miller. 2012. A taxonomically based ordinal estimate of soil productivity for landscape-scale analysis. Soil Science 177: 288–299.

Schlosser, E. 2001. Fast food nation. Mariner Books, Boston.

Schumacher, E. F. 1973. Small is Beautiful: Economics as if People Mattered. New York Perennial Library.

Seto, K.C., B. Guneralp, and L.R. Hutyra. 2012. Global forecasts of urban expansion to 2030 and direct impacts on biodiversity and carbon pools. Proceedings of the National Academy of Sciences of the USA 109: 16083–16088.

Shaffer, G., J. Day, S. Mack, P. Kemp, I. van Heerden, M. Poirrier, K. Westphal, D. FitzGerald, A. Milanes, C. Morris, R. Bea, and S. Penland. 2009. The MRGO navigation project: A massive human-induced environmental, economic, and storm disaster. Journal of Coastal Research, SI 54, 206–224.

Skrebowski C. 2004. Oil fields mega projects 2004. Petroleum Review (January), 18–20.

Smil, V. 2000. Phosphorus in the environment: natural flows and human interferences. Annual Review of Energy and the Environment, 25, 53–88.

Smil, V. 2011. Harvesting the biosphere: the human impact. Population and development review, 613–636.

Smil, V. 2013. *Harvesting the Biosphere: How Much We Have Taken from Nature*. The MIT Press, Cambridge, MA, 312 p.

Smit, J., A. Ratta, and J. Nasr. 1996. *Urban agriculture. Food, jobs, and sustainable cities*. United Nations Development Programme, Publication series for Habitat II, Volume 1. New York, USA.

Soddy, F. 1922. *Cartesian Economics: The Bearing of Physical Science upon State Stewardship*. Henderson, London.

Solow, R.M., 1994. Perspectives on growth theory. Journal of Economic Perspectives 8, 45–54.

Solow, RM. 1974. The economics of resources or the resources of economics. American Economic Review, 66, 1–14.

Sophocleous, M. 2012. Conserving and extending the useful life of the largest aquifer in North America: the future of the High Plains/Ogallala aquifer. Ground Water, 50, 831–839.

Srogi, K. 2007. Monitoring of environmental exposure to polycyclic aromatic hydrocarbons: a review. Environmental Chemistry Letters, 5, 169–195.

Steffen, William, Wendy Broadgate, Lisa Deutsch, Owen Gaffney, and Cronelia Ludwig. 2015. The trajectory of the Anthropocene: The great acceleration. The Anthropocene Review, 2, 81–98.

Steinhart, J.S., and C.E. Steinhart. 1974. Energy use in the US food system. Science, 184, 307–316.

Steward, D. R., Bruss, P. J., Yang, X., Staggenborg, S. A., Welch, S. M., & Apley, M. D. 2013. Tapping unsustainable groundwater stores for agricultural production in the High Plains Aquifer of Kansas, projections to 2110. Proceeding of the National Academy of Sciences. doi:10.1073/pnas.1220351110.

Stiglitz, Joseph. 2013. *The Price of Inequality: How Today's Divided Society Endangers our Future*. W.W. Norton, New York.

Strumsky, D., Lobo, J., & Tainter, J. A. 2010. Complexity and the productivity of innovation. *Systems Research and Behavioral Science, 27*(5), 496–509.

Sumaila, U., L. The, R. Watson, P. Tyedmers, and D. Pauly. 2008. Fuel price increase, subsidies, overcapacity, and resource sustainability. ICES Journal of Marine Science, 65, 832–840.

Surowiecki, James. The Puerto Rican problema. New Yorker April 6.

Tainter, J. A., & Taylor, T. G. 2014. Complexity, problem-solving, sustainability and resilience. *Building Research & Information, 42*(2), 168–181.

Tainter, Joseph. 2005. *The Collapse of Complex Societies*, Cambridge University Press.

Tao, B. et al. 2014. Increasing Mississippi river discharge throughout the 21st century influenced by changes in climate, land use, and atmospheric CO_2. Geophysical Research Letters. 21, 4978–4986. doi.10.1002/2014GL060361.

The Federal Writers' Project. 1938. *The WPA Guide to New Orleans: Guide to 1930s New Orleans*. Pantheon Books. New York City.

Tillery, A.C., Matherne, A.M., and Verdin K.L. 2012. Estimated probability of post-wildfire debris flows in the 2012 Whitewater–Baldy Fire burn area, southwestern New Mexico: U.S. Geological Survey Open-File Report 2012–1188, 11 p.

Tilman, D., K.G. Cassman, P.A. Matson, R. Naylor, and S. Polasky. 2002. Agricultural sustainability and intensive production practices. Nature, 418, 671–677.

Timoney, Kevin. 2013. *The Peace-Athabasca Delta – Portrait of a Dynamic Ecosystem*. The University of Alberta Press, Edmonton, 596 p.

Tumber, Catherine. 2012. *Small, Gritty, and Green*. The MIT Press, Cambridge, 211 p.

Turner, Graham. 2008. A Comparison of the Limits to Growth with Thirty Years of Reality. Socio-Economics and the Environment in Discussion. CSIRO Working Paper Series. Canberra Australia.

Turner, Graham. 2012. On the cusp of global collapse? Updated comparison of The Limits to Growth with historical data. Gaia. 21/2:116–124. www.oekom.de/gaia.

Tverberg, G. 2014. Oil price slide – no good way out. Our Finite World. http://ourfiniteworld.com/2014/11/05/oil-price-slide-no-good-way-out/

Vermeer, M., Rahmstorf, S. 2009. Global sea level linked to global temperature. Proc. Natl. Acad. Sci. U.S.A. 106, 21527–21532.

Vitousek PM, Ehrlich PR, Ehrlich AH, Matson PA. 1986. Human appropriation of the products of photosynthesis. BioScience, 36, 368–373.

Vitousek PM, Mooney HA, Lubchenco J, Melillo JM. 1997. Human domination of earth's ecosystems. Science, 277, 494–499.

Vitousek, P.M., J.D. Aber, R.W. Howarth, G.E. Likens, P.A. Matson, D.W. Schindler, W.H. Schlesinger, and D.G. Tilman.1997. Human alteration of the global nitrogen cycle: sources and consequences. Ecological Applications, 7, 737–750.

Vitousek, Peter M., Harold A. Mooney, Jane Lubchenco, and Jerry M. Melillo. 1997. "Human domination of Earth's ecosystems." *Science* 277(5325), 494–499.

Wackernagel, M., Schulz, N., Deumling, D., Callejas Linares, A., Jenkins, M., Kapos, V., Monfreda, C., Loh, J., Myers, N., Norgaard, R., & Randers, J., 2002. Tracking the ecological overshoot of the human economy. Proceedings of the National Academy of Science 99(14), 9266–9271.

Wackernagel, M., Schulz, N., Deumling, D., Callejas Linares, A., Jenkins, M., Kapos, V., Monfreda, C., Loh, J., Myers, N., Norgaard, R., & Randers, J., 2002. Tracking the ecological overshoot of the human economy. Proceedings of the National Academy of Science 99(14), 9266–9271.

Waldheim, C. 2010. Notes towards a history of agrarian urbanism. In: M. White and M. Przybylski (eds.), *Bracket 1: [on farming]*. Actar, Barcelona.

Walker, D. R. 1974. The metrology of the roman silver coinage. Part I: From Augustus to Domitian. Part II. From Verva to Commodus. Part III. From Pertinax to Uranius Antonius. British Archaeology Report. Oxford. Supplementary Series 5, 22, 40.

Webb, N.R. 1998. The traditional management of European heathlands. Journal of Applied Ecology, 35, 987–990.

Weber, C.L., and H.S. Matthews. 2008. Food-miles and the relative climate impacts of food choices in the United States. Environmental Science & Technology, 42, 3508–3513.

Weems, John Edward. 1957. *A Weekend in September*. Texas A&M University Press, College Station, 180 p.

Weißbach et al., 2013. Energy intensities, EROIs, and energy payback times of electricity generating power plants Energy 52, 210.

Weisz and Steinberger. 2010. Reducing energy and materials flows in cities. Current Opinion in Environmental Sustainability, 2, 185–192

Wilkinson, Richard and Kate Pickett. 2009. *The Spirit Level*. Bloomsbury Press, New York, 375 p.

Williams, R.C. 2006. *Horace Greeley: Champion of American Freedom*. New York University Press.

Wortman, S.E., and S.T. Lovell. 2013. Environmental challenges threatening the growth of urban agriculture in the United States. Journal of Environmental Quality, 42, 1283–1294.

Zhu, X. G., Long, S., and Ort, D. R. 2008. What is the maximum efficiency with which photosynthesis can convert solar energy into biomass? Current Opinion in Biotechnology, 19(2) 153–159.

Zumkehr, A., and J.E. Campbell. 2015. The potential for local croplands to meet US food demand. Frontiers in Ecology and the Environment, 13, 244–248.

Index

© Springer Science+Business Media New York 2016
J.W. Day, C. Hall, *America's Most Sustainable Cities and Regions*,
DOI 10.1007/978-1-4939-3243-6

Printed in the United States
By Bookmasters